T0336569

DESIGN OF BUILDINGS FOR WIND

Second Edition

DESIGN OF BUILDINGS FOR WIND

A Guide for ASCE 7-10 Standard Users and Designers of Special Structures

Second Edition

EMIL SIMIU, P.E., Ph.D.

JOHN WILEY & SONS, INC.

This book is printed on acid-free paper. ♾

Published by John Wiley & Sons, Inc., Hoboken, New Jersey
Published simultaneously in Canada

Library of Congress Cataloging-in-Publication Data:

Simiu, Emil.
Design of buildings for wind : a guide for ASCE 7-10 standard users and designers of special structures / Emil Simiu. – 2nd ed.
p. cm.
Includes index.
Originally published under title: Design of buildings and bridges for wind. 2006.
ISBN 978-0-470-46492-2 (hardback); 978-1-118-07735-1 (ebk); 978-1-118-07736-8 (ebk); 978-1-118-07737-5 (ebk); 978-1-118-08613-1 (ebk); 978-1-118-08622-3 (ebk); 978-1-118-08623-0 (ebk)
1. Wind-pressure. 2. Buildings–Aerodynamics. 3. Bridges–Aerodynamics. 4. Wind resistant design. I. Simiu, Emil. Design of buildings and bridges for wind. II. Title.
TA654.5.S54 2011
624.1′75–dc23

2011013566

Printed in the United States of America
10 9 8 7 6 5 4 3 2 1

CONTENTS

PREFACE

For most common types of structure, standard provisions on wind loads are in principle adequate for design purposes. The ASCE 7-10 Standard, among other standards, has incorporated a great deal of wind engineering knowledge accumulated within the last half-century. However, because the Standard has been developed by successive accretions, not always smoothly, its previous versions have been perceived by some practitioners as complicated and unwieldy. In an effort to respond to the demand for a clearer document, the ASCE 7-10 version of the Standard has been substantially expanded and revised. However, difficulties remain.

One of the main objectives of this book is to help the reader better understand the ASCE 7-10 Standard provisions for wind loads and apply them with confidence and ease. To this end, the book presents a guide to the Standard that explains the rationale of the provisions and illustrates their use through a large number of detailed numerical examples. Particular attention is given to the numerous changes made in the 2010 version of the ASCE Standard. Comparisons are presented between—or among—results of alternative provisions specified by the Standard for the same type of building. The comparisons show, for example, that for low-rise buildings, the so-called envelope procedure does not necessarily yield the lowest wind loads, as the Standard asserts. They also show that wind loads yielded by alternative regular methods, by regular and simplified methods, or by alternative simplified methods, can exhibit significant differences.

Wind loads and effects on special structures cannot in general be estimated by using standard provisions based on tables and plots. Rather, they need to be based on aerodynamic testing in the wind tunnel or larger facilities, and on information on the extreme wind speeds at the site. The Standard's provisions on the wind tunnel method are still vague and incomplete. In particular, they contain little or no material on dynamic analyses, the dependence of the response on wind directionality, and the calibration of the design mean recurrence intervals to account for larger-than-typical errors and uncertainties in the parameters that govern the wind-induced demand.

Some design offices have a policy of requiring wind effects estimates from more than one consultant. This is prudent, and justified by the fact that, owing to the lack of adequate standard provisions on the wind tunnel procedure, estimates by various laboratories can differ significantly. For example, recent estimates of the New York World Trade Center towers response to wind, performed by two well-known consultants, were found to differ by over 40%.

Consultant reports need, therefore, to be carefully scrutinized, and the need for transparency in their presentation cannot be overemphasized. The book includes as an appendix a report by Skidmore Owings & Merrill LLP, which presents a practitioner's perspective on the current state of the art in wind engineering and is a testimony to the need for transparent, traceable, and auditable procedures. Material presented in the book enables structural engineers to "ask the right questions," scrutinize effectively wind engineers' contributions to the determination of wind effects, and determine wind effects efficiently and accurately on their own, just as structural engineers do for seismic effects. This requires the use by the structural engineer of aerodynamic or aeroelastic data supplied by the wind engineer in standard, electronic form, and of directional extreme wind speed data obtained from wind climatological consultants. The wind effects of interest include internal forces, demand-to-capacity indexes for individual member design, as well as deflections and accelerations needed to check serviceability requirements. The book describes in detail modern, effective, and transparent methods for estimating such wind effects for any specified mean recurrence interval.

Wind effects on individual members are functions of influence coefficients that differ from member to member. In the past, the lack of sufficiently powerful computer resources did not allow this dependence to be taken into account accurately. The capability to do so is now routinely available. It has created a bridge between the wind engineer and the structural engineer that makes it possible to integrate the wind and structural phases of structural design more clearly and accurately than was heretofore possible. Public domain database-assisted design software referenced in the book allows the effective implementation of this capability.

Until about two decades ago, the time domain solution of large systems of differential equations posed insurmountable computational problems, and dynamics calculations were performed by using the spectral (frequency domain) approach, which transforms the differential equations of motion into algebraic equations. This approach is not always transparent and intuitive, and in practice suppresses phase information needed to correctly add wind effects from various sources (e.g., from two perpendicular lateral motions).

Computational capabilities and measurement technology developed in last two decades have made it possible to replace the frequency domain approach by the typically more effective time domain approach. The time domain approach, used in publicly available software referenced in the book, is not limited to the estimation of loads through the summation of pressures measured at large numbers of ports. Rather, in the integrated format known as database-assisted design, it can be directly and effectively applied to the design or checking of individual member strength, and can thus substantially improve structural design accuracy. In particular, the time domain approach eliminates the need for large numbers of cumbersome load combinations, based on guesswork—the method currently being used—and performs the requisite combinations through simple algebraic addition of time series of load effects. The improvements inherent in the use of time domain rather than frequency domain methods can be compared to those inherent in the calculation of structural response by finite element rather than by slide-rule based techniques.

The book contains material on structural reliability under wind loads that provides, among other matters, a perspective on the limitations of the Load and Resistance Factor Design (LRFD) approach, and a procedure for the calibration of design mean recurrence intervals. The calibration is required to ensure adequate safety levels if uncertainties in the determination of wind effects exceed typical uncertainties assumed in the ASCE 7 Standard. To date, the role of errors and uncertainties in the specification of design mean recurrence intervals has not been addressed by the ASCE 7 Standard provisions for the wind tunnel procedure, with the result that some designers and wind engineers resort to "magic numbers" that may be inadequate. For example, the same design mean recurrence interval is implied in the Standard for ordinary buildings and for tall buildings, whose response depends on dynamic parameters, including damping, which may exhibit large uncertainties. A reliability-based approach that takes into account such uncertainties can yield, for some tall buildings, longer mean recurrence intervals and hence larger wind effects than those specified in the ASCE Standard.

Structural reliability is also useful because it helps engineers design structures that do not consume more material and do not contain more embodied energy than necessary to ensure adequate safety levels. Inadequate safety levels can result in wind-induced losses, which are visible and costly. On the other hand, the cost of unnecessary materials—of "fat," as opposed to "muscle"—is much less visible to the public eye, but is nonetheless real, in monetary, energy consumption, and carbon footprint terms.

The book addresses incipient efforts to estimate ultimate capacities under fluctuating wind loads, aimed to achieve designs that are safer, more economical, and less demanding of embodied energy than those based on linear methods of analysis. The book also addresses wind-induced loss estimation, a topic fraught with difficulties, owing to the nonlinearities typically associated with the analysis of failures. Finally, in response to requests by students and practitioners of wind engineering, a number of theoretical developments are considered in some detail, mostly in appendices. On the other hand, it was decided in consultation with the editor that material on bridges be limited in this edition to fundamentals only.

I wish to express my sincere appreciation to the following contributors, who capably performed and checked calculations for Part II of the book: Dr. Girma Bitsuamlak, Dr. Arindam Gan Chowdhury, who also thoroughly reviewed the entire manuscript, and Dr. DongHun Yeo. I also wish to thank Professor Elena Dragomirescu, Dr. Dat Duthinh, Professor Mircea D. Grigoriu, Dr. Franklin T. Lombardo, Professor Jean-Paul Pinelli, Mr. Workamaw Warsido and Dr. Richard N. Wright, who provided helpful comments, suggestions, and criticism, and Professor Yuko Tamura, who kindly facilitated access to Tokyo Polytechnic University's extensive aerodynamics databases. Last but not least, I am indebted to Robert L. Argentieri, Executive Editor; Daniel Magers, Editorial Assistant; Doug Salvemini, Production Editor; and Holly Wittenberg, the talented designer of the book's cover, all of John Wiley and Sons; as well as Devra Kunin, copyeditor; for their capable contributions and gracious help.

The views expressed in this book do not necessarily represent those of the U.S. government or any of its agencies.

I dedicate this book gratefully and lovingly to my wife.

Emil Simiu
Rockville, MD, USA

PART I

INTRODUCTION

CHAPTER 1

OVERVIEW

The purpose of this book is to provide structural engineers with the knowledge and tools required for the proficient design of buildings for wind loads. The book is concerned with both ordinary and special structures.

Ordinary structures are typically designed by using standard provisions for wind loads. Owing in part to their development by successive and more or less disorderly accretions, the wind loading provisions of the ASCE 7 Standard have become increasingly difficult to apply. In an effort to respond to the demand for a clearer document, the ASCE 7-10 version of the Standard has been substantially expanded and revised. Nevertheless, difficulties remain. A main objective of this book is to provide clear and detailed guidance to the use of the ASCE 7-10 Standard, including information on the fact that alternative procedures specified in the Standard for buildings of the same type may yield significantly different results.

The design of special structures typically requires the use of aerodynamic data obtained in ad hoc tests conducted in wind tunnel and/or large-scale testing facilities, and of extreme wind speed data. The requisite aerodynamic and wind speed data are reflected in wind engineering consultant reports. However, such reports do not—or do not yet—have to conform to uniform standards of practice. For this reason, response estimates for the same building can differ by more than 40%, depending upon the wind engineering laboratories providing them. It is therefore in the structural engineers' interest to be able to scrutinize and evaluate consultant reports effectively. This book provides the wind engineering knowledge and tools required to do so. The book also enables structural engineers to perform, independently, detailed estimates of wind-induced response for both strength and serviceability, much as structural engineers do for seismic response. The estimates must use extreme wind speed data and aerodynamic or aeroelastic data provided in standardized formats by wind engineering consultants. The data are first applied to a preliminary structural design. Iterations of the calculations are then performed until the design is satisfactory. Such calculations, based on clear and transparent algorithms, can be performed routinely and efficiently by using public domain software referenced in the book.

Response calculations must allow for appropriate safety margins that reflect uncertainties in the parameters governing the wind-induced demand. These safety margins were provided in earlier versions of the Standard in the form of wind load factors. The book documents the limitations and shortcomings of the Load and Resistance Factor Design (LRFD) approach, in which wind load factors are used. In the ASCE 7-10 Standard, wind load factors are nominally equal to unity; however, values larger than unity are implicit in design wind speeds with mean recurrence intervals longer than those specified in the Standard's earlier versions. However, for some special structures, those mean recurrence intervals may not be adequate. This is the case if the uncertainties in the parameters affecting the demand are larger than the typical uncertainties inherent in the Standard provisions for ordinary, rigid structures. In particular, if uncertainties in the dynamic effects are significant, a calibration procedure is needed to calculate safe mean recurrence intervals of the design wind effects. Such a procedure was developed at the express request of structural engineering practitioners (see Appendix A5), and is discussed in the book's chapter on structural reliability.

Part II of this book is devoted to the ASCE 7-10 Standard, and is divided into eight chapters (Chapters 2 through 9) concerned with (1) general requirements (i.e., risk categories, basic design wind speeds, terrain exposure, enclosure classification, directional factors, topographic factors), and (2) the determination of wind effects on main wind force resisting systems and on components and cladding, by regular or simplified approaches. Part II illustrates the Standard provisions by means of a large number of calculation examples.

Part III is devoted to fundamentals. Chapter 10 is concerned with atmospheric circulations and the features of various types of storm. Chapter 11 provides descriptions of the atmospheric boundary layer, including the description of the wind velocity dependence on height above the surface, and of the turbulence within the atmospheric surface layer. Chapter 12 considers extreme wind speeds and extreme wind effects, their statistical estimation by parametric and non-parametric methods, estimation errors, wind speed simulations, and the dependence of wind effects on wind directionality. Chapter 13 provides fundamental notions of bluff body aerodynamics, and discusses modeling laws and aerodynamic measurements in the wind tunnel and large-scale aerodynamic testing facilities. Chapter 14 presents fundamentals of structural dynamics under stochastic loads for the general case of buildings with non-coincident mass and elastic centers. Chapter 15 is concerned with aeroelastic effects. Chapter 16 presents (1) a critique of conventional structural reliability approaches known as Load and Resistance Factor Design, (2) material on mean recurrence intervals calibrations as functions of parameter uncertainties, (3) an introduction to strength reserves in a wind engineering context, and (4) an innovative approach to multi-hazard design, which shows that ASCE Standard provisions on the design of structures in regions with strong earthquakes and wind storms can be unsafe. Chapter 17 is an introduction to wind-induced loss estimation.

Part IV is concerned with the determination of wind effects on rigid and flexible buildings (Chapters 18 and 19, respectively), and discusses database-assisted design (DAD) concepts and procedures. Pressure records can be used for the calculation of wind loads, or can be part of the more elaborate DAD approach, which allows combinations of wind effects to be developed conveniently and rigorously, and provides integrated loading and design calculations in one fell swoop.

Part V contains appendixes. Appendix A1 concerns fundamentals of the theory of stochastic processes. Appendix A2 presents elements of the theory of mean wind profiles in the atmospheric boundary layer. Appendix A3 presents elements of the theory of turbulence in the atmospheric boundary layer. Appendix A4 provides a description and critique of two commonly used but typically unsatisfactory approaches to the wind directionality problem. Appendix A5 provides an authoritative view by a prominent structural engineering firm on some important aspects of the state of the art in wind engineering.

PART II

GUIDE TO THE ASCE 7-10 STANDARD PROVISIONS ON WIND LOADS

CHAPTER 2

ASCE 7-10 WIND LOADING PROVISIONS

2.1 INTRODUCTION

The purpose of this Guide is to help the reader become familiar with and proficient in the use of the ASCE 7-10 Standard [2-1] provisions for wind loads. Because the provisions were largely developed by successive accretions, they have reached a level of complexity that has led some practitioners to perceive them as difficult to use. Largely for this reason, the Standard's ASCE 7-10 version of the wind load provisions has been reorganized, and comprises six chapters (ASCE Chapters 26 through 31)[1] instead of just one, as was the case for its predecessors. For a number of types of buildings or other structures, the Standard contains alternative provisions whose choice by the designer is optional. In spite of those changes, the Standard continues to exhibit problems in terms of its user-friendliness and internal consistency.

The main questions the Guide must answer are: For the building or other structure being considered, what are the steps required to determine the design wind loads? How are those steps implemented? How do results of alternative provisions compare with each other? Answers to these questions are provided in this chapter and in Chapters 3 to 9, which present material on the Standard provisions, Numerical Examples illustrating them, and comparisons

[1]Chapters, sections, figures, and tables preceded by the acronym ASCE belong to the ASCE 7-10 Standard (e.g., ASCE Sect. 26.2). Sections, figures, and tables not preceded by the acronym ASCE (e.g., Sect. 2.1) belong to this book. Users should note that in some cases there is no distinction in the Standard between tables and figures: some ASCE figures contain, or consist of, tables (e.g., Fig. 27.4.3), and vice versa. Numbers in brackets (e.g., [2-1]) refer to references listed at the end of the book.

among results obtained by alternative Standard procedures applicable to the same buildings. Table 2.3.1, List of Numerical Examples, is found in Sect. 2.3.

The Guide is not intended to be a substitute for the ASCE 7-10 Standard. For this reason, most of the Standard's figures and tables are not reproduced in the Guide. However, clear reference is made to those figures and tables for use, as needed, in conjunction with the Guide. Errata to the ASCE 7-10 Standard are posted periodically on the site www.SEInstitute.org.

Section 2.2 provides a brief overview of the Standard. Section 2.3 describes the contents of the Guide.

2.2 ASCE 7-10 STANDARD: AN OVERVIEW

This section notes the types of procedure for determining wind loads specified by the Standard (Sect. 2.2.1), lists the buildings and other structures covered by those procedures (Sect. 2.2.2), summarizes provisions on minimum design wind loads (Sect. 2.2.3), discusses the pressure sign convention, the definition of net pressures, and the representation of the pressures in the Standard (Sect. 2.2.4), and defines the Standard's regular and simplified approach for determining wind loads (Sect. 2.2.5).

2.2.1 ASCE 7-10 Standard Procedures for Determining Wind Loads

The ASCE 7-10 Standard specifies two basic types of procedure for determining wind loads: (1) procedures that use aerodynamic data listed in tables and/or plots (ASCE Chapters 27 through 30), and (2) the wind tunnel procedure (ASCE Chapter 31).

The procedures based on tabulated and/or plotted data applied to main wind force resisting systems (MWFRS, see ASCE Sect. 26.2, p. 243) are referred to in the Standard as the *directional procedure* (ASCE Chapters 27 and 29) and the *envelope procedure* (ASCE Chapter 28). These procedures, as well as those based on tabulated and/or plotted data applied to components and cladding (C&C, see ASCE Sect. 26.2, p. 243; ASCE Chapter 30),[2] are also referred to in the Standard as *analytical* (ASCE Sect. 28.2, User Note; ASCE Sect. 26.1.2.2). The term "analytical" may therefore be used to designate both

[2]Any component of an MWFRS must be designed for (1) the global demand it experiences as part of the MWFRS *and* (2) the simultaneous local demand it experiences as specified for C&C. The relative magnitude of the global and local demands on a component depends upon the structural system. As an example, wind normal to a building face induces stresses in the side walls. Each side wall, as a whole, is part of the MWFRS insofar as those stresses are due to pressures acting on more than one surface (in this case, to pressures on the windward and leeward walls). But components of the side walls also experience, simultaneously, stresses due to internal and external pressures acting locally on the side walls and on the roof. The pressures acting locally are specified in the Standard as C&C pressures.

the "directional" and the "envelope" procedures—that is, all of the Standard's procedures other than the wind tunnel procedure.[3]

The wind tunnel procedure is applied primarily for special, one-of-a-kind structures, and is not discussed in this Guide; it is discussed, however, in Chapters 13, 18, and 19. See also ASCE Sect. 31.4.3.1

2.2.2 Buildings and Other Structures Covered by the Standard

With the exception of the wind tunnel procedure, the ASCE 7 procedures are applicable only to regular-shaped buildings and to structures not subjected to across-wind loading,[4] galloping, flutter, channeling effects, or buffeting from upwind obstructions.

ASCE Chapter 26 contains definitions and notations, and covers, among others, the following topics applicable to the determination of wind loads on all buildings and other structures: risk category, basic wind speeds, enclosure classification, exposure category, and topographic factors.

ASCE Chapter 27 ("Directional Procedure") covers MWFRS of:

(a) *Enclosed and partially enclosed buildings of all heights*, including rigid and flexible buildings, with flat, gable, hip, monoslope, mansard, domed, or arched roofs.

(b) *Open buildings* with monoslope, pitched, and troughed free roofs.

(c) *Enclosed simple diaphragm buildings* with mean roof height $h \leq 160$ ft and flat, gable, hip, monoslope, and mansard roofs.[5]

ASCE Chapter 28 ("Envelope Procedure") covers MWFRS of:

(a) *Enclosed and partially enclosed low-rise buildings*, that is, buildings with mean roof height h (i) less than or equal to 60 ft *and* (ii) less than the building's least horizontal dimension, with flat, gable, or hip roofs.

[3]In fact, both the directional procedure and the envelope procedure are based on envelopes of pressures obtained in wind tunnel tests for more than two flow directions. The directional procedure uses envelopes of measured pressures, whereas the envelope procedure, developed in the late 1970s, uses envelopes of "pseudo-pressures," that is, fictitious pressures purported to induce in selected structural members the same forces or moments that would be induced by actual pressures (ASCE Sect. 26.2, item "Envelope Procedure").

[4]An exception not mentioned explicitly in the Standard is tall flexible buildings, for which ASCE Sect. 26.9.5 specifies the use of gust factors G_f, and which typically experience across-wind loads.

[5]The term "enclosed" is defined in Sect. 3.3. *Simple diaphragm buildings* are defined as enclosed buildings in which both windward and leeward wind loads are transmitted to the MWFRS by rigid or flexible diaphragms (i.e., roof, floors, or other membrane or bracing systems). Diaphragms that may be considered rigid typically consist of untopped or concrete-filled steel decks and concrete slabs with span-to-depth ratio of two or less. Diaphragms constructed of wood structural panels may be considered flexible (ASCE Sect. 26.2).

(b) *Enclosed low-rise buildings of the simple diaphragm building type* with flat, gable, or hip roofs. The tables and plots of ASCE Chapter 28 are largely based on measurements performed after the late 1970s.[6]

ASCE Chapter 29 covers MWFRS of

(a) Solid attached signs.
(b) Rooftop structures and equipment on buildings.
(c) Solid freestanding signs or solid freestanding walls.
(d) Chimneys, tanks, rooftop equipment, open signs, lattice frameworks, and trussed towers.

ASCE Chapter 30 covers C&C.

2.2.3 Minimum Design Wind Loads

Minimum design wind loads are specified as follows.

For *MWFRS of enclosed and partially enclosed buildings*, the design wind load shall not be less than 16 psf times the building's wall area, and 8 psf times the building's roof area projected onto a vertical plane normal to the assumed wind direction. Wall and roof loads shall be applied simultaneously (ASCE Sects. 27.6.1 and 28.4.4). For simple diaphragm low-rise buildings, it is also required that these loads be applied while vertical loads on the roof are assumed to be zero (ASCE Sect. 28.6.4).

For *MWFRS of open buildings*, the design wind load shall not be less than 16 psf times the area A_f of the structure (ASCE Sect. 27.4.7).

For *MWFRS of chimneys, tanks, rooftop equipment and similar structures; open signs; lattice frameworks; and trussed towers*, the design wind load shall not be less than 16 psf times the vertical projection of the area A_f of the structure (ASCE Sect. 29.8).

For C&C design wind pressures shall not be less than 16 psf acting in either direction normal to the surface (ASCE Sect. 30.2.2).

2.2.4 Pressure Sign Convention, Net Pressures, Single and Double Arrow Representation of Pressures

By aerodynamic convention, positive and negative pressures are directed, respectively, toward and away from the surface on which they act. The *net pressure* is the *vector* sum of the external and internal pressures acting on a surface; for roof overhangs, it is the vector sum of the external pressures acting on the upper and lower sides of the overhang; for parapet loads on

[6]The User Note in ASCE Sect. 28.2 states that *the provisions on low-rise buildings generally yield "the lowest wind pressure of all of the analytical methods specified in this standard."* This statement is *not* generally valid; see, for example, Sect. 6.2.

MWFRS, it is the vector sum of the external pressures acting on the outer and inner parts of the parapet. (If the overhang or parapet is permeable, internal pressures must also be accounted for.) The term "vector sum," as opposed to "algebraic sum," should be used because the directions of the pressures being added must be referred to a common oriented coordinate axis.

For example, assume that the external pressure acting on the surface of a cladding panel is 40 psf (toward the exterior surface, since in accordance with the aerodynamic convention the external pressure is positive), and the internal pressure acting on that panel is 10 psf (toward the interior surface, since the internal pressure is positive). The net pressure on the panel is then $40 + (-10) = 30$ psf (Fig. 2.2.1a). This is because, on a coordinate axis for which the positive direction is defined as the direction of the positive external pressure, the positive internal pressure is negative in a vectorial sense. Assuming that the internal pressure were -10 psf (away from the internal surface of the panel), the net pressure would be $40 - (-10) = 50$ psf (Fig. 2.2.1b). Assuming that the external pressure was -40 psf and the internal pressure were 10 psf, the net pressure on the panel would be -50 psf (Fig. 2.2.1c).

A *single arrow* representation of pressures indicates that the pressure on a surface is only positive or only negative. A *double arrow* representation of pressures indicates that the pressure on a surface can be either positive or negative (see, e.g., Fig. 8.5.1).

2.2.5 Regular Approach and Simplified Approach to Determining Wind Loads

The *regular approach* is applicable to both *main wind force resisting systems* (MWFRS) and *components and cladding* (C&C), and is represented in the flowchart of Fig. 2.2.2. Alternatively, for restricted sets of enclosed buildings, the designer has the option of determining design wind loads on MWFRS and C&C by using a *simplified approach* (Fig. 2.2.3).

2.2.5.1 Regular Approach. Tables 2.2.1, 2.2.2, and 2.2.3 list buildings and other structures covered by the regular approach for determining *pressures on MWFRS*, *forces on MWFRS*, and *pressures on C&C*, respectively. They

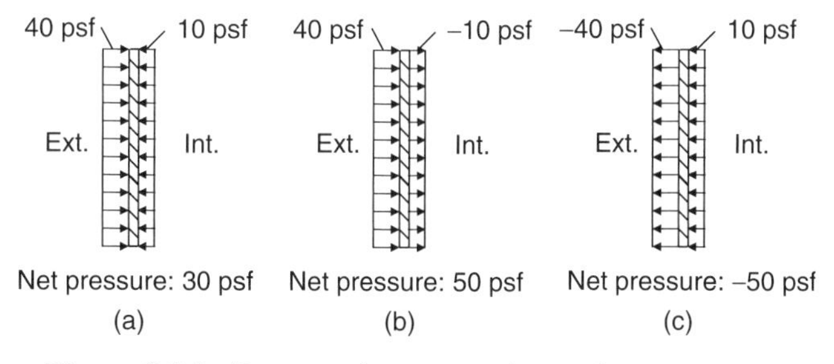

Figure 2.2.1. Pressure sign convention and net pressures.

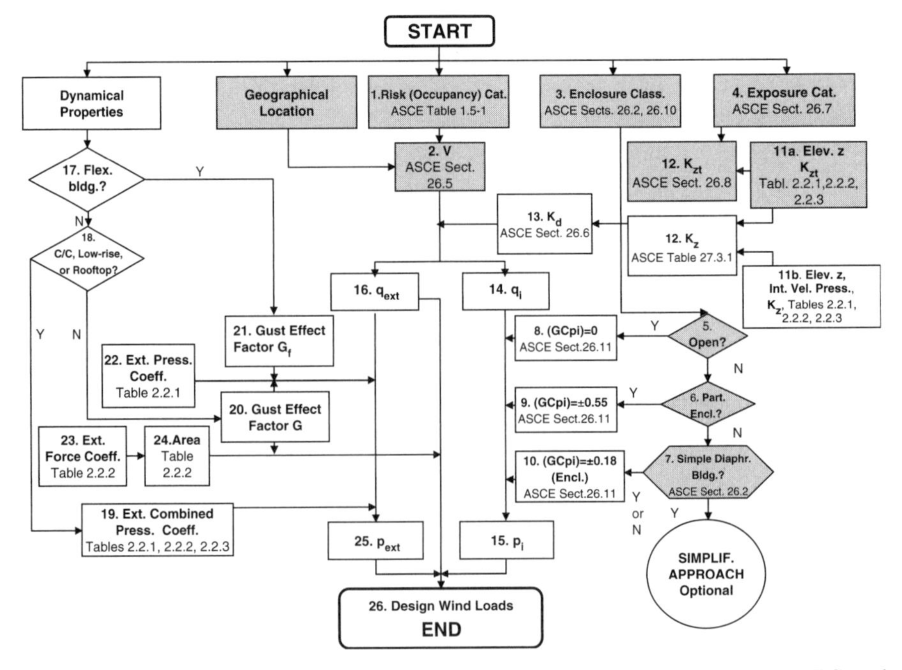

Figure 2.2.2. Flowchart for determining design wind loads or forces on MWFRS and C&C, regular approach. Shaded boxes are used in both the regular and the simplified approach. For design wind load cases, see ASCE Sects. 27.4.6 and ASCE Appendix D.

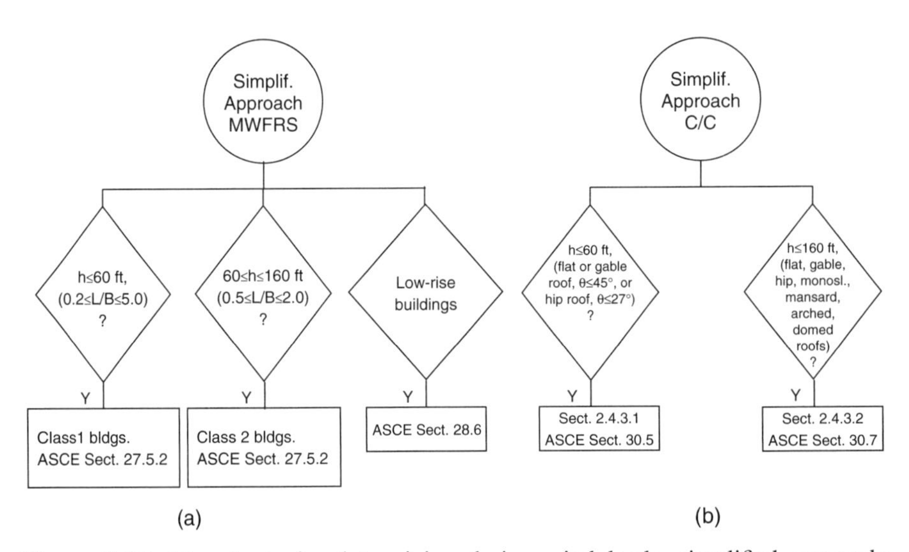

Figure 2.2.3. Flowcharts for determining design wind loads, simplified approach: (a) MWFRS (applicable only for buildings of the simple diaphragm type); (b) C&C.

TABLE 2.2.1. Summary of Requirements, Wind Pressures on MWFRS of Buildings, Regular Approach

	(1)	(2)	(3)	(4)	(5)	(6)	(7)
	External pressure p_{ext}	Internal pressure p_i	External press. coeff.	Height at which ext. velocity pressure is evaluated[a]	Height at which int. velocity pressure is evaluated	Gust effect factor G or G_f	Exposure reqt.[b, c]
A. Enclosed, part. encl., and open bldgs. with flat, gable, hip, mono-slope, or mansard roofs; domed or arched roofs (directional proced.). *See ASCE Table 27.2-1.* *Design Wind Load Cases: ASCE Sect. 27.4.6*[e]	qGC_p (rigid bldgs.) qG_fC_p (flexible bldgs.)	$\pm q_i(GC_{pi})$: ASCE Table 26.11-1	C_p: ASCE Figs. 27.4.1-1 to 27.4.1-3	Height z (windward wall); Height h (leewd./side walls; roofs); Height of top of dome	Mean roof height h^f	$G = 0.85$ or ASCE Sect. 26.9.4 G_f: ASCE Sect. 26.9.5	1 ASCE Sects. 26.7.1, 26.7.4.1
B. Enclosed and partially enclosed low-rise bldgs. (envelope procedure). *See ASCE Table 28.2-1.* *Design Wind Load Cases: ASCE Fig. 28.4-1*	$q_h(GC_{pf})$	$\pm q_i(GC_{pi})$ ASCE Table 26.11-1	(GC_{pn}): ASCE Fig. 28.4-1	Mean roof height h	Mean roof height h	—	2 ASCE Sects. 26.7.1, 26.7.4.2

(*continued overleaf*)

TABLE 2.2.1. ***(Continued)***

	(1) External pressure p_{ext}	(2) Internal pressure p_i	(3) External press. coeff.	(4) Height at which ext. velocity pressure is evaluated[a]	(5) Height at which int. velocity pressure is evaluated	(6) Gust effect factor G or G_f	(7) Exposure reqt.[b,c]
C. Open bldgs. with monoslope, pitched, or troughed free roofs (directional procedure). *See ASCE Table 27.2-1.* *Design Wind Load Cases: ASCE Sect. 27.4.6*	$q_h GC_N$	0	C_N: ASCE Figs. 27.4-4 to 27.4-7	Mean roof height h	—	$G = 0.85$ or ASCE Sect. 26.9.4	2 ASCE Sects. 26.7.1, 26.7.4.1

[a] Velocity pressures are evaluated at the heights that apply for K_z and K_{zt} (Steps 11, 12, 13, Fig. 2.2.2).
[b] Exposure Reqt. 1: Wind loads based on appropriate exposure for each wind direction considered (ASCE Sects. 26.7.1, 26.7.4).
[c] Exposure Reqt. 2: Wind loads based on exposure category resulting in highest wind loads for any direction at the site (ASCE Sects. 26.7.1, 26.7.4).
[d] Design wind load cases requirements for torsion do not apply to 1-story buildings with $h \leq 30$ ft and 1- or 2-story buildings with light frame constr. or flexible diaphragms.
[e] Design wind load cases requirements for torsion do not apply to 1-story buildings with $h \leq 30$ ft and 1- or 2-story buildings with light frame constr. or flexible diaphragms.
[f] Except for positive internal pressures in partially encl. buildings, for which internal velocity pressures may be evaluated at level of highest opening affecting those pressures.

TABLE 2.2.2. Summary of Requirements, Wind Forces on MWFRS of Building Appurtenances and Other Structures, Regular Approach (see also ASCE Table 29.1-1)

	(1) Force	(2) Force coeff.	(3) Height at which velocity pressure is evaluated	(4) Gust effect factor G or G_f	(5) Exposure Reqt. [a,b]
A. Solid free-standing walls, solid freestanding signs. *A_s = gross area of solid wall or sign.*	$q_h G C_f A_s$	C_f: ASCE Sect. 29.4-1	See ASCE Fig. 29.4-1	$G = 0.85$ or ASCE Sect. 26.9.4	1 ASCE Sects. 26.7.1, 26.7.4.3
B. Rooftop structures and equipment.	See Recommendation Sect. 7.5.1				
C. Chimneys & tanks, open signs, lattice frame-works, trussed towers. *A_f = projected area normal to wind, except if C_f is specified for actual surface area.*	$q_h G C_f A_f$	C_f: ASCE Figs. 29.5-1 to 29.5-3	Height of centroid of area A_f	$G = 0.85$ or ASCE Sect. 26.9.4	1 ASCE Sects. 26.7.1, 26.7.4.3

[a]Exposure Reqt. 1: Wind loads based on appropriate exposure for each wind direction considered (ASCE Sects. 26.7.1, 26.7.4.3).

[b]Exposure Reqt. 2: Wind loads based on exposure category resulting in highest wind loads for any direction at the site (ASCE Sects. 26.7.1, 26.7.4.3).

reference relevant ASCE provisions covered and illustrated subsequently in this Guide.

2.2.5.2 Simplified Approach. *For wind loads on MWFRS*, the simplified process is restricted to enclosed simple diaphragm buildings. *For wind loads on C&C*, one simplified method (ASCE Chapter 30, Part 2) is applied to enclosed buildings with mean roof height $h \leq 60$ ft and flat, gable, or hip roofs. A second simplified method (ASCE Chapter 30, Part 4) is applied to enclosed buildings with mean roof height $h \leq 160$ ft and flat, gable, hip, monoslope, and mansard roofs. Since if $h \leq 60$ ft, it is also the case that

TABLE 2.2.3. Summary of Requirements, C&C Design Wind Pressures, Regular Approach

	(1) External pressures p_{ext}	(2) Intern. press. p_i	(3) External combined press. coeff.	(4) Heights at which ext. velocity press. is evaluated	(5) Height at which int. velocity press. is evaluated
A. Enclosed or partially enclosed buildings, $h \le 60$ ft, including low-rise buildings (walls; flat, gable, multispan gable, hip, monosl., stepped, sawtooth, arched, domed roofs). *See ASCE Table 30.4-1.* (ASCE Chapter 30, Part 1)	$q_h(GC_p)$	$\pm q_i(GC_{pi})$	(GC_p): ASCE Figs. 30.4-1 to 30.4-6, 27.4-3 (Note 4), 30.4-7	Mean roof height h	Mean roof height h
B. Enclosed or partially enclosed buildings w/$h > 60$ ft (walls; flat, gable, hip, pitched, mansard, arched, domed roofs). *See ASCE Table 30.6.1.* (ASCE Chapter 30, Part 3)	$q(GC_p)$	$\pm q_i(GC_{pi})$	(GC_p): ASCE Fig. 30.6-1 (roofs, flat or with slopes $\le 10°$), Fig. 27.4-3, Note 4 (arched roofs), Fig. 30.4-7 (domed roofs), Fig. 30.6-1, Note 6 (other roof angles & geom.)	Height z for windward wall Mean roof height h for leeward/ side walls/roofs	Mean roof height h
C. Open buildings of all heights with free monoslope, pitched, troughed roofs. *See ASCE Table 30.8.-1.* (ASCE Chapter 30, Part 5)	$q_h GC_N$	0.0	C_N: ASCE Figs. 30.8-1–30.8-3	—	N/A
D. Rooftop structures and equipment for bldgs. $h \le 60$ ft. *See ASCE Sect. 30.9-1.* (ASCE Chapter 30, Part 6)	ASCE Sects. 30.11 and 29.6				

Notes:

1. Exposure Requirement 2 for all cases (ASCE Sects. 26.7.1, 26.7.4.4).
2. For parapets and roof overhangs, see ASCE Chapter 30, Part 6, ASCE Sects. 30.9 and 30.10

$h < 160$ ft, for enclosed buildings with mean roof height $h \leq 60$ ft and flat, gable, or hip roofs, the user has the option of using the simplified provisions of either ASCE Chapter 30, Part 2, or ASCE Chapter 30, Part 4.

2.3 ORGANIZATION OF THE GUIDE: CHAPTERS 3 TO 9

Chapter 3 contains material on steps common to the regular and the simplified approach, as applied to both the MWFRS and C&C of buildings and other structures. These steps define the *risk category* of the structure; the structure's *wind environment* (represented by the *basic wind speed*, the *exposure classification*, and the *topographic factor*); and the building *enclosure category*, which affects the extent to which wind induces pressures in the interior of the building.

Chapter 4 presents basic aerodynamic quantities used within the framework of the *regular approach* for determining wind loads on both MWFRS and C&C of any building or other structure. These quantities are: the *combined internal pressure coefficient*, the *velocity pressure exposure coefficient*, the *wind directionality factor*, the *velocity pressure*, and the *gust effect factor*. (For flexible structures, the gust effect factor has a dynamic component as well, which is also described in Chapter 4.)

Additional aerodynamic information is required for determining wind pressures or forces on MWFRS or C&C of specific types of building or structure. Such information is provided as follows, in:

Chapter 5, on the *regular approach* applied to *MWFRS of enclosed, partially enclosed, and open buildings of all heights, roof overhangs, and parapets* (ASCE Sect. 27.4).

Chapter 6, on the *regular* approach applied to *MWFRS of low-rise buildings* (ASCE Sect. 28.4).

Chapter 7, on the *regular* approach applied to *MWFRS of structures other than buildings* (ASCE Chapter 29).

Chapter 8, on the *simplified* approach applied to *MWFRS* of simple diaphragm buildings (ASCE Sects. 27.6 and 28.6).

Chapter 9, on the *regular* and *simplified* approaches applied to *C&C* (ASCE Chapter 30).

Numerical Examples presented throughout the Guide are listed in Table 2.3.1.

Comparisons between or among pressures or pressure coefficients obtained by using alternative Standard provisions applicable to the same structure are based on results of Numerical Examples (see Sects. 6.2, 8.1.2, 8.4.2, 8.5.2, and 9.3.2). Cases where more than one set of provisions are applicable to the same types of building are pointed out in subsequent chapters.

TABLE 2.3.1. List of Numerical Examples

Numerical Example	Topic	MWFRS or C&C	Approach
3.3.1 3.4.1 3.5.1	Enclosure classification Exposure category Topographic factors	MWFRS and C&C	Regular and Simplified
5.2.1, 5.2.2, 5.3.1, 5.3.2 5.2.3 5.4.1	Enclosed or partially enclosed rigid or flexible buildings of all heights, including parapets and overhangs Domed roofs Open buildings with monoslope free roof	MWFRS	Regular
6.1.1	Enclosed or partially enclosed low-rise buildings ($h \leq 60$ ft; h/least horiz. dim. ≤ 1.0).		
7.1.1 7.1.2 7.2.1 7.2.2 7.2.3 7.3.1 7.4.1 7.5.1 7.5.2	Solid freestanding walls Solid freestanding signs Open signs Trussed towers Lattice frameworks Chimneys Solid attached signs Rooftop equipment $h > 60$ ft Rooftop equipment $h \leq 60$ ft	MWFRS	Regular
8.1.1 8.2.1 8.3.1 8.4.1 8.5.1	Enclosed simple diaphragm buildings with mean roof height $h \leq 60$ ft Parapets for enclosed simple diaphragm buildings Roof overhangs for enclosed simple diaphragm buildings Enclosed simple diaphragm buildings with 60 ft $< h \leq 160$ ft Enclosed simple diaphragm low-rise buildings ($h \leq 60$ ft)	MWFRS	Simplified
9.2.1 9.2.2 9.2.3	Enclosed buildings with $h \leq 60$ ft Open buildings of all heights w/ free roofs Parapets and roof overhangs	C&C	Regular
9.3.1 9.3.2	Enclosed buildings with $h \leq 60$ ft Enclosed buildings with $h < 160$ ft	C&C	Simplified

CHAPTER 3

REGULAR AND SIMPLIFIED APPROACH: RISK CATEGORY, BASIC WIND SPEED, ENCLOSURE, EXPOSURE, TOPOGRAPHIC FACTOR

The following steps, indicated in Fig. 2.2.2, define the building's risk category, enclosure classification, and environmental conditions (basic wind speed, exposure category, topographic factor):

- Risk category for the building or other structure (Sect. 3.1)
- Basic wind speed as a function of risk category (Sect. 3.2)
- Enclosure classification (Sect. 3.3)
- Exposure category (Sect. 3.4)
- Topographic factor K_{zt} (Sect. 3.5).

These steps are required for determining design wind loads for MWFRS and C&C of all buildings and other structures, by either the regular or the simplified approach.

3.1 RISK CATEGORY (ASCE TABLE 1.5-1)

Structures are divided into four risk categories, depending upon the hazard to human life in the event of failure, and upon whether the structure is designated as an essential facility. Risk Category I includes, among others, agricultural facilities, minor storage facilities, and certain temporary facilities. Risk Category III includes, among others, structures where more than 300 people congregate in one area.[1] Risk Category IV is assigned to structures designated as essential facilities. Structures not listed in Categories I, III, and IV are classified as Category II structures.

For example, consider a typical office building. ASCE Table 1.5-1 does not include structures with this function in Risk Categories I, III, or IV. Hence, the building is classified as belonging to Risk Category II. Consider now a manufacturing facility for hazardous chemicals. In ASCE Table 1.5-1, this type of facility is assigned Risk Category III.

3.2 BASIC WIND SPEED *V* (ASCE SECT. 26.5, ASCE FIGS. 26.5.-1a, b, c)

The basic wind speed is determined, for any geographical location, as a function of risk category.

For Risk Category II, V is taken from ASCE Fig. 26.5-1A, and corresponds to a probability of being exceeded of approximately 7% in an average 50-yr period, that is, to an average 0.07/50 = 0.14% probability of being exceeded in any one year, or a 1/0.0014 $\cong$ 700-yr mean recurrence interval[2] (MRI). This basic speed V corresponds approximately to the speed with a 50-yr MRI times the square root of the wind load factor 1.6 specified in earlier versions of the Standard for strength design. The quantity V^2 is therefore approximately equal to the square of the 50-yr speed times 1.6, and is used in the ASCE 7-10 Standard for strength design in conjunction with a wind load factor equal to 1.0.

For Risk Categories III and IV, V is taken from ASCE Fig. 26.5-1B; the probability of exceedance is 3% in an average 50-yr period, that is, the MRI is approximately 1,700 years. The increase in the MRI of the basic wind speed for Risk Categories III and IV with respect to Risk Category II corresponds approximately to the multiplication of the wind pressures specified in earlier versions of the Standard by an importance factor larger than 1.0. (ASCE 7-10 no longer specifies an importance factor for either Risk Category III or I.)

For Risk Category I, V is taken from ASCE Fig. 26.5-1C; the probability of exceedance is 15% in an average 50-yr period, that is, the MRI is approximately 300 years.

As of this writing an Applied Technology Council Web site (www.atcouncil.org/windspeed.html) is planned, which will provide, free of charge, 3-s peak gust speeds for Category I, II, III, and IV buildings shown in ASCE 7-10 wind maps; 10-yr, 25-yr, 50-yr, and 100-yr wind speeds provided in the ASCE 7-10

[1]This criterion is usually interpreted as referring to places of assembly such as, e.g., auditoria. A criterion for classification as Category III buildings that is applicable to buildings whose overall occupancy exceeds 5,000 people is specified in Table 1604.5 of the 2003 International Building Code, which is otherwise mostly based on ASCE 7-02. We are indebted to William F. Baker of Skidmore Owings & Merrill for drawing our attention to this criterion.

[2]For a definition of and explanation of mean recurrence intervals, see Sect. 12.2.

Commentary for serviceability design; and ASCE 7-05 3-s basic wind speeds, as well as fastest mile wind speeds.

State of Florida authorities are considering the specification of wind speeds for High Velocity Hurricane Zones (HVHZ),[3] such that the pressures corresponding to those speeds are at least as large as the pressures specified in the ASCE 7-05 Standard.

In certain cases, it is necessary to determine basic design wind speeds from regional climatic data (ASCE Sects. 26.5.2 and 26.5.3). For details, see Chapter 12 and Sect. 16.7.

3.3 ENCLOSURE CLASSIFICATION (ASCE SECTS. 26.2 AND 26.10)

The enclosure classification of a building controls the internal pressures specified for design. Internal pressures develop (1) if air blown into a space cannot freely leave that space, in which case the internal pressures are positive, or (2) if air sucked away from a space cannot be freely replaced, in which case the internal pressures are negative (see Fig. 13.2.9). The internal pressures—and the enclosure classification—thus depend on the way and the degree to which the building is enclosed, that is, on the size and distribution of openings in the building envelope.

For purposes of enclosure classification, glazing (i.e., glass or translucent plastic sheet in windows, doors, skylights, or curtain walls) is *not* defined as an opening. In *wind-borne debris regions*, defined in ASCE Sect. 26.10.3.1, glazing must be protected as specified in ASCE Sect. 26.10.3.2, except that no protection is required for (1) Risk Category I buildings, and (2) glazing located over 60 ft above the ground and over 30 ft above aggregate surface roofs, including roofs with gravel or stone ballast, located within 1,500 ft of the building.

A building or other structure is *open* if each of its walls is at least 80% open.

A building is *partially enclosed* if (1) in at least one wall that experiences positive external pressure, the total area of openings exceeds 1.1 times the sum of the areas of openings in the balance of the building envelope (walls and roof), *and* (2a) the total area of openings in that wall exceeds 4 sq ft or 1% of the wall's gross area, whichever is smaller, *and* (2b) the percentage of openings in the balance of the building envelope does not exceed 20%. A structure that complies with the definitions of open and partially enclosed buildings is classified as *open* (ASCE 26.10.4).

A building is *enclosed* if it is not open or partially enclosed.

Numerical Example 3.3.1. *Enclosure classification.* Consider a building with rectangular shape in plan (60 ft $\times$ 30 ft), flat roof, and eave height h = 20 ft. The exterior walls have gross areas $60 \times 20 = 1{,}200$ sq ft and $30 \times 20 =$ 600 sq ft. Assume that the area of the openings in each of the 1,200-sq-ft

[3]HVHZ consist of Florida's Miami Dade and Broward counties.

walls is 240 sq ft, the 600-sq-ft walls have no openings, and the roof has two 1-sq-ft openings.

If one of the 1,200-sq-ft walls is considered as the windward wall receiving positive pressure, the sum of the areas of openings in the balance of the building envelope is 242 sq ft. Condition 2a for the classification of the building as partially enclosed is satisfied; that is, the total area of openings in the 1,200-sq-ft wall (i.e., 240 sq ft) exceeds the smaller of the areas 4 sq ft and $0.01 \times 1{,}200 = 12$ sq ft. Condition 2b is also satisfied; that is, the percentage of openings in the balance of the building envelope is $242/(1{,}200 + 2 \times 600 + 60 \times 30) < 20\%$. However, condition 1 is *not* satisfied; that is, 240 sq ft $< 1.1(240 + 2 \times 1) = 266.2$ sq ft, so the building is not classified as partially enclosed. The building is not classified as open; that is, it does not satisfy the condition that, for each wall, the ratio of the area of the wall's openings to the wall's gross area is at least 80%. Since the building is neither open nor partially enclosed, it is classified as enclosed.

3.4 EXPOSURE CATEGORY (ASCE SECT. 26.7)

The exposure category is based on the roughness characteristics of the terrain or water surface upwind of the structure. The exposure category governs the changes in wind speeds from the standard conditions for which they are defined in the Standard (i.e., 3-s peak gust speed at 10 m above ground in terrain with open exposure) to the conditions prevailing at the structure's site at any specified elevation.

The Standard distinguishes between *surface roughness* and *exposure*. For the atmospheric flow to acquire properties associated with a specified exposure (these properties include the wind speed variation with height above ground and the characteristics of the flow turbulence), the surface roughness that prevails upwind of the structure must extend over a sufficiently long fetch.

Section 3.4.1 presents information on *surface roughness categories*. Section 3.4.2 presents information on *exposure categories*. Sect. 3.4.3 discusses *exposure requirements* specified in the Standard for situations in which the exposure depends upon direction.

3.4.1 Surface Roughness Categories (ASCE Sect. 26.7.2)

Surface roughness categories are based on measurements by meteorologists and assessments by wind engineers. They are defined as follows:

Surface Roughness B: Urban and suburban areas, wooded areas, or other terrain with numerous closely spaced obstructions having the size of single-family dwellings or larger.

Surface Roughness C: Open terrain with scattered obstructions generally less than 30 ft high, flat open country, and grasslands.

Surface Roughness D: Flat, unobstructed areas and water surfaces, smooth mud flats, salt flats, and unbroken ice.

3.4.2 Exposure Categories (ASCE Sect. 26.7.3)

The Standard specifies the following exposure categories.

Exposure B: For buildings with mean roof height $h \leq 30$ ft, Exposure B applies where Surface Roughness B prevails in the upwind direction for at least 1,500 ft. For buildings with $h > 30$ ft, Exposure B applies where that distance is greater than (a) 2,600 ft and (b) 20 times the building height (ASCE Commentary Fig. C26.7-1).

Exposure D: Exposure D applies where Surface Roughness D prevails in the upwind direction for a distance greater than (a) 5,000 ft and (b) 20 times the building height. Exposure D also applies wherever the site is within less than 600 ft or 20 times the building height from an Exposure D condition, whichever is greater (ASCE Commentary Fig. C26.7-2).

Exposure C: Exposure C applies where Exposures B and D do not apply, and is commonly referred to as open terrain exposure.

For a site located in the transition zone between exposure categories, the category resulting in the largest wind loads must be used, except where an intermediate exposure can be determined by a rational method (see ASCE Commentary Sect. C26.7).

In early versions of the Standard, Exposure A was defined for centers of large cities. However, the variety of roughness conditions in the center of a large city is such that Exposure A was deemed not to be useful for design purposes and was not included in the ASCE 7-10 Standard and recent versions thereof.

Aerial photographs provided to help designers identify various types of exposure are included on pp. 546, 546a, and 546b of the Standard's Commentary.

3.4.3 Exposure Requirements (ASCE Sects. 26.7.1 and 26.7.4)

For each direction in which wind loads are determined, the exposure category is determined for two upwind sectors extending 45° on either side of that direction. The exposure category for the 45° sector that results in the largest wind loads (see ASCE Commentary Fig. C26.7-5) is then used for that direction.

The Standard specifies two exposure requirements, that is, two approaches to the application of the method just described. In the first approach, referred to as *Exposure Requirement 1*, the wind loads are determined for each direction by using the method described. This approach is required for the design of the MWFRS of: (1) all enclosed and partially enclosed buildings designed by using the regular or simplified approach of ASCE Chapter 27, including all diaphragm type buildings with $h \leq 160$ ft, parapets, and roof overhangs, and (2) structures and building appurtenances designed by using the provisions of ASCE Chapter 29.

The second approach, henceforth called *Exposure Requirement 2*, requires that the design be based on the exposure resulting in the highest wind loads for any direction, and is applied for the design of: (1) all C&C (ASCE Chapter 30), (2) the MWFRS of enclosed and partially enclosed low-rise buildings (ASCE Chapter 28), and (3) the MWFRS of open buildings with free monoslope, pitched, and troughed roofs (ASCE Chapter 27.4.3).

Numerical Example 3.4.1. *Wind directionality and exposure requirements.* A building has the directional exposures shown in Fig. 3.4.1. For Exposure Requirement 1, the wind loads are to be determined for eight wind directions at 45° intervals (see ASCE Fig. C26.7-5). For each of the eight directions, the upwind exposure is to be determined for each of two 45° sectors, one on each side of the wind direction being considered. The sector with the exposure resulting in the highest loads is to be used to define wind loads for that direction. (Note that, in a given region, wind speeds at any given elevation are greater for Exposure D than for Exposure C, and greater for Exposure C than for Exposure B.)

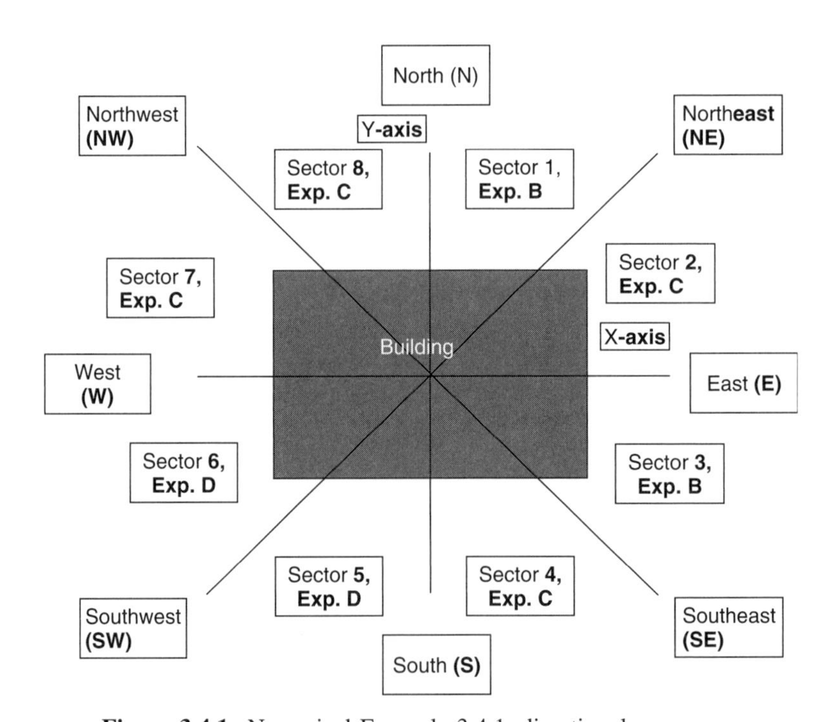

Figure 3.4.1. Numerical Example 3.4.1, directional exposures.

Exposure Requirement 1. For the given building, the exposures that define the highest wind loads for each of the eight directions are shown here:

Wind from North: Exposure C (based on Sectors 8 and 1)
Wind from Northeast: Exposure C (based on Sectors 1 and 2)
Wind from East: Exposure C (based on Sectors 2 and 3)
Wind from Southeast: Exposure C (based on Sectors 3 and 4)
Wind from South: Exposure D (based on Sectors 4 and 5)
Wind from Southwest: Exposure D (based on Sectors 5 and 6)
Wind from West: Exposure D (based on Sectors 6 and 7)
Wind from Northwest: Exposure C (based on Sectors 7 and 8)

Exposure Requirement 2. The design is based on the exposure resulting in the highest wind loads for any direction. For this example, Exposure D is used for determining wind loads for all directions.

3.5 TOPOGRAPHIC FACTOR K_{zt} (ASCE SECT. 26.8, ASCE FIG. 26.8-1)

Unless the surrounding terrain is flat for a sufficiently long distance upwind, the basic wind speeds are modified not only by the upwind terrain roughness, but also by the structure's surrounding topography. Over the surface of a rising slope, speeds are larger, for any given height above ground, than their counterparts over the horizontal terrain upwind of the slope. The increase in the wind speeds due to the topography is called *speed-up*, and is reflected in the exposure-dependent factor K_{zt}.

The heights above ground at which the factor K_{zt} is evaluated depend upon type of structure and portion thereof, and upon whether K_{zt} is used in the evaluation of internal or external pressures. Those heights are the same as for the corresponding velocity pressure exposure coefficients K_z and are listed for the regular approach in: Tables 2.2.1 and 2.2.1 for MWFRS; Table 2.2.1 for C&C; and in Chapters 5, 6, 8, and 9 for parapets and roof overhangs. For the simplified approach, K_{zt} is evaluated at each height z, although for enclosed simple diaphragm buildings covered by ASCE Sect. 27.5, it may be evaluated instead for height $h/3$, while for low-rise simple diaphragm buildings it is determined at height h.

The Standard provides speed-up models applicable to two-dimensional (2-D) ridges, 3-D isolated hills, and 2-D escarpments, provided that all the following conditions are satisfied (ASCE Sect. 26.8.1; see Fig. 3.5.1 for notations):

1. No topographic features of comparable height exist for a horizontal distance of 100 times the height of the hill H or 2 miles, whichever is less, from the point at which the height H is determined.

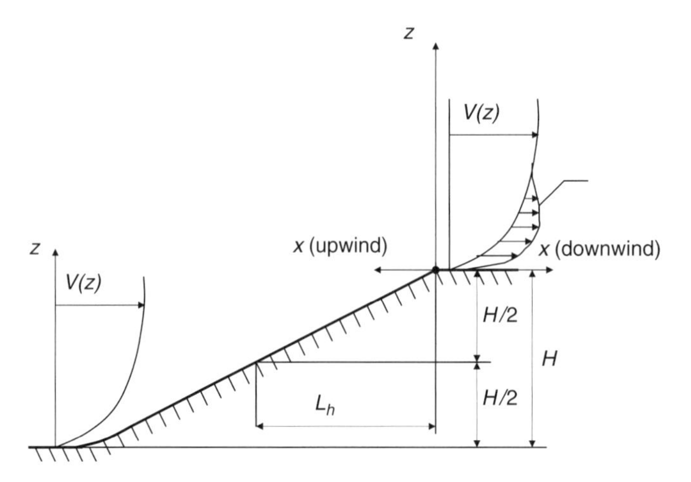

Figure 3.5.1. Notations.

2. The topographic feature protrudes above the height of upwind terrain features within a 2-mile radius by a factor of two or more in any quadrant.
3. The structure is located in the upper half of a hill or ridge or near the crest of an escarpment.
4. $H/L_h \geq 0.2$.
5. The height of the hill H exceeds 15 ft for Exposures C and D, and 60 ft for Exposure B.

If any of conditions 1–5 is not satisfied, $K_{zt} = 1$.

The Standard topographic factor is defined as $K_{zt} = [V(z,x)/V(z)]^2$, where $V(z) =$ 3-s peak gust speed at height z above ground in horizontal terrain with no topographic feature. The expression for K_{zt} is:

$$K_{zt} = (1 + K_1\ K_2\ K_3)^2, \tag{3.5.1}$$

where the factor K_1 accounts for the shape of the topographic feature, K_2 accounts for the variation of the speed-up as a function of distance from the crest, and K_3 accounts for the variation of the speed-up as a function of height above the surface of the topographic feature. Values of and expressions for K_1, K_2, K_3 are given in ASCE Fig. 26.8-1. For example, for $H/L_h \leq 0.5$,

$$K_1 = aH/L_h; \quad K_2 = 1 - \frac{|x|}{\mu L_h}; \quad K_3 = \exp(-\gamma z/L_h), \qquad \text{(3.5.2a, b, c)}$$

where for 2-D escarpments, $\gamma = 2.5$; $\mu = 1.5$ (upwind of crest), $\mu = 4.0$ (downwind of crest); $a = 0.75$ (Exposure B), $a = 0.85$ (Exposure C), $a = 0.95$ (Exposure D).

Numerical Example 3.5.1. *Topographic factor for a 2-dimensional escarpment.* A 2-dimensional escarpment has Exposure B and dimensions $H = 100$ ft, $Lh = 400$ ft. The topography upwind of the escarpment is assumed to satisfy conditions 1 and 2. The building is located at the top of the escarpment, and the downwind distance (see Fig. 3.5.1) between the crest and the building's windward face is x = 40 ft. (In Fig. 3.5.1, the building would be located to the right of the crest.) We seek the quantity K_{zt} for elevation $z = 25$ ft above ground at $x = 40$ ft.

Condition 4 is satisfied, since $H/L_h = 100/400 = 0.25 > 0.2$, as is condition 5, since $H = 100$ ft > 60 ft.

Since $H/L_h < 0.5$, Eqs. 3.5.2 a, b, c yield:

$$K_1 = 0.75 \times 100/400 = 0.1875$$

$$K_2 = 1 - 40/(4.0 \times 400) = 0.975$$

$$K_3 = \exp(-2.5 \times 25/400) = 0.855$$

From Eq. 3.5.1, the topographic factor is

$$K_{zt} = (1 + 0.1875 \times 0.975 \times 0.855)^2 = 1.16^2 = 1.35$$

This result implies that at $x = 40$ ft downwind of the crest and $z = 25$ ft above ground, the increased peak 3-s gust is 1.16 times larger than the peak 3-s gust at 25 ft above ground upwind of the escarpment, and the corresponding pressures are $(1.16)^2 = 1.35$ times larger than upwind of the escarpment.

CHAPTER 4

REGULAR APPROACH: STEPS COMMON TO ALL BUILDINGS/OTHER STRUCTURES (MWFRS AND C&C)

4.1 INTRODUCTION

The following steps applicable to both the regular and the simplified approach were described in Chapter 3: determining the risk category, basic wind speed, enclosure classification, exposure category, and topographic factor (see shaded boxes of Fig. 2.2.2). Additional steps applicable to the regular approach represented in Fig. 2.2.2 consist of determining the following aerodynamic quantities:

- Combined internal pressure coefficient (Sect. 4.2.1)
- Velocity pressure exposure coefficient (Sect. 4.2.2)
- Wind directionality factor (Sect. 4.2.3)
- Velocity pressure (Sect. 4.2.4)
- Gust effect factor (Sect. 4.2.5)[1]

These steps are common for MWFRS of all buildings and other structures for which wind loads are determined by the regular approach.

Following these steps, the regular approach is completed by determining the appropriate external pressure or force coefficients and calculating the net pressures or forces for:

- MWFRS of various types of buildings and other structures (see Chapter 5 for buildings of all heights, including parapets and roof overhangs;

[1]For flexible structures, the gust factors account not only for aerodynamic effects but for dynamic effects as well; see Sect. 4.2.5.2.

Chapter 6 for low-rise buildings, including parapets and roof overhangs; and Chapter 7 for other structures)
- C&C, see Chapter 9

4.2 REGULAR APPROACH: STEPS COMMON TO ALL BUILDINGS AND OTHER STRUCTURES (MWFRS AND C&C)

This section covers the following steps of Fig. 2.2.2: combined internal pressure coefficient (Sect. 4.2.1), velocity pressure exposure coefficient (Sect. 4.2.2), wind directionality factor (Sect. 4.2.3), velocity pressure (Sect. 4.2.4), and gust effect factor (Sect. 4.2.5).

4.2.1 Combined Internal Pressure Coefficients (ASCE Sect. 26.11 and ASCE Table 26.11-1)

As shown in Fig. 2.2.2, if the building's enclosure classification is open, the combined internal pressure coefficient[2] $(GC_{pi}) = 0$, meaning that there are no internal pressures.

If the enclosure is not open, it must be ascertained whether it is partially enclosed, in which case $(GC_{pi}) = \pm 0.55$, or enclosed, in which case $(GC_{pi}) = \pm 0.18$.

If the building is enclosed and, in addition, it is a *simple diaphragm building* (defined in ASCE Sect. 26.2; see also Sect. 2.2.2, footnote 5), the designer has the option of using the simplified approach for determining wind loads on MWFRS (see Fig. 2.2.3), instead of the regular approach represented in Fig. 2.2.2.

4.2.2 Velocity Pressure Exposure Coefficient K_z (ASCE Table 27.3.1)

The values of the exposure-dependent coefficient K_z at elevation z can be taken for all buildings and other structures from ASCE Table 27.3.1[3] or can be calculated as follows:

$$K_z = 2.01\ (z/z_g)^{2/\alpha} \quad \text{for} \quad z_1 \le z \le z_g \qquad (4.2.1\text{a})$$

$$K_z = 2.01\ (z_1/z_g)^{2/\alpha} \quad \text{for} \quad z < z_1 \qquad (4.2.1\text{b})$$

For *Exposure B*, $z_g = 1{,}200$ ft, $\alpha = 7.0$
For *Exposure C*, $z_g = 900$ ft, $\alpha = 9.5$
For *Exposure D*, $z_g = 700$ ft, $\alpha = 11.5$

[2]As used in the Standard, the term "combined pressure coefficient" means that the two factors between the parentheses, one of which is a gust factor while the other is a pressure coefficient, are specified as a product, rather than individually, that is, they cannot be separated.
[3]To facilitate the user's task, the Standard also reproduces ASCE Table 27.3-1 in ASCE Chapters 28, 29, and 30 as ASCE Tables 28.3-1, 29.3-1, and 30.3.1.

(ASCE Table 27.3-1). In Eqs. 4.2.1a and b, $z_1 = 15$ ft, *except* that, for Exposure B, $z_1 = 30$ ft for C&C of all buildings and for MWFRS of low-rise buildings designed by using ASCE Fig. 28.4-1.

Heights above the ground or water surface used for the evaluation of K_z depend upon type of structure and portion thereof, and upon whether K_z is used in the evaluation of internal or external pressures. Heights are listed for the regular approach in Tables 2.2.1 and 2.2.2 for MWFRS, and in Table 2.2.3 for C&C.

4.2.3 Wind Directionality Factor K_d (ASCE Sect. 26.6, ASCE Table 26.6-1)

The absence in the Standard of directional information on wind speeds and wind pressures typically results in a conservative envelope of the pressures or forces. To make up for this conservatism, the Standard specifies wind directionality factors K_d, which reduce the pressures calculated in the absence of directional information.

The following values are specified in the Standard:

$K_d = 0.85$ for all buildings and structures, except as follows:
$K_d = 0.90$ for square chimneys, tanks, and similar structures.
$K_d = 0.95$ for hexagonal and round chimneys, tanks, and similar structures, and for trussed towers with other than triangular, square, and rectangular cross sections.

The wind directionality factor K_d shall be applied only when used in conjunction with the load combinations of ASCE Sect. 2.3 (see ASCE Sect. 26.6).

4.2.4 Velocity Pressures

The expression for the velocity pressure at height z above ground is

$$q_z = 0.00256\, K_z\, K_{tz}\, K_d\, V^2 \text{ (psf; } V \text{ in mph)} \quad (4.2.2a)$$

$$\text{[In SI units: } q_z = 0.613\, K_z\, K_{tz}\, K_d\, V^2 \text{ (N/m}^2\text{; } V \text{ in m/s)]}^4 \quad (4.2.2b)$$

(ASCE Sects. 27.3.2, 28.3.2, 29.3.2, 30.3.2), where K_z and K_{tz} are evaluated at the heights specified in Tables 2.2.1, 2.2.2, and 2.2.3, and V is specified as a function of risk category (Sect. 3.2.1). The corresponding values of the velocity pressures are evaluated for internal pressures and, separately, for external pressures. No reductions in velocity pressure are permitted on account of shielding by other structures or by terrain features (see, e.g., ASCE Sect. 27.1.4). The Standard uses several notations for the velocity pressure

[4]For some geographical regions, a different value for the coefficient 0.00256 (0.613 in SI units) may be selected if sufficient climatic data are available to justify that value.

q_z (e.g., q, q_h, q_i, q_z), depending upon the height at which it is evaluated and on whether it pertains to external or internal pressures. For brevity, in Fig. 2.2.2 and Tables 2.2.1, 2.2.2, and 2.2.3, the notations used are q_i for internal pressures and q_h or q for external pressures.

4.2.5 Gust Effect Factor (ASCE Sect. 26.9)

The gust effect factor affects external pressures or forces. It is denoted by G for rigid buildings, and by G_f for flexible buildings. Whether a building is rigid or flexible is determined in accordance with ASCE Sects. 26.2 and 26.9.

The gust effect factor G or G_f is used only where indicated in Tables 2.2.1 and 2.2.2. For all other cases, the Standard specifies combined pressure coefficients denoted by (GC), where C denotes generically the pressure coefficient. (Subscripts are added to the factor C that depend on the application.) It is the *products* (GC), rather than individual G and C values, that are specified; the two factors G and C may not be separated.

4.2.5.1 Rigid Structures (ASCE Sects. 26.9.1 and 26.9.3). At any one point in time, the difference between the fluctuating velocities at two distinct points in space increases as the distance between those points increases. This spatial variation is referred to as *imperfect spatial coherence.* (Should the fluctuations at each instant be exactly the same over an entire area, they would be called perfectly coherent spatially.) Consider the peak external aerodynamic pressure $p_{pk}(P)$ measured at a point P on the surface of the structure, and an area A of that surface surrounding point P. If A is very small, it can be assumed approximately that the peak aerodynamic force acting on A is equal to $p_{pk}(P)A$. However, over an area $A_1 > A$, at the instant at which the pressure $p_{pk}(P)$ occurs, the pressures at points of the area A_1 other than point P are smaller than $p_{pk}(P)$, so the peak aerodynamic force acting on the area is *less* than $p_{pk}(P)A_1$. Therefore, the *average* pressure on area A_1 is less than $p_{pk}(P)$. This fact explains the decrease of the average design pressures specified in the Standard as the tributary area increases. For rigid structures, this decrease is accounted for by area-dependent *gust factors* $G < 1$ that multiply the peak pressures and depend upon the tributary area A_1.

The Standard permits the calculation of gust effect factors for rigid structures by an equation that assumes that pressure fluctuations are proportional to the along-wind velocity fluctuations of the oncoming turbulent wind. This assumption is approximately valid in the particular case of pressures on a building face induced by wind normal to that face. However, it is also used for convenience, even though it is physically unwarranted, where the imperfect spatial coherence of the pressures (e.g., on roofs or side walls) is inherent in the aerodynamics of the body, rather than being related to the oncoming along-wind velocity fluctuations.

In view of the fact that a high degree of precision in calculating G is typically unwarranted, the Standard permits the use of the simple assumption $G = 0.85$.

4.2.5.2 Flexible Structures (ASCE Sects. 26.9.4 and 26.9.5). Dynamic effects experienced by flexible structures, defined in ASCE Sects. 26.2 and 26.9, are associated with *resonance*, a phenomenon wherein the structural response to a periodic or nearly periodic force is amplified if the period of the forcing and the natural period of vibration of the structure are nearly the same. Resonance can occur, for example, if soldiers march in step over a bridge whose natural period of vibration is nearly the same as the time between successive steps.

The dynamic response is affected by the structure's damping ratio, fundamental natural period of vibration, and fundamental modal shape, which the Standard assumes to vary linearly with height. The gust effect factor, denoted for flexible structures by G_f, accounts for both the dynamic response and the imperfect spatial coherence. As specified in the Standard, G_f is a measure of the total along-wind response induced by wind normal to a face of the structure.

Wind induces response not only in the mean wind velocity direction (i.e., *along-wind response*), but also in the direction *normal* to the mean wind velocity (i.e., the *across-wind response*). For tall structures, the across-wind response is due primarily to vorticity effects induced by wind in the structure's wake, and can be considerably larger than the response in the mean velocity direction, especially for long mean recurrence intervals of the wind speeds. Note, however, that the Standard provisions specifically exclude vorticity effects from consideration, except when the wind tunnel procedure is used. For this reason, the gust effect factor is used only for preliminary design purposes, except for structures with width-to-depth ratios for which the response at high speeds is typically larger in along-wind than in across-wind directions. Nevertheless, the Commentary contains material on across-wind response that may be used for rough preliminary, qualitative estimates.

CHAPTER 5

REGULAR APPROACH: BUILDINGS, PARAPETS, OVERHANGS (“DIRECTIONAL” PROCEDURE), MWFRS

5.1 INTRODUCTION

For the regular approach applied to MWFRS of buildings of all heights, the pressures and the design cases are based on ASCE Sect. 27.4.

Section 5.2 presents material on *enclosed or partially enclosed buildings of all heights*, including: Numerical Examples 5.2.1a, b (*rigid and flexible enclosed buildings with mean roof height* $h = 95\,ft$), Numerical Example 5.2.2 (*enclosed building with mean roof height* $h = 20\,ft$), and Numerical Example 5.2.3 (*building with domed roof*). The building considered in Numerical Example 5.2.2 is also considered, as a low-rise building, in Chapter 6 (Numerical Example 6.1.1) and in Chapter 8 (Numerical Example 8.5.1).

Section 5.3 presents material on *roof overhangs and parapets for buildings of all heights*, including Numerical Examples 5.3.1 and 5.3.2.

Section 5.4 presents material on *open buildings with free monoslope, pitched, or troughed free roofs*, including Numerical Example 5.4.1.

5.2 REGULAR APPROACH: ENCLOSED OR PARTIALLY ENCLOSED BUILDINGS OF ALL HEIGHTS, MWFRS

The material in this section covers enclosed or partially enclosed buildings of all heights (Numerical Examples 5.2.1 and 5.2.2), including buildings with domed roofs (Numerical Example 5.2.3).

In accordance with ASCE Sect. 27.4.1, design wind pressures are determined by the equation

$$p = qGC_p - q_i(GC_{pi}) \tag{5.2.1}$$

(ASCE Eq. 27.4.1); see Table 2.2.1 for the definition of q, q_i, G, C_p, and (GC_{pi}).

Numerical Example 5.2.1. *95 ft high (a) rigid and (b) flexible building.* The building being considered is an enclosed office building with rectangular shape in plan (60 ft × 125 ft), eave height 95 ft, and flat roof (Fig. 5.2.1), located at the southern tip of the Florida peninsula in flat terrain with Exposure B from all directions.[1] The calculations that follow demonstrate the sequence of steps required to determine the design wind pressures.

Design wind pressures. The following steps defined in Fig. 2.2.2 (an expanded version of the steps listed in ASCE Table 27.2-1) are required to determine the design wind pressures:

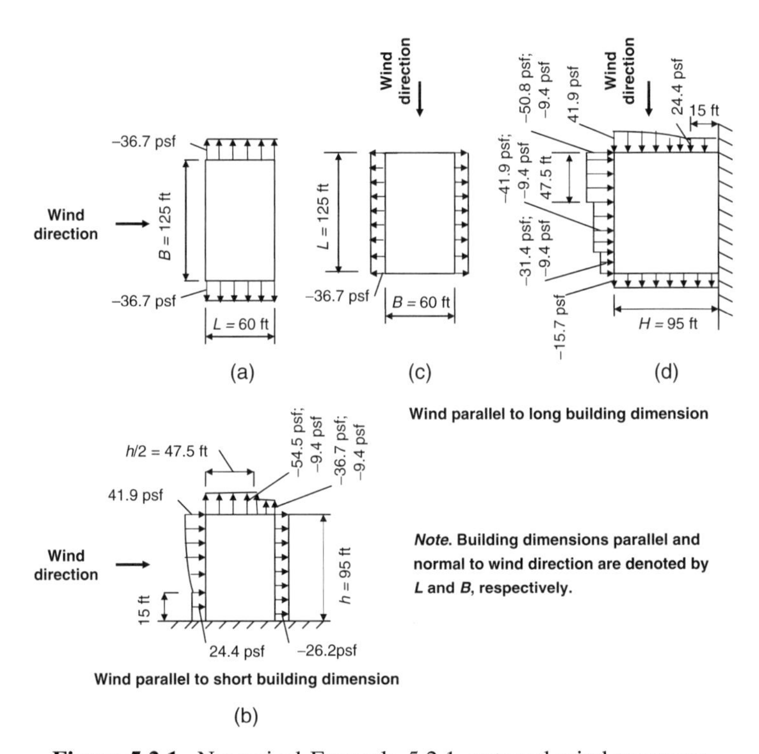

Figure 5.2.1. Numerical Example 5.2.1, external wind pressures.

[1]For Florida's High Velocity Hurricane Zones (HVHZ) (see Sect. 3.2), Sect. 1620.3 of the 2007 Florida Building Code (FBC 2007) requires that: "All buildings and structures shall be considered to be in Exposure Category C as defined in Section 6.5.6.3 of ASCE 7-05." This restriction is not specified in the ASCE 7 Standard, and in this example Exposure B was used.

Step 1. ***Risk category*** (ASCE Table 1.5-1). In accordance with ASCE Table 1.5-1, office buildings are assigned *Risk Category II*; see example in Sect. 3.1.

Step 2. ***Basic wind speed*** (ASCE Sect. 26.5). For Risk Category II, use Fig. 26.5-1a. For the southern tip of Florida, the basic wind speed is $V = 170$ *mph*.

Step 3. [2]

- ***Wind directionality factor*** K_d (ASCE Table 26.6-1). For buildings, $K_d = 0.85$.
- ***Enclosure classification*** (ASCE Sect. 26.10; see Chapter 3, Numerical Example 3.3.1). For the building considered here, it is assumed that the building is *enclosed*. (Given the location of the building, this requires that the glazing be protected or that the conditions listed in Sect. 3.3 for wind-borne debris regions be satisfied.)
- ***Exposure category*** (ASCE Sect. 26.7). In this example, it is assumed that the site has *Exposure B in all directions.* (For an example of the application of the Standard provisions if the Exposure Category depends on direction, see Chapter 3, Numerical Example 3.4.1.) As indicated in Table 2.2.1, for the type of building considered in this example, Exposure Requirement 1 applies; that is, the wind loads are based on the appropriate exposure for each wind direction (ASCE Sect. 26.7.4.1).
- ***Is the building an enclosed simple diaphragm building?*** (ASCE Sect. 26.2). We assume that the building has simple diaphragms as defined in ASCE Sect. 26.2. However, since $h > 60$ ft, the building is not a Class 1 building (ASCE Sect. 27.6.1). Since the ratio of along-wind to across-wind dimension is $60/125 < 0.5$, the building is not a Class 2 building (ASCE 27.5.2). Hence, the building is *not a simple diaphragm building* as defined in the Standard. (As is indicated in Fig. 2.2.2, had the building been a simple diaphragm building, the designer would have had the option of determining the loads either by the regular approach or by the simplified approach.)
- ***Is the building a low-rise building?*** (ASCE Sects. 26.2 and ASCE Chapter 28). Since $h > 60$ ft, *the building is not a low-rise building.*
- ***Combined internal pressure coefficient*** (GC_{pi}) (ASCE Sect. 26.11). Since the building is neither open nor partially enclosed, the flowchart (Fig. 2.2.2) leads to $(GC_{pi}) = \pm 0.18$.
- ***Elevations at which the velocity pressures are evaluated*** (ASCE 27.4.1; Table 2.2.1, item A, cols. 4 and 5). *For external pressures*

[2]Step 3 corresponds, in an expanded form, to the nomenclature of ASCE Table 27.2-1.

on windward wall, the elevations z at which velocity pressures are calculated are chosen as follows. (Different intervals may be chosen as deemed reasonable by the designer.)

$$z(\text{ft}) \leq 15 \quad 35 \quad 55 \quad 75 \quad 95$$

For internal pressures on windward wall and for external and internal pressures on all other walls and on roof, $z = 95$ ft (Table 2.2.1).

- ***Gust Effect Factor*** (ASCE Sect. 26.9)

  Is the building rigid or flexible? (ASCE Sects. 26.2 and 26.9).

  (a) ***Rigid building, G*** (ASCE Sect. 26.9.4; Table 2.2.1, item A, col. 6). Assume that the building's main wind force resisting system consists of sufficiently stiff concrete shear walls (i.e., shear walls such that ASCE Eq. 26.9-5 yields $n_a \geq 1$ Hz). In accordance with ASCE Sect. 26.2 ("building and other structure: rigid"), the building is then classified as *rigid*.

  In this example, the value $G = 0.85$ will be employed, as permitted for rigid buildings by ASCE Sect. 26.9.4. It is of interest to also obtain G by using Eqs. 26.9-6 to 26.9-9. The results are as follows.

  Wind parallel to short building dimension (i.e., $B =$ 125 ft)

$$\overline{z} = 0.6h = 57 \text{ ft}; \quad c = 0.3 \text{ (Table 26.9-1)};$$

$$\ell = 320 \text{ ft}; \; I_{\overline{z}} = 0.3(33/\overline{z})^{1/6} = 0.274; \; \overline{\varepsilon} = 1/3;$$

$$L_{\overline{z}} = \ell \left(\frac{\overline{z}}{33}\right)^{\overline{\varepsilon}} = 383.9 \text{ ft}$$

$$Q = \sqrt{\frac{1}{1 + 0.63\left(\dfrac{B+h}{L_{\overline{z}}}\right)^{0.63}}}$$

$$= 1/\{1 + 0.63[(125 + 95)/383.9]^{0.63}\}^{1/2} = 0.83$$

$$G = 0.925\left(\frac{1 + 1.7 \times 3.4 I_{\overline{z}} Q}{1 + 1.7 \times 3.4 I_{\overline{z}}}\right)$$

$$= 0.925\{(1 + 1.7 \times 3.4 \times 0.274 \times 0.83)/$$

$$(1 + 1.7 \times 3.4 \times 0.274)\}$$

$$= 0.83.$$

Wind parallel to long building dimension (i.e., $B = 60$ ft),

$$Q = 1/\{1 + 0.63[(60 + 95)/383.9]^{0.63}\}^{1/2} = 0.86$$

$$G = 0.85.$$

It is seen that the use of the value $G = 0.85$ is reasonable for both wind directions.

(b) ***Flexible building, G_f*** (ASCE Sect. 26.9.5; Table 2.2.1, item A, col. 6). We now consider the assumption that the MWFRS consists of structural steel moment-resistant frames. The approximate natural frequency is determined by ASCE Eq. 26.9-2:

$$n_a = 22.2/h^{0.8}$$

where h is in feet, so $n_a = 22.2/95^{0.8} = 22.2/38.2 = 0.58$ Hz < 1 Hz, and in accordance with ASCE Sect. 26.2 (Building and other structure: flexible), the building is classified as *flexible*.

Since the approximate fundamental natural frequency for the building is $n_a = 0.58$ Hz < 1 Hz along both principal axes, the building is classified as *flexible*. The damping ratio is assumed to be $\beta = 2\%$. The gust effect factor G_f (Table 2.2.1, item A, col. 1; ASCE Sect. 27.9.5) is determined using ASCE Eqs. 26.9-10 to 26.9-15. Assuming the fundamental natural frequency $n_1 \cong n_a$, these equations yield:

$$g_R = [2 \ln (3600 \times 0.58)]^{1/2} + 0.577/[2 \ln (3600 \times 0.58)]^{1/2}$$

$$= 3.91 + 0.577/3.91 = 4.06.$$

$$\overline{V}_{\overline{z}} = \overline{b} \left(\frac{\overline{z}}{33} \right)^{\overline{\alpha}} \frac{88}{60} V = 0.45 \times (57/33)^{1/4} (88/60)\ 170$$

$$= 128.6 \text{ ft/s (ASCE Table 26.9-1)}$$

$$N_1 = \frac{n_1 L_{\overline{z}}}{V_{\overline{z}}} = 0.58 \times 383.9/128.6 = 1.73$$

$$R_n = \frac{7.47 N_1}{(1 + 10.3 N_1)^{5/3}} = 7.47 \times 1.73/(1 + 10.3 \times 1.73)^{5/3}$$

$$= 0.097$$

To calculate R_h, R_B evaluate the quantities $\eta = 4.6 n_a \ell / \overline{V}_{\bar{z}}$ for $\ell = h, B$. To calculate R_L evaluate $\eta = 15.4 n_a \ell / \overline{V}_{\bar{z}}$ for $\ell =$ L. Then $R_\ell = 1/\eta - 2/\eta^2 (1 - e^{-2\eta})$,

$$R = \sqrt{\frac{1}{\beta} R_n R_h R_B (0.53 + 0.47 R_L)}$$

and

$$G_f = 0.925 \left(\frac{1 + 1.7 I_{\bar{z}} \sqrt{(3.4Q)^2 + g_R^2 R^2}}{1 + 1.7 \times 3.4 I_{\bar{z}}} \right)$$

For ***wind parallel to short horizontal dimension of building***, $B = 125$ ft, $L = 60$ ft. The results are, respectively, $\eta = 1.97$, $\eta = 2.59$, and $\eta = 4.17$; $R_h = 0.38$, $R_B = 0.32$, $R_L = 0.21$; $R = 0.60$; $G_f = 0.98$.

For ***wind parallel to long horizontal dimension of building***, $B = 60$ ft, $L = 125$ ft and $\eta = 1.97$, $\eta = 1.24$, and $\eta = 8.68$; $R_h = 0.38$, $R_B = 0.51$, $R_L = 0.11$; $R = 0.737$; $G_f = 1.06$.

The balance of the calculations for this Numerical Example is performed by assuming that the building is rigid and $G = 0.85$.

Step 4. *Note:* Topographic factor and velocity pressure coefficient K_z are evaluated at the same heights.

- ***Velocity pressure exposure coefficient K_z*** (ASCE Table 27.3-1; Table 2.2.1). See Table in Step 5.
- ***Topographic factor K_{zt}*** (ASCE Sect. 26.8; Sect. 2.3.5). In this example, it is assumed that the terrain is flat, so $K_{zt} = 1.0$. For an application to the case where topographic features are present, see Chapter 3, Numerical Example 3.5.1.

Step 5. Velocity pressures q (ASCE Sect. 27.3.2, Eq. 27.3-1).

For the windward wall, the velocity pressures q_z are:

z (ft)	K_z	K_{zt}	K_d	V (mph)	q_{ext} (psf)
0–15	0.57	1	0.85	170	35.8
35	0.73	1	0.85	170	45.9
55	0.83	1	0.85	170	52.2
75	0.91	1	0.85	170	57.2
95	0.98	1	0.85	170	61.6

For leeward walls, side walls, and roof:

$$q_h \text{ (psf)} = 61.6 \text{ (all heights; based on } K_z \text{ at } z = h = 95 \text{ ft)}$$

Step 6. Internal pressures (ASCE Sect. 27.4; Table 2.2.1, item A, col. 2). Since $(GC_{pi}) = \pm 0.18$ and $q_i = 61.6$ psf, $p_i = (GC_{pi})\, q_i = \pm 0.18 \times 61.6$, that is, $q_i = \pm 11.0$ psf.

Step 7. External pressure coefficients (ASCE Fig. 27.4-1; Table 5.2.1). The coefficients are obtained for two wind directions: parallel to short building dimension, and parallel to long building dimension. As was mentioned earlier, for the purpose of calculating MWFRS pressures, the building is assumed to be rigid, and the value $G = 0.85$ will be employed.

Wind direction parallel to short building dimension (i.e., wind blowing normal to ridge, Fig. 5.2.1a), $L = 60$ ft, $L/B = 60/125 < 1.0$, and $h/L = 95/60 > 1$.

Windward wall, $C_p = 0.8$.

Leeward wall, $C_p = -0.5$.

Side walls, $C_p = -0.7$.

Roof: Since $\theta < 10^\circ$ and $h/L > 1$:

For distance $h/2 = 47.5$ ft from windward edge: $C_p = -1.3$ (before reduction) and $C_p = -0.18$. The corresponding roof area is $Bh/2 = 125 \times 47.5 = 5938$ sq ft, so a reduction factor (Table 5.2.1, footnote #) is applied to the coefficient $C_p = -1.3$. Therefore, $C_p = -1.3 \times 0.8 = -1.04$, and $C_p = -0.18$.

For roof area from 47.5 ft from windward edge to leeward edge, $C_p = -0.7$, and $C_p = -0.18$.

Wind direction parallel to long building dimension (i.e., wind blowing in longitudinal direction, Fig. 5.2.1c). The calculations are similar, but $L = 125$ ft and $B = 60$ ft (L = building dimension parallel to wind direction, Fig. 5.2.1 and Table 5.2.1, Note 1). $L/B = 125/60 = 2.08$ (note that $2 < 2.08 < 4$), and $h/L = 95/125 = 0.76$ (note that $0.5 < 0.76 < 1.0$).

Windward wall, $C_p = 0.8$.

Leeward wall, $C_p = -0.3 + [-0.2 - (-0.3)] \times (2.08 - 2.0)/(4 - 2). = -0.296 \approx -0.3$

Side walls, $C_p = -0.7$.

Roof: $\theta < 10^\circ$ and $h/L = 0.76$:

TABLE 5.2.1. External Pressure Coefficients C_p, Enclosed and Partially Enclosed Buildings of All Heights with Roof Types of ASCE Fig. 27.4-1

Walls (use with q_h [except for windward wall, use q_z])	
Windward wall	$C_p = 0.8$ (use with q_z)
Leeward wall	$0 < L/B \leq 1$: $C_p = -0.5$; $L/B = 2$: $C_p = -0.3$; $L/B \geq 4$: $C_p = -0.2$
Side walls	$C_p = -0.7$

Roof (use with q_h)

		Windward side						Leeward side	
Normal to ridge $\theta \geq 10^\circ$	θ°	10	15	20	25	30	10	15	≥ 20
	$h/L \leq 0.25$	−0.7	−0.5	−0.3	−0.2	−0.2	−0.3	−0.5	−0.6
		−0.18	0.0	0.2	0.3	0.3			
	$h/L = 0.5$	−0.9	−0.7	−0.4	−0.3	−0.2	−0.5	−0.5	−0.6
		−0.18	−0.18	0.0	0.2	0.2			
	$h/L \geq 1.0$	$-1.3^{\#}$	−1.0	−0.7	−0.5	−0.3	−0.7	−0.6	−0.6
		−0.18	−0.18	−0.18	0.0	0.2			

		Horizontal distance u from windward edge	C_p
Normal to ridge for $\theta < 10^\circ$, and parallel to ridge for all θ	$h/L \leq 0.5$	$0 < u \leq h$	−0.9,−0.18
		$H < u \leq 2h$	−0.5,−0.18
		$u > 2h$	−0.3,−0.18
	$h/L \geq 1.0$	$0 < u \leq h/2$	$-1.3^{\#}$,−0.18
		$u > h/2$	−0.7,−0.18

#May be reduced via multiplication by factor m obtained as follows: $Bh/2 \leq 100$ sq ft: $m = 1.0$; $Bh/2 = 200$ sq ft: $m = 0.9$; $Bh/2 \geq 1000$ sq ft: $m = 0.8$ (linear interpolation permitted).

Notes:

1. L and B = horizontal dimension of building parallel and normal to wind direction, respectively.
2. For values of θ, L/B, and h/L not shown, linear interpolations are permitted between entries of the same sign; for h/L, if the two entries have opposite signs, use 0.0 as one of the entries.
3. The horizontal component of the roof pressures shall not be used to reduce the total horizontal shear on the building, but shall be taken into account for members and connections near eaves.
4. For monoslope roofs, entire surface is either a windward or a leeward surface.

Windward edge to $h/2 = 47.5$ ft from windward edge. The roof area is $Bh/2 = 60 \times 47.5 = 2850$ sq ft, so a 0.8 reduction factor (Table 5.2.1, footnote #) is applied to the coefficient $C_p = -1.3$, that is, $C_p = -1.04$. Interpolation between the values −0.9 (for $h/L = 0.5$) and −1.04 (for $h/L = 1.0$) yields $C_p = -0.9 + [-1.04 - (-0.9)](0.76 - 0.5)/0.5 = -0.97$. Therefore, $C_p = -0.97$, and $C_p = -0.18$.

From 47.5 ft to 95 ft from windward edge, $C_p = -0.9 + [-0.7 - (-0.9)](0.76 - 0.5)/0.5$, that is, $C_p = -0.8$, and $C_p = -0.18$.

Between 95 ft from the windward edge to leeward edge,

$$C_p = -0.5 - (0.76 - 0.5)(0.2/0.5), \text{ that is,}$$
$$C_p = -0.60, \text{ and } C_p = -0.18.$$

Step 8. Wind pressures (ASCE Eq. 27.4-1). The external and net wind pressures are:

MWFRS Pressures: Wind Parallel to the Short Building Dimension

Surface	z (ft)	q_z (psf)	G	C_p	$p_{ext} = q_z GC_p$ (psf)	$p_{net} = p_{ext} - p_i$ (psf) $(GC_{pi}) =$ 0.18	$(GC_{pi}) =$ −0.18
Windward wall	0–15	35.8	0.85	0.8	24.4	35.4	13.4
	35	45.9	0.85	0.8	31.2	42.2	20.2
	55	52.2	0.85	0.8	35.5	46.5	24.5
	75	57.2	0.85	0.8	38.9	49.9	27.9
	95	61.6	0.85	0.8	41.9	52.9	30.9
Leeward wall	All	61.6	0.85	−0.5	−26.2	−15.2	−37.2
Side walls	All	61.6	0.85	−0.7	−36.7	−25.7	−47.7
Roofs*	0 to $0.5h$*	61.6	0.85	−1.04	−54.5	−43.5	−65.5
		61.6	0.85	−0.18	−9.4	1.6	−20.4
	$>0.5h$*	61.6	0.85	−0.7	−36.7	−25.7	−47.7
		61.6	0.85	−0.18	−9.4	1.6	−20.4

*Distance from windward edge ($h = 95$ ft)

MWFRS Pressures: Wind Parallel to the Long Building Dimension

Surface	z (ft)	q_z (psf)	G	C_p	$p_{ext} = q_z GC_p$ (psf)	$p_{net} = p_{ext} - p_i$ (psf) $(GC_{pi}) =$ 0.18	$(GC_{pi}) =$ −0.18
Windward wall	0–15	35.8	0.85	0.8	24.4	35.4	13.4
	35	45.9	0.85	0.8	31.2	42.2	20.2
	55	52.2	0.85	0.8	35.5	46.5	24.5
	75	57.2	0.85	0.8	38.9	49.9	27.9
	95	61.6	0.85	0.8	41.9	52.9	30.9
Leeward wall	All	61.6	0.85	−0.3	−15.7	−4.7	−26.7
Side walls	All	61.6	0.85	−0.7	−36.7	−25.7	−47.7
Roofs*	0 to $0.5h$*	61.6	0.85	−0.97	−50.8	−39.8	−61.8
		61.6	0.85	−0.18	−9.4	1.6	−20.4
	$0.5h$ to h*	61.6	0.85	−0.8	−41.9	−30.9	−52.9
		61.6	0.85	−0.18	−9.4	1.6	−20.4
	$>h$*	61.6	0.85	−0.6	−31.4	−20.4	−42.4
		61.6	0.85	−0.18	−9.4	1.6	−20.4

*Distance from windward edge ($h = 95$ ft)

Note on roof pressures for flexible buildings. For flexible buildings, the roof pressures are calculated by using for the velocity pressures the gust factor G_f. Physically, this is not warranted, because the resonant amplification effects implicit in the gust factor effect G_f do not affect the effective aerodynamic pressures on the roof. For this reason, in calculating roof pressures for flexible buildings, it is reasonable to assume $G = 0.85$.

Design wind load cases (ASCE Sect. 27.4.6; ASCE Fig. 27.4-8; ASCE Appendix D, Sect. D1.1). The MWFRS of enclosed and partially enclosed buildings for which wind loads are determined by the regular approach shall be designed for all wind load cases of ASCE Sect. 27.4-6 (ASCE Fig. 27.4-8), except that no cases involving torsional moments need to be considered for buildings with mean roof height $h \leq 30$ ft, or for buildings with one or two stories framed with light frame construction, and buildings with one or two stories designed with flexible diaphragms (ASCE Appendix D, Sect. D1.1). This exception does not apply to the building considered in this example, for which all four cases of ASCE Fig. 27.4-8 need to be considered.

The windward and leeward external pressures calculated at $h = 95$ ft were shown to be, respectively, $p_{ext} = 41.7$ psf, $p_{ext} = -15.6$ psf for wind direction parallel to the long sides of the building (denoted the x-direction), and $p_{ext} = 41.7$ psf, $p_{ext} = -26.1$ psf for wind parallel to the short sides of the building (y-direction). For the elevation $h = 95$ ft, the loading cases of ASCE Fig. 27.4-8 are as follows:

Case 1a. Horizontal loading: $41.7 - (-15.6) = 57.3$ psf (x-direction).

Case 1b: Horizontal loading: $41.7 - (-26.1) = 67.8$ psf (y-direction).

Case 2a: Horizontal loading: $0.75 \times 57.3 = 43.0$ psf (x-dir.) and, simultaneously, torsional moment $M_T = \pm 0.75 \times 57.3 \times 60(0.15 \times 60)/1000 = \pm 23.2$ kip-ft/ft.

Case 2b: Horizontal loading: $0.75 \times 67.8 = 50.9$ psf (y-dir.) and, simultaneously, torsional moment $M_T = \pm 0.75 \times 67.8 \times 125(0.15 \times 125)/1000 = \pm 119.2$ kip-ft/ft.

Note: For wind in x direction, the dimension normal to the wind direction, denoted by B in Table 5.2.1 (ASCE Fig. 27.4-1), is denoted for clarity in ASCE Fig. 27.4-8 by $B_x (B_x = 125$ ft); similarly, for wind in y direction, the dimension normal to the wind direction is denoted by $B_y (B_y = 60$ ft).

Case 3: Horizontal loading: $0.75 \times 57.3 = 43.0$ psf (x-direction) and, simultaneously, $0.75 \times 67.8 = 50.9$ psf (y-direction).

Case 4: Horizontal loading: $0.563 \times 57.3 = 32.3$ psf (x-dir.) and, simultaneously, 0.563×67.8 psf $= 38.2$ psf (y-dir.), torsional moment $M_T = \pm 0.75(M_{Tx} + M_{Ty}) = \pm 106.8$ kip-ft/ft.

These results correspond to elevation $h = 95$ ft. Similar results must be obtained for other elevations of the windward and leeward walls. They depend upon elevation, owing to the variation of the windward wall pressures with height. The resulting overall forces and moments are used for the design of the MWFRS. Internal pressures do not contribute to the *overall* transverse and longitudinal forces and torsional moments. However, they can affect members belonging to the MWFRS that carry wind loads on walls and on the roof.

Numerical Example 5.2.2. *Commercial building with 15-ft eave height.* The building being considered is rectangular in plan with dimensions 45 ft × 40 ft, eave height 15 ft, gable roof with slope $\theta = 26.6^{\circ}$, and mean roof height $h = 20$ ft. It is assumed that the building is located in Arkansas in flat terrain with exposure B in all directions and that it is enclosed. The determination of the design wind pressures follows the steps of Fig. 2.2.2. The wind loading is determined in accordance with the provisions for "buildings of all heights" (ASCE Sect. 27.4), the use of which is optional.

***Step 1. Risk category*:** II (ASCE Table 1.5-1).

***Step 2. Basic wind speed*:** $V = 115$ mph (ASCE Fig. 26.5-1a).

Step 3. [3]

- ***Wind directionality factor K_d*** (ASCE Table 26.6-1). For buildings $K_d = 0.85$.
- ***Enclosure classification*** (ASCE Sect. 26.10; see Chapter 3, Numerical Example 3.3.1). For the building considered here, it is assumed that the building is *enclosed*. (Given the location of the building, this requires that the glazing be protected or that the two conditions listed in Sect. 3.3 for wind-borne debris regions be satisfied.)
- ***Exposure category*** (ASCE Sect. 26.7). In this example, it is assumed that the site has *Exposure B in all directions.* For an example of the application of the Standard provisions on exposure category, see Chapter 3, Numerical Example 3.4.1. As indicated in Table 2.2.1, if the exposure category depends on direction, Exposure Requirement 1 applies; that is, the wind loads are based on the appropriate exposure for each wind direction (ASCE Sect. 26.7.4.1).
- ***Is the building an enclosed simple diaphragm building?*** (ASCE Sect. 26.2). We assume that the building has simple diaphragms as defined in Sect. 26.2. Since $h < 60$ ft and the ratio L/B is nearly unity, the building is a Class 1 building (ASCE Sect. 27.5.2). The building may be classified as *a simple diaphragm building*. (As is indicated in Fig. 2.2.2, the designer has the option

[3] Step 3 corresponds, in an expanded form, to the nomenclature of ASCE Table 27.2-1.

of determining the loads either by the regular approach or by the simplified approach.) The choice made in this example is to use the regular approach.

- ***Is the building a low-rise building?*** (ASCE Sects. 26.2 and ASCE Chapter 28). Since $h < 60$ ft, and the ratios h/B and h/L are less than 1, the building may be classified as a *low-rise building* (ASCE Sect. 27.5.2). The wind loads may be determined by using the provisions for the "buildings of all heights" approach, or those for "low-rise buildings." The choice made in this example is to use the regular, "buildings of all heights" approach.
- ***Combined internal pressure coefficient*** (GC_{pi}) (ASCE Sect. 26.11). Since the building is neither open nor partially enclosed, the flowchart (Fig. 2.2.2) leads to $(GC_{pi}) = \pm 0.18$.
- ***Elevations at which the velocity pressures are evaluated*** (ASCE 27.4.1; Table 2.2.1, item A, cols. 4 and 5).

  For external pressures on windward wall, the elevations z at which velocity pressures are calculated are $z = 15$ ft, and $z = 20$ ft.

  For internal pressures on windward wall and for external and internal pressures on leeward and side walls and on roof, $z = 20$ ft (Table 2.2.1).
- **Gust effect factor** (ASCE Sect. 26.9).
- ***Is the building rigid or flexible?*** (ASCE Sects. 26.2 and 26.9). The building's MWFRS is assumed to consist of sufficiently stiff shear walls (i.e., shear walls such that ASCE Eq. 26.9-5 yields $n_a \geq 1$ Hz) that, in accordance with ASCE Sect. 26.2 ("building and other structure: rigid"), the building is classified as *rigid.* The gust response factor G may therefore be assumed to be $G = 0.85$.

Step 4. *Note:* Topographic factor and velocity pressure coefficient K_z are evaluated at the same heights.

- ***Velocity pressure exposure coefficient*** K_z (ASCE Table 27.3-1; Table 2.2.1). Using ASCE Table 27.3-1 for Exposure B, the following results are obtained:

z (ft)	15	20
K_z	0.57	0.62

- ***Topographic factor*** K_{zt} (ASCE Sect. 26.8; Sect. 2.3.5). In this example, it is assumed that the terrain is flat, so $K_{zt} = 1.0$. (For an application to the case where topographic features are present, see Chapter 3, Numerical Example 3.5.1.)

Step 5. Velocity pressures q (ASCE Sect. 27.3.2, Eq. 27.3-1). For $V =$ 115 mph,

For windward walls:

$$q_{15\text{ ft}} = 0.00256 \times 1.0 \times 0.575 \times 0.85 \times 1.0 \times 115^2 = 16.4 \text{ psf.}$$

For leeward walls, side walls, and roof:

$$q_h = 0.00256 \times 1.0 \times 0.62 \times 0.85 \times 1.0 \times 115^2 = 17.8 \text{ psf.}$$

Step 6. Internal pressures (ASCE Sect. 27.4; Table 2.2.1, item A, col. 2). Since $(GC_{pi}) = \pm 0.18$ and $q_i = 17.8$ psf, $p_i = (GC_{pi})q_i = \pm 0.18 \times 17.8$, that is, $q_i = \pm 3.2$ psf.

Step 7. External pressure coefficients (ASCE Fig. 27.4-1; Table 5.2.1). The coefficients are obtained for two wind directions: parallel to short building dimension, and parallel to long building dimension.

Wind direction parallel to short building dimension $L =$ 40 ft (normal to ridge) (Table 2.2.1, Note 1), $L/B = 40/45 < 1$, and $h/L = 20/40 < 1$.

Windward wall, $C_p = 0.8$.

Leeward wall, $C_p = -0.5$.

Side walls, $C_p = -0.7$.

Roof windward: $\theta = 26.6^\circ$ and $h/L = 20/40 = 0.5$;

$$C_p = -0.27 \text{ or } C_p = 0.2.$$

Wind direction parallel to long building dimension $L = 45$ ft, $L/B = 45/40 = 1.125$, and $h/L = 20/45 < 0.5$.

Windward wall, $C_p = 0.8$.

Leeward wall, $C_p = -0.48$.

Side walls, $C_p = -0.7$.

Roof, 0 to h = 20 ft from windward wall, $C_p = -0.9$ or $C_p = -0.18$.

Roof, 20 ft to 40 ft from windward wall, $C_p = -0.5$ or $C_p = -0.18$.

Roof, 40 ft to 45 ft from windward wall, $C_p = -0.3$ or $C_p = -0.18$.

Step 8. External and net pressures are:

MWFRS Pressures: Wind Parallel to the Short Building Dimension (Normal To Ridge)

Surface	z (ft)	q_z (psf)	G	C_p	$p_{ext} = q_z GC_p$ (psf)	$p_{net} = p_{ext} - p_i$ (psf) $(GC_{pi}) =$ 0.18	$(GC_{pi}) =$ −0.18
Windward wall	0–15	16.4	0.85	0.8	11.2	14.4	8.0
Leeward wall	All	17.8	0.85	−0.5	−7.6	−4.4	−10.8
Side walls	All	17.8	0.85	−0.7	−10.6	−7.4	−13.8
Windward roof	–	17.8	0.85	−0.27	−4.1	−0.9	−7.3
		17.8	0.85	0.2	3.0	6.2	−0.2
Leeward roof	–	17.8	0.85	−0.6	−9.1	−5.9	−12.3

MWFRS Pressures: Wind Parallel to Long Building Dimension

Surface	z (ft)	q_z (psf)	G	C_p	$p_{ext} = q_z GC_p$ (psf)	$p_{net} = p_{ext} - p_i$ (psf) $(GC_{pi}) =$ 0.18	$(GC_{pi}) =$ −0.18
Windward wall	0–15	16.4	0.85	0.8	11.2	14.4	8.0
Leeward wall	All	17.8	0.85	−0.48	−7.3	−4.1	−10.5
Side walls	All	17.8	0.85	−0.7	−10.6	−7.4	−13.8
Roofs*	0 to h*	17.8	0.85	−0.9	−13.6	−10.4	−16.8
		17.8	0.85	−0.18	−2.7	0.5	−5.9
	h to $2h$*	17.8	0.85	−0.5	−7.6	−4.4	−10.8
		17.8	0.85	−0.18	−2.7	0.5	−5.9
	$>2h$*	17.8	0.85	−0.3	−4.5	−1.3	−7.7
		17.8	0.85	−0.18	−2.7	0.5	−5.9

*Distance from windward edge ($h = 20$ ft)

Design wind load cases (ASCE Sect. 27.4.6; ASCE Fig. 27.4-8; ASCE Appendix D, Sect. D1.1). The MWFRS of enclosed and partially enclosed buildings for which wind loads are determined by the regular approach shall be designed for all wind load cases of ASCE Sect. 27.4-6 (ASCE Fig. 27.4-8), except that no cases involving torsional moments need to be considered for buildings with mean roof height $h \leq 30$ ft, or for buildings with one or two stories framed with light frame construction or designed with flexible diaphragms (ASCE Appendix D, Sect. D1.1). This exception will not be applied to the building considered in this example, for which all four cases of ASCE Fig. 27.4-8 will be considered, for illustration purposes.

The external pressures determined at $h = 15$ ft on the windward and leeward walls were shown to be, respectively, $p_{ext} = 11.2$ psf, $p_{ext} = -7.3$ psf for the wind direction parallel to the long sides of the building (denoted the x-direction), and $p_{ext} = 11.2$ psf, $p_{ext} = -7.6$ psf for wind parallel to the short sides of the building (y-direction). The loading cases of ASCE Fig. 27.4-8 are as follows:

Case 1a. Horizontal loading: $11.2 - (-7.3) = 18.5$ psf (x-direction).

Case 1b: Horizontal loading: $11.2 - (-7.6) = 18.8$ psf (y-direction).

Case 2a: Horizontal loading: $0.75 \times 18.5 = 13.9$ psf (x-dir.) and, simultaneously, torsional moment $M_T = 0.75 \times 18.5 \times 40(0.15 \times 40)/1000 = \pm 3.3$ kip-ft/ft.

Case 2b: Horizontal loading: $0.75 \times 18.8 = 14.1$ psf (y-dir.) and, simultaneously, torsional moment $M_T = \pm 0.75 \times 18.8 \times 45(0.15 \times 45)/1000 = \pm 4.3$kip-ft/ft.

Note: For wind in x direction, the dimension normal to the wind direction, denoted by B in Table 5.2.1 (ASCE Fig. 27.4-1), is denoted for clarity in ASCE Fig. 27.4-8 by $B_x (B_x = 45$ ft); similarly, for wind in y direction, the dimension normal to the wind direction is denoted by $B_y (B_y = 40$ ft).

Case 3: Horizontal loading: $0.75 \times 18.5 = 13.9$ psf (x-direction) and, simultaneously, $0.75 \times 18.8 = 14.1$ psf (y-direction).

Case 4: Horizontal loading: $0.563 \times 18.5 = 10.4$ psf (x-dir.) and, simultaneously, 0.563×18.8 psf$= 10.6$ psf (y-dir.), torsional moment $M_T = \pm 0.75(M_{Tx} + M_{Ty}) = \pm 0.75(3.3 + 4.3) = \pm 5.7$ kip-ft/ft.

These results are valid for all elevations, since for this building the eave height is 15 ft, and the design pressures do not vary with height. Internal pressures do not contribute to the overall transverse and longitudinal forces or to the torsional moments.

Numerical Example 5.2.3. *External pressure coefficients on domed roofs.* Design wind pressures on domed roofs are determined in accordance with ASCE Sect. 27.4.1. However, the velocity pressure q associated with the external pressure coefficient is calculated at the top of the dome (i.e., at elevation $h_D + f$, Fig. 5.2.2), as specified in Note 2 of ASCE Fig. 27.4-2 (see also Table 2.2.1, item A, col. 4).

This section is concerned only with the determination of the pressure coefficients C_p. All other calculations are similar to those performed for buildings of all heights, except for the height at which q is evaluated, as was just indicated.

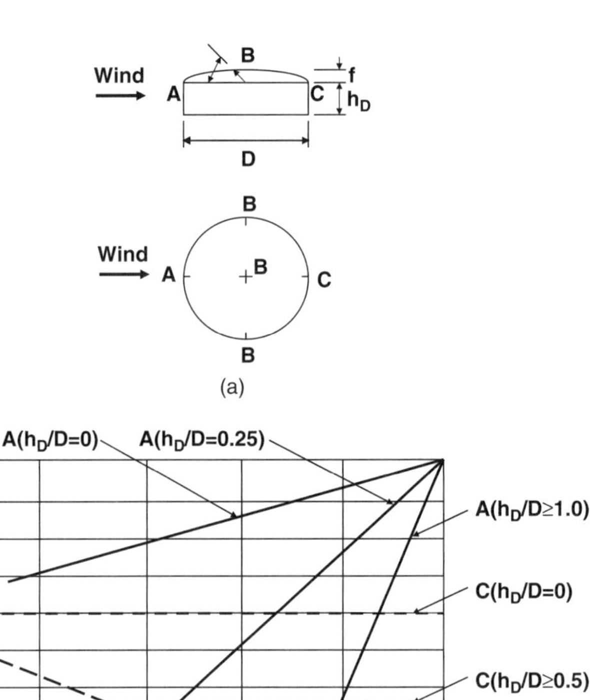

Figure 5.2.2. Enclosed and partially enclosed buildings of all heights with domed roofs: (a) elevation and plan of a dome with a circular base; (b) plot for the estimation of the pressure coefficient C_p. From ASCE Standard 7-10, *Minimum Design Loads for Buildings and Other Structures*. Copyright 2010 by the American Society of Civil Engineers. With permission from ASCE.

To determine C_p, two cases, referred to as Case 1 and Case 2, are considered:[4]

Case 1 is used to check maximum uplift. C_p values are obtained from Fig. 5.2.2b by linear interpolation between lines *A* and *B* for the

[4]Case A and Case B of ASCE Fig. 27.4-2 are denoted here as Case 1 and Case 2, respectively. This avoids confusion with the notation of the points A and B in ASCE Fig. 27.4-2 (Fig. 5.2.2).

windward side of the domed roof, and between lines B and C for the leeward side.

Case 2 is used to check maximum base shear. For $0 \leq \theta \leq 25^\circ$ (θ is defined in Fig. 5.2.2a), C_p is constant and equal to the value at point A. For $25^\circ < \theta \leq 90^\circ$, C_p is determined by linear interpolation between lines A and B, and for $90^\circ < \theta \leq 180^\circ$, by linear interpolation between lines B and C. Interpolations may be based on horizontal distances. For ratios h_D/D not shown in Fig. 5.2.2b, linear interpolation between lines A, between lines B, and between lines C is permitted. C_p is constant on the domed surface on arcs normal to the wind direction (on the arc B-B-B and all arcs parallel to B-B-B in Fig. 5.2.2a).

The total horizontal force may not be reduced by accounting for wind forces on the roof. For $f/D < 0.05$, Table 5.2.1 (ASCE Fig. 27.4-1) is used.

Let $D = 150$ ft, $h_D = 37.5$ ft, $f = 30$ ft; $h_D/D = 0.25$, $f/D = 0.2$. For $A(h_D/D = 0.25)$, $C_p = -0.67$. For $B(h_D/D = 0)$, $C_p = -0.58$, for $B(h_D/D = 0.5)$, $C_p = -1.07$, so for $B(h_D/D = 0.25)$, $C_p = -0.83$. For $C(h_D/D = 0.25)$, $C_p = -0.25$.

Case 1: Obtain pressure coefficients by linear interpolation, (1) between points A and B: Point A, $C_p = -0.67$; point B, $C_p = -0.83$; point at, for example, 30-ft horizontal distance downwind from point A, $C_p = -0.67 - [30/(150/2)](0.83 - 0.67) = -0.73$; (2) between points B and C: point B, $C_p = -0.83$; point C, $C_p = -0.25$; point at, e.g., 30-ft distance downwind from line B-B-B, $C_p = -0.25 - [(150/2 - 30)/(150/2)](0.83 - 0.25) = -0.60$.

Case 2: Horizontal distance downwind of A over which pressures are the same as at A: $x = 44.2$ ft (x depends upon the curvature of the domed surface and was determined graphically). Pressure coefficients: For A and entire domed surface downwind of A up to horizontal coordinate x: $C_p = -0.67$. For B: $C_p = -0.83$. For point at 20 ft downwind of coordinate x: $C_p = -0.67 - [20/(150/2 - 44.2)](0.83 - 0.67) = -0.77$. Between B and C, pressures are the same as for Case 1.

5.3 REGULAR APPROACH: ROOF OVERHANGS AND PARAPETS, MWFRS

Roof overhangs and parapets for which wind loads on MWFRS are determined by the regular approach are specified in the Standard for buildings of all heights (ASCE Chapter 27), which are considered in this chapter, and for low-rise buildings (ASCE Chapter 28), which are considered in Chapter 6. ASCE Sects. 29.6 and 29.7, in which parapets and overhangs are considered as building appurtenances, limit themselves to referring the user to ASCE Chapters 27 and 28, without providing any additional information.

5.3.1 Regular Approach: Roof Overhangs for Buildings of All Heights, MWFRS (ASCE Sect. 27.4.4; ASCE Fig. 27.4-1)

Under ASCE Sect. 27.4.4, roof overhangs are designed for the top surface pressures on the roof (ASCE Fig. 27.4-1), in combination with an upward pressure on the bottom surface,

$$p_{bottom} = q_h C_p \tag{5.3.1}$$

where $C_p = 0.8$, and the velocity pressure q_h is evaluated at the mean roof height h. For details, see Numerical Example 5.3.1.

Numerical Example 5.3.1. *Roof Overhangs, Windward Building Faces (ASCE Sect. 27.4.4).* The building of Numerical Example 5.2.1 is assumed to have a roof with overhangs projecting 3 ft outside of the longitudinal exterior walls. For that building, it was shown that the largest external pressure coefficients for the roof near the exterior walls are -1.04 and -0.97 for wind parallel to the short building dimension and wind parallel to the long building dimension, respectively; and that the velocity pressure at eave height is 61.6 psf. The net pressure coefficients for the roof overhangs along the long and short building dimensions are then, respectively, $0.8 + 1.04 = 1.84$ and $0.8 + 0.97 = 1.77$, directed upwards, and the uplift forces are, respectively, 113 psf and 109 psf.

5.3.2 Regular Approach: Fascia for Open Buildings with Monoslope, Pitched, or Troughed Free Roofs, MWFRS (ASCE Sect. 27.4.3)

Fascia panels on free roofs with slope less than or equal to 5° shall be considered inverted parapets subjected to pressures determined as

$$p_p = q_h (GC_{pn}) \tag{5.3.2}$$

where q_h is the velocity pressure evaluated at the mean roof height, and the combined net pressure coefficient is $(GC_{pn}) = +1.5$ for windward fascia, and -1.0 for leeward fascia.

5.3.3 Regular Approach: Parapets for Enclosed, Partially Enclosed, and Open Buildings, MWFRS (ASCE Sects. 27.4.5, 28.4.2)

Design wind pressures on parapets for rigid or flexible buildings of all heights and low-rise buildings, with flat, gable, or hip roofs, are determined as

$$p_p = q_p (GC_{pn}) \tag{5.3.3}$$

where q_p is the velocity pressure evaluated at the top of the parapet, and the combined net pressure coefficient is $(GC_{pn}) = +1.5$ for windward parapets, and -1.0 for leeward parapets.

Numerical Example 5.3.2. *Parapets (ASCE 27.4.5).* The building of Numerical Example 5.2.1 is assumed in this example to have a 3-ft-high parapet on all roof sides. At the top of the parapet (98-ft elevation), the velocity pressure exposure coefficient for exposure B is $K_z = 2.01(98/1200)^{2/7} = 0.98$ (ASCE Table 27.3.1), and the velocity pressure is $q_p = 0.00256 \times 0.984 \times 1.0 \times 0.85 \times 170^2 = 61.9$ psf (ASCE Eq. 27.3.1; $V = 170$ mph; the wind directionality factor is $K_d = 0.85$). The total pressures on the windward and leeward parapets are $p_p = 61.9 \times 1.5 = 92.9$ psf and $p_p = 61.9 \times (-1.0) = -61.9$ psf (directed toward the windward parapet and away from the leeward parapet), respectively.

5.4 REGULAR APPROACH: OPEN BUILDINGS WITH MONOSLOPE, PITCHED, OR TROUGHED FREE ROOFS, MWFRS

The net design pressure (ASCE Sect. 27.4.3) is

$$p = q_h G C_N \tag{5.4.1}$$

where q_h is the velocity pressure at the mean height of the roof h (ASCE Figs. 27.4-4, 27.4-5, 27.4-6), and the gust effect factor is $G = 0.85$ (ASCE 26.9.3). For wind directions $\gamma = 0^\circ$ and $\gamma = 180^\circ$ (i.e., normal to the horizontal edges of the roof), roof slope $\theta \le 45^\circ$, and $0.25 \le h/L \le 1.0$, the net pressure coefficient (i.e., the sum of pressure coefficients for the top and bottom surfaces) C_N is taken from ASCE Figs. 27.4-4, 27.4-5, and 24.4-6, in which $C_N = C_{NW}$ for the windward half of the roof, and $C_N = C_{NL}$ for the leeward half; both Cases A and B shall be considered in design. Linear interpolation is permitted, except that for $\theta < 7.5^\circ$ the values listed for monoslope roofs for $\theta = 0^\circ$ shall be used. Clear flow occurs if obstructions block 50% or less of the area between the ground and the windward edge of the roof. Obstructed flow occurs if the obstructions block more than 50% of the area.

For all roofs with slopes $\theta \le 45^\circ$, the net pressure coefficients C_N for wind parallel to the direction of the roof's horizontal edges ($\gamma = 90^\circ$ or $\gamma = 270^\circ$) depend on the horizontal distance u from the windward roof edge, as shown in ASCE Fig. 27.4-7.

As noted earlier, *fascia panels* of roofs with $\theta \le 5^\circ$ shall be considered as inverted parapets, for which q_h is used instead of q_p (ASCE Sect. 27.4.3).

Numerical Example 5.4.1. *Net pressure coefficients for MWFRS of open monoslope building with free roof.* Assume that for a monoslope roof $h = 15$ ft, $L = 24$ ft, the length (dimension transverse to the horizontal eave)

TABLE 5.4.1. Pressure Coefficients for Open Building with Monoslope Roof

Wind dir. γ	Load case	C_{NW}	C_{NL}
0°	A	−0.9	−1.3
	B	−1.9	0.0
180°	A	1.3	1.6
	B	1.8	0.6

Wind dir. γ	Dist. u from windward edge	Load case	C_N
90°	$u \leq 15$ ft	A	−0.8
		B	0.8
270°	15 ft $< u \leq 21$ ft	A	−0.6
		B	0.5

is 21 ft, and $\theta = 15^\circ$. The site has Exposure B for all directions, and is located in flat terrain near Toledo, Ohio. It is determined from ASCE Table 1.5-1 that the building belongs to Risk Category II.

The basic wind speed is $V = 115$ mph. The topographic factor is $K_{zt} = 1.0$, and the velocity pressure exposure coefficient is $K_h = 0.57$ (ASCE Table 27.3.1). The wind directionality factor is 0.85 (ASCE Table 26.6-1). The wind flow is assumed to be clear. From ASCE Figs. 27.4-1 and 27.4-4, the net pressure coefficients are shown in Table 5.4.1.

The velocity pressure is $q_h = 0.00256\, K_z K_{zt} K_d V^2$ (ASCE Sect. 27.3.2), that is, $q_h = 16.4$ psf. Equation 5.4.1 yields design wind pressures $p = 16.4 \times 0.85\, C_N = 13.9\, C_N$ (psf). Recall that C_N is a generic notation, and that for angles 0° and 180° the specific notations are C_{NW} and C_{NL} for the windward and leeward area of the roof, respectively. For example, for angle 0°, Load Case A, the windward design wind pressure coefficient is $C_{NW} = -0.9$, so $p = -13.9 \times 0.9 = -12.5$ psf.

CHAPTER 6

REGULAR APPROACH: LOW-RISE BUILDINGS, PARAPETS, OVERHANGS ("ENVELOPE" PROCEDURE), MWFRS

6.1 NET PRESSURES ON WALLS AND ROOF

Low-rise buildings (ASCE Sect. 28.4.1) are defined as buildings with mean roof height $h \leq 60$ ft and ratio $h/L \leq 1$, where L is the least horizontal building dimension. The design wind pressures are determined as follows:

$$p = q_h[(GC_{pf}) - (GC_{pi})] \qquad (6.1.1)$$

(ASCE Eq. 28.4-1), where q_h = velocity pressure, (GC_{pf}) = combined external pressure coefficient from Table 6.1.1, and (GC_{pi}) = combined internal pressure coefficient (ASCE Fig. 26.11-1) for the zones indicated in Fig. 6.1.1 (ASCE Fig. 28.4-1). In Fig. 6.1.1a, is 10% of the least horizontal building dimension or $0.4h$, whichever is smaller, but not less than 4% of the least horizontal dimension or 3 ft. For roof angles not shown in Table 6.1.1, linear interpolation is permitted.

If for Zone 2 or Zone 2E the combined pressure coefficient (GC_{pf}) is negative, it is applied (1) from the edge of the roof for a distance equal to half the dimension of the building parallel to the direction of the MWFRS being designed, or (2) from the edge of the roof for a distance equal to 2.5 times the windward wall eave height, whichever is smaller. If the latter distance is smaller, then the remaining portion of Zone 2 or Zone 2E extending to the ridge line is to be treated as Zone 3 or 3E, respectively (Note 8, ASCE Fig. 28.4-1).

TABLE 6.1.1. External Combined Pressure Coefficients (GC_{pf}), Low-Rise Buildings*

	Load Case A Zone (Fig. 6.1.1)							
Roof slope	1	2	3	4	1E	2E	3E	4E
$0 \leq \theta \leq 5^\circ$	0.40	0.69	0.37	0.29	0.61	1.07	0.53	0.43
$\theta = 20^\circ$	0.53	0.69	0.48	0.43	0.80	1.07	0.69	0.64
$30 \leq \theta \leq 45^\circ$	0.56	0.21	−0.43	−0.37	0.69	0.27	−0.53	−0.48
$\theta = 90^\circ$	0.56	0.56	−0.37	−0.37	0.69	0.69	−0.48	−0.48

	Load Case B Zone (Fig. 6.1.1)											
All Roof Slopes	1	2	3	4	5	6	1E	2E	3E	4E	5E	6E
	−0.45	−0.69	−0.37	−0.45	0.40	−0.29	−0.48	−1.07	−0.63	−0.48	0.61	−0.43

*From ASCE Standard 7-10, *Minimum Design Loads for Buildings and Other Structures*. Copyright 2010 by the American Society of Civil Engineers. With permission from ASCE.

For the MWFRS in the direction parallel to the ridge, or for flat roofs, C_{pf} is based on an angle $\theta = 0^\circ$. The boundary between Zones 2 and 3, and between Zones 2E and 3E, must be located at mid-width of building (Note 7, ASCE Fig. 28.4-1). This requirement governs even if the windward wall eave height is sufficiently small for requirement (2) of Note 8, just noted, to be applicable.

Except for one-story buildings with $h \leq 30$ ft, and one- or two-story buildings with light-framed construction or flexible diaphragms, torsional loading shall be considered in design by applying to Zones 1T, 2T, 3T, 4T, 5T, and 6T (Fig. 6.1.2; ASCE Fig. 28.4-1) 25% of the design pressures calculated by Eq. 6.1.1 for Zones 1, 2, 3, and 4, 5, and 6 respectively, while the design pressures for the balance of the building are given by Eq. 6.1.1 (Note 5, ASCE Fig. 28.4-1).

The load patterns of Figs. 6.1.1 and 6.1.2 for winds acting transversely and longitudinally are associated with the reference corner shown therein. The load patterns shall be applied by considering in turn each building corner as reference corner. Therefore, four load cases need to be considered for each of Figs. 6.1.1a and 6.1.1b, 6.1.2a, and 6.1.2b. As noted in Table 2.2.1, item B, if the building exposure depends on direction, Exposure Requirement 2 applies; that is, the wind loads are based on the exposure category resulting in the highest wind loads from any direction.

The total horizontal shear on the building shall not be less than that determined by neglecting the wind forces on the roof. This provision does not apply to buildings using moment frames for MWFRS (Note 6, ASCE Fig. 28.4-1).

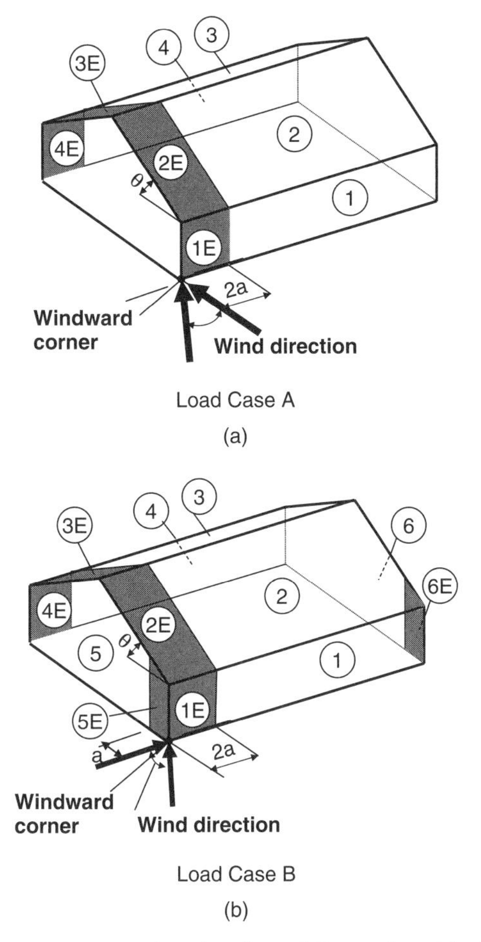

Figure 6.1.1. Pressure zones, enclosed and partially enclosed buildings, $h \leq 60$ ft, low-rise buildings, "envelope procedure," regular approach, MWFRS, Load Case A and Load Case B.

Numerical Example 6.1.1. *Commercial building with 15 ft eave height*. A rectangular office building has dimensions in plan 45 ft × 40 ft, eave height 15 ft, gable roof with slope $\theta = 26.6^\circ$ and mean roof height $h = 15 \text{ ft} + {}^1\!/_2({}^1\!/_2 \times 40 \text{ ft}) \tan 26.6^\circ = 20 \text{ ft} < 40$ ft. Since $h < 60$ ft and $20/40 < 1$, the building is a low-rise building. The building is located in Arkansas in flat terrain with Exposure B in all directions, and has been determined to be enclosed (for an example of how the enclosure is determined, see Numerical Example 3.1.1). We follow the flowchart of Fig. 2.2.2.

Risk category: II (ASCE Table 1.5-1)

Basic wind speed: $V = 115$ mph (ASCE Fig. 26.5-1a).

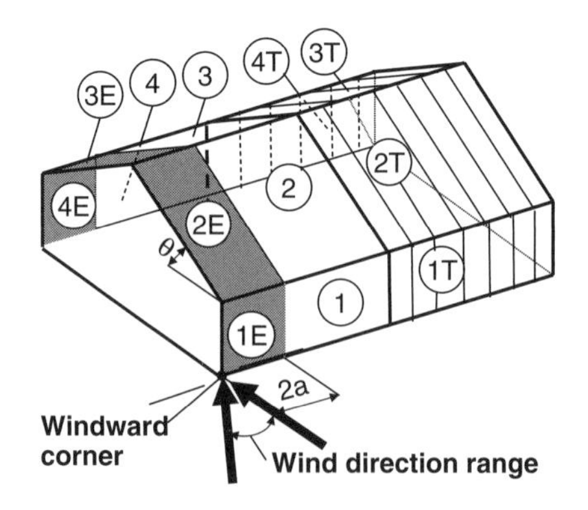

Case A Torsion (Transverse Direction)

(a)

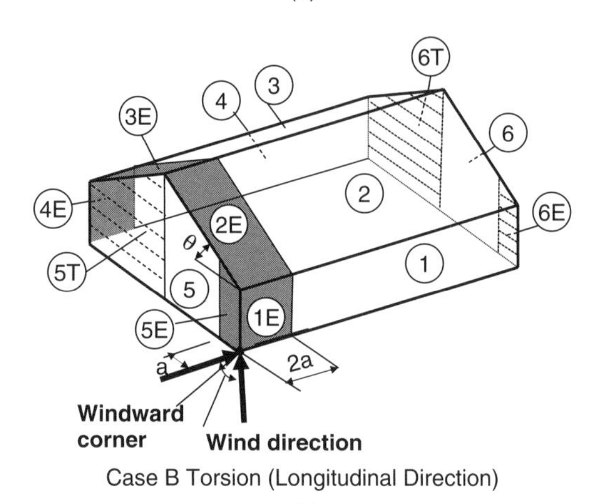

Case B Torsion (Longitudinal Direction)

(b)

Figure 6.1.2. Pressure zones, torsional loadings, enclosed and partially enclosed buildings, $h \leq 60$ ft, low-rise buildings, "envelope procedure," regular approach, MWFRS, Case A and Case B.

Exposure category: It is assumed that the building has Exposure Category B for all directions. (For an example of how the exposure category is determined if the exposure depends on direction, see Numerical Example 3.2.1.)

Enclosure classification: It is assumed that the building is enclosed.

Combined internal pressure coefficient for enclosed building: $(GC_{pi}) = \pm 0.18$ (ASCE Sect. 26.11).

***Elevation at which external and internal velocity pressures are evaluated*:** Mean roof height $h = 20$ ft (Table 2.2.1, item A, cols. 4 and 5; ASCE Sect. 28.4.1).

Velocity pressure exposure coefficient: $K_z = 0.70$ (ASCE Table 28.3-1 or Eqs. 4.2.1, $z_1 = 30$ ft for Exposure B).

Topographic factor $K_{zt} = 1.0$ (ASCE Sect. 26.11). This value corresponds to flat terrain surrounding the building. For an example of how K_{zt} is determined, see Numerical Example 3.4.1.

Wind directionality factor $K_d = 0.85$ (ASCE Sect. 26.8).

Velocity pressure (ASCE Sect. 28.3.2):

$$q_h = 0.00256 \times 0.70 \times 1.0 \times 0.85 \times 115^2 = 20.1 \text{ psf}.$$

Design pressures are given by Eq. 6.1.1 for the zones of Fig. 6.1.1.

Wind parallel to short horizontal dimension of building

Load Case A (Transverse direction):

Zone 1. $p = 20.1[0.53 + (0.56 - 0.53)(26.6 - 20.0)/(30 - 20.0)$ $(\pm 0.18)]$ $= 20.1[0.55 - (\pm 0.18)] = 7.4$ psf or 14.7 psf.

Zone 2. $p = 20.1[-0.69 + (0.21 - (-0.69))(26.6-20.0)/(30 - 20)$ $(\pm 0.18)]$ $= 20.1[-0.1-(\pm 0.18)] = -5.6$ psf or 1.6 psf.

Zone 3. $p = 20.1[-0.45 - (\pm 0.18)] = -12.7$ psf or -5.4 psf.

Zone 4. $p = 20.1[0.39 - (\pm 0.18)] = -11.5$ psf or -4.2 psf.

Similarly,

Zone 1E. $p = 11.0$ psf or 18.2 psf.

Zone 2E. $p = -7.3$ psf or -0.1 psf.

Zone 3E. $p = 15.4$ psf or -8.1 psf.

Zone 4E. $p = -14.4$ psf or -7.1 psf.

As indicated in ASCE Sect. 27.4-6, torsion need not be considered for buildings meeting the conditions listed in ASCE Appendix D (ASCE Sect. D1.1). However, for illustration purposes, the design pressures are calculated for Zones 1T, 2T, 3T, and 4T (Fig. 6.1.2). These are specified to be 25% of the pressures for Zones 1, 2, 3, and 4, respectively, that is:

Zone 1T. $p_T = 0.25 \times 7.4 = 1.9$ psf or $0.25 \times 14.7 = 3.7$ psf.

Zone 2T. $p_T = -1.4$ or 0.4 psf.

Zone 3T. $p_T = -3.2$ psf or -1.4 psf.

Zone 4T. $p_T = -2.9$ psf or -1.1 psf.

Load Case B: Longitudinal direction

Selected calculations are as follows:

Zone 2. $p = 20.1[-0.69 - (\pm 0.18)] = -17.5$ psf or -10.3 psf.

Zone 3. $p = 20.1[-0.37 - (\pm 0.18)] = -11.1$ psf or -3.8 psf.

Zone 2E. $p = 20.1[-1.07 - (\pm 0.18)] = -25.1$ psf or -17.9 psf.

Zone 3E. $p = 20.1[-0.53 - (\pm 0.18)] = -14.3$ psf or -7.0 psf.

Note: Wind loads on MWFRS of enclosed or partially enclosed low-rise buildings may be determined by the regular approach, as specified in ASCE Sect. 27.4 (buildings of all heights) or in ASCE Sect. 28.4 (low-rise buildings). ASCE Sect. 28.4 was added to an earlier version of the Standard to provide for an improved set of wind loads. As is shown by comparisons between the results of this Numerical Example and those of Numerical Example 5.2.2, the statement in the User Note of ASCE Sect. 28.2 that the provisions of ASCE Sect. 28.4.1 "generally yield the lowest wind pressures of all the analytical methods specified in this standard" is not necessarily warranted.

6.2 COMPARISON BETWEEN RESULTS BASED ON ASCE SECTS. 27.4.1 AND 28.4.1

Table 6.2.1 shows results of pressure calculations for a building with mean roof height $h = 20$ ft, based on the provisions of ASCE Sect. 27.4.1 on the one hand (Numerical Example 5.2.2) and ASCE Sect. 28.4.1 on the other (Numerical Example 6.1.1). The areas on the building exterior surface defined by ASCE Fig. 27.4-1 correspond approximately to the zones defined in ASCE Fig. 28.4-1 and listed in Table 6.2.1. Pressures in several instances, including in particular those shown in bold type, are larger in magnitude for the "low-rise buildings" (ASCE Sect. 28.4) than for the "buildings of all heights" (ASCE Sect. 27.4) calculations, by as much as 62%. While in some instances the differences are reduced on account of the Standard's minimum pressure requirements (Sect. 2.2.3), this would not be the case for buildings in higher wind zones.

TABLE 6.2.1. Pressures on Building with $h = 20$ ft Based on Sect. 27.4.1 (Buildings of All Heights) and Sect. 28.4.1 (Low-Rise Buildings), Regular Approach (psf)

	Zone	ASCE Sect. 27.4.1 (Buildings of all heights)	ASCE Sect. 28.4.1 (Low-rise bldgs.)
Wind parallel to short building dimension (Case A)	1	14.4 or 8.0	14.7 or 7.4
	2	6.2 or −7.3	1.6 or −5.6
	3	−12.3 or −5.9	−12.7 or −5.4
	4	−10.8 or −4.4	−11.5 or −4.2
	1E	14.4 or 8.0	**18.2** or 11.0
	2E	6.2 or −7.3	−0.1 or −7.3
	3E	−12.3 or −5.9	**−15.4** or −8.1
Wind parallel to long building dimension (Case B)	2	−10.8 or −5.9	**−17.5** or −10.3
	2E	−16.8 or −10.4	**−25.1** or −17.9

6.3 REGULAR APPROACH: PARAPETS AND ROOF OVERHANGS, MWFRS

Provisions for design wind pressures for parapets are identical for low-rise buildings (ASCE Sect. 28.4.2) and for buildings of all heights (ASCE Sect. 27.4.5). For an application to buildings of all heights, see Numerical Example 5.3.2.

The calculations of design wind pressures for roof overhangs are similar to those for buildings of all heights (see Numerical Example 5.3.1), except that (1) for low-rise buildings, the external pressures on the roof are calculated in accordance with ASCE Sect. 28.4.1, and (2) the upward pressure on the bottom surface of windward overhangs for low-rise buildings is:

$$p_{bottom} = q_h C_p \tag{6.3.1}$$

where $C_p = 0.7$ (ASCE Sect. 28.4.3). For buildings of all heights it is $C_p = 0.8$.

CHAPTER 7

REGULAR APPROACH: STRUCTURES OTHER THAN BUILDINGS, MWFRS

This chapter presents material, including Numerical Examples, on:

Solid freestanding walls (Numerical Example 7.1.1)
Solid freestanding signs (Numerical Example 7.1.2)
Open signs (Numerical Example 7.2.1)
Trussed towers (Numerical Example 7.2.2)
Lattice frameworks (Numerical Example 7.2.3)
Chimneys (Numerical Example 7.3.1)
Solid attached signs (Numerical Example 7.4.1)
Rooftop equipment $h > 60$ ft (Numerical Example 7.5.1)
Rooftop equipment $h \leq 60$ ft (Numerical Example 7.5.2)

7.1 SOLID FREESTANDING WALLS AND SOLID SIGNS

Signs and walls are defined as solid if the ratio of solid area to gross area is $\varepsilon > 30\%$. The design force is

$$F = q_h G\, C_f A_g \tag{7.1.1}$$

where G = gust factor (ASCE 26.9), A_g = gross area of wall or sign, and C_f is given in ASCE Fig. 29.4-1.

Signs with ratios $s/h < 1$ are designed for Cases A, B depicted in ASCE Fig. 29.4-1, and, for ratios of width to depth $B/s \geq 2$, also for Case C. If $s/h = 1$, the forces in Cases A, B, and C shall be applied at an elevation $h/2 + 0.05h$.

For Case C, if $s/h > 0.8$, C_f shall be multiplied by the reduction factor $\{1.8 - s/h\}$.

For walls with ratios $\varepsilon < 1$, the coefficients C_f shall be multiplied by the factor $\{1 - (1 - \varepsilon)^{1.5}\}$. When a return corner with length L_r is present, tabulated values marked with an asterisk (*) in ASCE Fig. 29.4-1 shall be reduced as indicated therein.

Numerical Example 7.1.1. *Solid freestanding wall.* A freestanding concrete wall with horizontal dimension $B = 45$ ft extends from the ground to height $h = 15$ ft (i.e., vertical dimension of the wall $s = h = 15$ ft). The wall has openings whose total area is 25% of the gross area of the wall. The wall is located near Miami, Florida, in flat terrain with Exposure C in all directions. We follow the steps of ASCE Table 29.1-1.

Step 1. Risk category: II (ASCE Table 1.5-1)

Step 2. Basic wind speed: $V = 170$ mph (ASCE Fig. 26.5-1A).[1]

Step 3.
- ***Wind directionality factor***: $K_d = 0.85$ (ASCE Table 26.6-1, for solid freestanding wall).
- ***Exposure category***: The building has Exposure Category C for all directions.
- ***Topographic factor***: $K_{zt} = 1.0$ (ASCE Sect. 26.8). This value corresponds to flat terrain surrounding the building. For an example of how K_{zt} is determined, see Numerical Example 3.4.1.
- ***Gust effect factor***: It is assumed that the wall was determined to be rigid, so $G = 0.85$ (ASCE Sect. 26.9). For an example of the calculation of the gust effect factor G_f for flexible structures, see Numerical Example 5.2.1, Step 3.

Step 4. Velocity pressure exposure coefficient: $K_h = 0.85$ (ASCE Table 29.3-1, for $h = 15$ ft and Exposure C).

Step 5. Velocity pressure: $q_h = 0.00256 \times 0.85 \times 1.0 \times 0.85 \times 170^2 = 53.5$ psf (ASCE Sect. 29.3.2).

Step 6. Force coefficient (ASCE Fig. 29.4-1, $s/h = 1.0$, $B/s = 3.0$): Based on Note 2 of ASCE Fig. 29.4-1, signs (or freestanding walls) with total area of openings not exceeding 30% of the gross area are classified as solid signs (or freestanding walls); the wall in this example is therefore classified as a *solid* wall. However, according to Note 2 of ASCE Fig. 29.4-1, the force coefficients for solid signs (or freestanding walls) with openings may be multiplied by a reduction factor of $(1 - (1\text{-}\varepsilon)^{1.5})$, where ε is the ratio of solid to gross area. In this case, the reduction factor is $(1 - (1 - 0.75)^{1.5}) = 0.875$.

Load CASES A and B: $C_f = 1.38 \times 0.875 = 1.21$.

[1]As of this writing, consideration is given to the possibility that the Florida Building Code will require higher wind speeds for Miami-Dade and Broward counties.

As per Note 3 of ASCE Fig. 29.4-1, since $B/s \geq 2.0$, CASE C must also be considered. As per Note 4 of ASCE Fig. 29.4-1, since $s/h > 0.8$, force coefficients shall be multiplied by the reduction factor $(1.8 - s/h) = 0.8$.

$$\textit{Load CASE C}: \; C_f = 2.60 \times 0.875 \times 0.8 = 1.82 \; (0 \text{ to } s)$$
$$1.70 \times 0.875 \times 0.8 = 1.19 \; (s \text{ to } 2s)$$
$$1.15 \times 0.875 \times 0.8 = 0.81 \; (2s \text{ to } 3s).$$

Step 7. Wind force (ASCE Sect. 29.4, using A_s the gross area of the solid freestanding wall):

CASE A (see sample illustration for CASE A in ASCE Fig. 29.4-1):

$$F = 53.5 \times 0.85 \times 1.21 \times (45 \times 15)/1000 = 37.0 \text{ kips},$$

acting normal to the face of the wall through the vertical centerline and at a height $0.55h = 8.25$ ft above the ground (see Note 3 of ASCE Fig. 29.4-1 for $s/h = 1$).

CASE B (see sample illustration for CASE B in ASCE Fig. 29.4-1):

$$F = 53.5 \times 0.85 \times 1.21 \times (45 \times 15)/1000 = 37.0 \text{ kips},$$

acting normal to the face of the wall at a distance of $0.2B = 9$ ft from the vertical centerline, closer to the windward edge, and at a height $0.55h = 8.25$ ft above the ground (see Note 3 of ASCE Fig. 29.4-1 for $s/h = 1$). For CASE B, each corner of the wall should be considered separately as the windward corner for cornering winds, as shown in the sample illustration for CASE B in ASCE Fig. 29.4-1.

CASE C (see sample illustration for CASE C in ASCE Fig. 29.4-1):

$F_1 = 53.5 \times 0.85 \times 1.82 \times (15 \times 15)/1000 = 18.6$ kip, acting normal to the face of the wall and along a vertical line at $0.5s = 7.5$ ft from the windward edge.

$F_2 = 53.5 \times 0.85 \times 1.19 \times (15 \times 15)/1000 = 12.2$ kip, acting normal to the face of the wall and along a vertical line at $1.5s = 22.5$ ft from the windward edge.

$F_3 = 53.5 \times 0.85 \times 0.81 \times (15 \times 15)/1000 = 8.2$ kip, acting normal to the face of the wall and along a vertical line at $2.5s = 37.5$ ft from the windward edge.

All forces are assumed to act at a height $0.55h = 8.25$ ft above the ground (see Note 3 of ASCE Fig. 29.4-1 for $s/h = 1$).

Numerical Example 7.1.2. *Solid freestanding sign with return corner.* A solid freestanding sign has horizontal dimension $B = 50$ ft, vertical dimension $s = 10$ ft, and height $h = 20$ ft. The sign has a return corner ($L_r = 3$ ft) at one end and is located in Melbourne, Florida, in flat terrain with Exposure B in all directions. We follow the flowchart of ASCE Table 29.1-1.

Step 1. ***Risk category***: II (ASCE Table 1.5-1)

Step 2. ***Basic wind speed***: $V = 150$ mph (ASCE Fig. 26.5-1A).

Step 3.
- ***Wind directionality factor***: $K_d = 0.85$ (ASCE Table 26.6-1, for solid freestanding sign)
- ***Exposure category***: The building has Exposure Category B for all directions.
- ***Topographic factor***: $K_{zt} = 1.0$ (ASCE Sect. 26.8). This value corresponds to flat terrain surrounding the building. For an example of how K_{zt} is determined, see Numerical Example 3.4.1.
- ***Gust effect factor***: It is assumed that the sign was determined to be rigid, so $G = 0.85$ (ASCE Sect. 26.9).

Step 4. ***Velocity pressure exposure coefficient***: $K_h = 0.62$ (ASCE Table 29.3-1, for $h = 20$ ft and Exposure B).

Step 5. ***Velocity pressure***: $q_h = 0.00256 \times 0.62 \times 1.0 \times 0.85 \times 150^2 = 30.4$ psf (ASCE Sect. 29.3.2).

Step 6. ***Force coefficient*** (ASCE Fig. 29.4-1, $s/h = 0.5$, $B/s = 5.0$, $L_r/s = 0.3$):

$$\text{Load CASES A and B: } C_f = 1.7.$$

As per Note 3 of ASCE Fig. 29.4-1, since $B/s \geq 2.0$, CASE C must also be considered.

Load CASE C: $C_f = 3.10 \times 0.9 = 2.79$ (reduction factor of 0.9 is applied for $L_r/s = 0.3$ when the corner *with* the return is the windward corner); 3.10 (when the corner *without* the return is the windward corner) (0 to s).

$$2.00\ (s \text{ to } 2s)$$

$$1.45\ (2s \text{ to } 3s)$$

$$1.05\ (3s \text{ to } 5s)$$

Step 7. ***Wind force*** (ASCE Sect. 29.4, using A_s the gross area of the solid freestanding sign):

CASE A (see sample illustration for CASE A in ASCE Fig. 29.4-1):

$$F = 30.4 \times 0.85 \times 1.7 \times (10 \times 50)/1000 = 21.9 \text{ kips}$$

acting normal to the face of the sign through the geometric center (see Note 3 of ASCE Fig. 29.4-1 for $s/h < 1$).

CASE B (see sample illustration for CASE B in ASCE Fig. 29.4-1):

$$F = 30.4 \times 0.85 \times 1.7 \times (10 \times 50)/1000 = 21.9 \text{ kips}$$

acting normal to the face of the sign at a distance of $0.2B = 10$ ft from the geometric center, closer to the windward edge (see Note 3 of ASCE Fig. 29.4-1 for $s/h < 1$). For CASE B, each corner of the wall should be considered separately as the windward corner for cornering winds, as shown in the sample illustration for CASE B in ASCE Fig. 29.4-1.

CASE C (see sample illustration for CASE C in ASCE Fig. 29.4-1):

$$F_1 = 30.4 \times 0.85 \times 2.79 \times (10 \times 10)/1000 = 7.2 \text{ kips}$$

(when the corner *with* the return is the windward corner) or $F_1 = 30.4 \times 0.85 \times 3.10 \times (10 \times 10)/1000 = 8.0$ kip (when the corner *without* the return is the windward corner), acting normal to the face of the sign through the geometric center of the region (at $0.5s = 5$ ft from the windward edge) (Note 3 of ASCE Fig. 29.4-1 for $s/h < 1$).

$$F_2 = 30.4 \times 0.85 \times 2.0 \times (10 \times 10)/1000 = 5.2 \text{ kips}$$

acting normal to the face of the sign through the geometric center of the region (at $1.5s = 15$ ft from the windward edge).

$$F_3 = 30.4 \times 0.85 \times 1.45 \times (10 \times 10)/1000 = 3.7 \text{ kips}$$

acting normal to the face of the sign through the geometric center of the region (at $2.5s = 25$ ft from the windward edge).

$$F_4 = 30.4 \times 0.85 \times 1.05 \times (10 \times 20)/1000 = 5.4 \text{ kips}$$

acting normal to the face of the sign through the geometric center of the region (at $4s = 40$ ft from the windward edge).

7.2 OPEN SIGNS, LATTICE FRAMEWORKS, TRUSSED TOWERS

The design force for *open signs* is

$$F = q_z G\, C_f A_f \tag{7.2.1}$$

where q_z = velocity pressure at the height z of the area A_f, and G is specified for rigid structures in ASCE 26.9.4 and for flexible structures in ASCE 26.9.5.

For *open signs* (defined as signs with ratio of area of openings to gross area equal to or larger than 0.3) and *lattice frameworks*, A_f = area of all exposed members and elements projected on a plane normal to the wind direction, C_f = net force coefficient from ASCE Fig. 29.5.2, and D = diameter of typical round member (two entries are listed for rounded members for each ratio ε in the tower segment under consideration, owing to the dependence of the drag force on wind speed for cylindrical members; see Sect. 13.2.5). The force F shall be assumed to act in the wind direction.

For *trussed towers*, A_f = solid area of a tower projected on the plane of the face under consideration, and C_f = net force coefficient for towers with structural angles or similar flat-sided members (ASCE Fig. 29.5.3). For trusses with *rounded members*, the forces on those members may be multiplied by the reduction factor $\{0.51\ \varepsilon^2 + 0.57\}$, where ε = ratio of solid area to gross area of one tower face for the tower segment being considered, provided that factor does not exceed unity. For square towers, the force along a tower diagonal is obtained through multiplication of the force based on ASCE Sect. 29-5.3 by the factor $\{1 + 0.75\varepsilon\}$ or 1.2, whichever is smaller. Wind forces on tower appurtenances such as ladders, conduits, lights, elevators, and so forth shall be calculated using appropriate force coefficients (see, e.g., [7-1, p. 420]). Loads due to ice accretion shall be accounted for in accordance with ASCE Chapter 10.

Numerical Example 7.2.1. *Open sign.* A freestanding sign has horizontal dimension B = 50 ft, vertical dimension s = 10 ft, and height h = 20 ft. The sign has flat-sided members and openings with total area equal to 30% of the gross area. The sign is located in Melbourne, Florida, in flat terrain with Exposure B in all directions. We follow the steps of ASCE Table 29.1-1.

Step 1. ***Risk category***: II (ASCE Table 1.5-1)

Step 2. ***Basic wind speed***: V = 150 mph (ASCE Fig. 26.5-1A).

Step 3.
- ***Wind directionality factor***: $K_d = 0.85$ (ASCE Table 26.6-1, for open signs)
- ***Exposure category***: The building has Exposure Category B for all directions.
- ***Topographic factor***: K_{zt} = 1.0 (ASCE Sect. 26.8). This value corresponds to flat terrain surrounding the building. For an example of how K_{zt} is determined, see Numerical Example 3.4.1.
- ***Gust effect factor***: It is assumed that the sign has been determined to be rigid, so G = 0.85 (ASCE Sect. 26.9).

Step 4. ***Velocity pressure exposure coefficient***: K_h = 0.62 (ASCE Table 29.3-1, for h = 20 ft and Exposure B).

Step 5. ***Velocity pressure***: $q_h = 0.00256 \times 0.62 \times 1.0 \times 0.85 \times 150^2 =$ 30.4 psf (ASCE Sect. 29.3.2).

Step 6. Force coefficient (ASCE Fig. 29.5-2, solidity ratio $\varepsilon = 0.7$, flat-sided members):

Based on Note 1 of ASCE Fig. 29.5-2, signs with openings comprising 30% or more of the gross area are considered *open* signs, and $C_f = 1.6$.

Step 7. Wind force (ASCE Sect. 29.5, using A_f the projected area normal to the wind):

The area of all exposed members and elements projected on a plane normal to the wind direction is the gross area multiplied by the solidity ratio, i.e., $A_f = 10 \times 50 \times 0.7 = 350$ sq ft.

$$F = 30.4 \times 0.85 \times 1.6 \times 350/1000 = 14.4 \text{ kips}$$

acting normal to the face of the sign through the geometric center.

Note. The wind load on an open sign is almost 52% less than that for a solid sign (CASE A of Example 7.1.2) with similar dimensions, even though the porosities are similar (30% vs. 25%). Research is recommended to eliminate such differences between design wind loads on sign structures with similar porosities.

The design forces for *trussed towers* are determined as indicated in Table 2.3.3.

Numerical Example 7.2.2. *Trussed tower.* A trussed tower with tapered square cross section has rounded members and horizontal dimensions $B =$ 20 ft and 10 ft at the bottom and top, respectively, and height $h = 200$ ft. The tower has openings with total area equal to 80% of the gross area and is located near Miami Beach, Florida, in flat terrain with Exposure C in all directions. We follow the steps of Table ASCE 29.1-1.

Step 1. Risk category: II (ASCE Table 1.5-1)

Step 2. Basic wind speed: $V = 170$ mph (ASCE Fig. 26.5-1A). See Numerical Example 7.1.1, Step 2.

Step 3.
- ***Wind directionality factor***: $K_d = 0.85$ (ASCE Table 26.6-1, for trussed tower with square cross section)
- ***Exposure category***: The building has Exposure Category C for all directions.
- ***Topographic factor***: $K_{zt} = 1.0$ (ASCE Sect. 26.8). This value corresponds to flat terrain surrounding the building. For an example of how K_{zt} is determined, see Numerical Example 3.4.1.
- ***Gust effect factor***: For the purposes of this example, it is assumed that the tower is rigid, so $G = 0.85$ (ASCE Sect. 26.9). For flexible structures, the gust effect factor, denoted by G_f, can be calculated as shown in Numerical Example 5.2.1 (Step 3).

TABLE 7.2.1. Wind Loads on a Trussed Tower

z (ft)	K_z	K_{zt}	K_d	V (mph)	q_z (psf)	G	C_f	C_f'	B (ft)	A_f (sq ft)	F (kip)
10	0.85	1	0.85	170	53.5	0.85	2.98	1.76	19.5	55.86	5.1
20	0.90	1	0.85	170	56.6	0.85	2.98	1.76	19	54.45	5.3
30	0.98	1	0.85	170	61.6	0.85	2.98	1.76	18.5	53.03	5.6
40	1.04	1	0.85	170	65.4	0.85	2.98	1.76	18	51.62	5.8
50	1.09	1	0.85	170	68.5	0.85	2.98	1.76	17.5	50.20	5.9
60	1.13	1	0.85	170	71.1	0.85	2.98	1.76	17	48.79	6.0
70	1.17	1	0.85	170	73.6	0.85	2.98	1.76	16.5	47.38	6.0
80	1.21	1	0.85	170	76.1	0.85	2.98	1.76	16	45.96	6.0
90	1.24	1	0.85	170	78.0	0.85	2.98	1.76	15.5	44.55	6.0
100	1.26	1	0.85	170	79.2	0.85	2.98	1.76	15	43.13	5.9
110	1.29	1	0.85	170	80.8	0.85	2.98	1.76	14.5	41.72	5.8
120	1.31	1	0.85	170	82.4	0.85	2.98	1.76	14	40.31	5.7
130	1.34	1	0.85	170	84.0	0.85	2.98	1.76	13.5	38.89	5.6
140	1.36	1	0.85	170	85.5	0.85	2.98	1.76	13	37.48	5.5
150	1.38	1	0.85	170	86.5	0.85	2.98	1.76	12.5	36.06	5.4
160	1.39	1	0.85	170	87.4	0.85	2.98	1.76	12	34.65	5.2
170	1.41	1	0.85	170	88.7	0.85	2.98	1.76	11.5	33.23	5.1
180	1.43	1	0.85	170	89.9	0.85	2.98	1.76	11	31.82	4.9
190	1.45	1	0.85	170	90.9	0.85	2.98	1.76	10.5	30.41	4.7
200	1.46	1	0.85	170	91.8	0.85	2.98	1.76	10	28.99	4.6

Step 4. Velocity pressure exposure coefficient: K_z (ASCE Table 29.3-1, for Exposure C): See Table 7.2.1.

Step 5. Velocity pressure: $q_z = 0.00256 \times K_z \times K_{zt} \times K_d \times V^2$ (ASCE Sect. 29.3.2): See Table 7.2.1.

Step 6. Force coefficient: C_f (ASCE Fig. 29.5-3, $\varepsilon = 0.2$):
For square cross-sectioned tower, force coefficient $C_f = 4.0 \times 0.2^2 - 5.9 \times 0.2 + 4.0 = 2.98$.

However, as per Note 3 of ASCE Fig. 29.5-3 for towers containing rounded members, it is acceptable to multiply the specified force coefficients by: $0.51 \times 0.2^2 + 0.57 = 0.59$ (≤ 1.0). The modified force coefficient is then $C_f' = 1.76$.

Step 7. Wind force: $F = q_z \times G \times C_f \times A_f$ (ASCE Sect. 29.5, Eq. 29.5-1)

A_f is the solid area of the tower face projected on the plane of that face for the tower segment under consideration. As per Note 4 of ASCE Fig. 29.5-3, for towers with square or rectangular cross sections, wind forces shall be multiplied by: $1 + 0.75 \times 0.2 = 1.15$

($\leq$ 1.2) if the wind is directed along the tower diagonal. For wind directed along the tower diagonal, the loads will be greater than for wind normal to the tower face because of the greater area A_f and the 1.15 factor. Since the tower is tapered, we use for the wind load F_i on a tower segment i the average projected solid area between the two heights (top and bottom of the segment) and the velocity pressure q_i at the top of the segment:

$$F_i = q_i \times G \times C_f' \times 1.15 \times [s \times (0.5 \times (B_i + B_{i-1}) \times 1.41) \times 0.2]$$

where $s = 10$ ft is the height of each tower segment under consideration. See Table 7.2.1.

Note. In the Standard, ASCE Table 29-2.1 is erroneously named Table 29-3.1.

Numerical Example 7.2.3. *Lattice framework.* A tower with square cross section is built of lattice framework with rounded members of 2 in. diameter and has horizontal dimension $B = 20$ ft and height $h = 80$ ft. The lattice framework has openings comprising 75% of the gross area and is located in Galveston, Texas, in flat terrain with Exposure D in all directions. We follow the flowchart of ASCE Table 29.1-1.

Step 1. ***Risk category***: II (ASCE Table 1.5-1)

Step 2. ***Basic wind speed***: $V = 150$ mph (ASCE Fig. 26.5-1A).

Step 3.
- ***Wind directionality factor***: $K_d = 0.85$ (ASCE Table 26.6-1, for lattice framework)
- ***Exposure category***: The building has Exposure Category D for all directions.
- ***Topographic factor***: $K_{zt} = 1.0$ (ASCE Sect. 26.8). This value corresponds to flat terrain surrounding the building. For an example of how K_{zt} is determined, see Numerical Example 3.4.1.
- ***Gust effect factor***: $G = 0.85$ (ASCE Sect. 26.9, assuming the structure to be rigid).

Step 4. ***Velocity pressure exposure coefficient***: K_z (ASCE Table 29.3-1, for Exposure D): See Table 7.2.2.

Step 5. ***Velocity pressure***: $q_z = 0.00256 \times K_z \times K_{zt} \times K_d \times V^2$ (ASCE Sect. 29.3.2): See Table 7.2.2.

Step 6. ***Force coefficient***: C_f (ASCE Fig. 29.5-2, $\varepsilon = 0.25$, $D \times q_z^{1/2} < 2.5$ for all z): See Table 7.2.2.

Step 7. ***Wind force***: $F = q_z \times G \times C_f \times A_f$ (ASCE Sect. 29.5, Eq. 29.5-1)
The area of all exposed members and elements projected on a plane normal to the wind direction (wind along diagonal will produce the

TABLE 7.2.2. Wind Loads on a Lattice Tower

z (ft)	K_z	K_{zt}	K_d	V (mph)	q_z (psf)	G	$D \times D\sqrt{q_z}$	C_f	A_f (ft^2)	F (kip)
10	1.03	1	0.85	150	50.4	0.85	1.2	1.3	70.7	3.9
20	1.08	1	0.85	150	52.9	0.85	1.2	1.3	70.7	4.1
30	1.16	1	0.85	150	56.8	0.85	1.3	1.3	70.7	4.4
40	1.22	1	0.85	150	59.7	0.85	1.3	1.3	70.7	4.7
50	1.27	1	0.85	150	62.2	0.85	1.3	1.3	70.7	4.9
60	1.31	1	0.85	150	64.1	0.85	1.3	1.3	70.7	5.0
70	1.34	1	0.85	150	65.6	0.85	1.3	1.3	70.7	5.1
80	1.38	1	0.85	150	67.6	0.85	1.4	1.3	70.7	5.3

highest wind load) is the gross area of each segment ($s = 10$ ft high) multiplied by the solidity ratio, i.e., $A_f = 10 \times 20 \times 1.41 \times 0.25 = 70.7$ ft^2. We use for wind load (F_i) calculation on a segment i the velocity pressure q_i at the top of the segment. See Table 7.2.2 for wind load results.

7.3 CHIMNEYS, TANKS, ROOFTOP EQUIPMENT, AND SIMILAR STRUCTURES

The design force is

$$F = q_z G\, C_f A_f \qquad (7.3.1)$$

(ASCE Figure 29.5.1), where q_z is the velocity pressure at height z, G is the gust effect factor (ASCE Sect. 26.9), C_f is taken from ASCE Figs. 29.5-1, D = diameter of circular cross section, side of square cross section, or depth of hexagonal or octagonal cross section (i.e., cross-sectional dimension normal to two parallel sides), and D' = depth of protruding elements such as ribs and spoilers.

For comments on the adequacy of ASCE Fig. 29.5-1 with regard to rooftop equipment, see Sect. 7.5 and the Recommendation of Sect. 7.5.1.

Numerical Example 7.3.1. *Chimney.* A chimney with circular cross section, 60 ft high with a uniform diameter of 12 ft, is moderately smooth. The chimney is located near Miami Beach, Florida, in flat terrain with Exposure D in all directions (the Florida Building Code requirement that for High Velocity Hurricane Zones only Exposure C be assumed would in this case be unconservative; see footnote, Example 5.2.1). We follow the flowchart of ASCE Table 29.1-1.

TABLE 7.3.1. Wind Forces on Chimney

z (ft)	K_z	K_{zt}	K_d	V (mph)	q_z (psf)	G	$D\sqrt{q_z}$	C_f	A_f (sq ft)	F (kips)
10	1.03	1	0.95	170	72.4	0.85	102.1	0.57	120	4.2
20	1.08	1	0.95	170	75.9	0.85	104.5	0.57	120	4.4
30	1.16	1	0.95	170	81.5	0.85	108.4	0.57	120	4.7
40	1.22	1	0.95	170	85.7	0.85	111.1	0.57	120	5.0
50	1.27	1	0.95	170	89.3	0.85	113.4	0.57	120	5.2
60	1.31	1	0.95	170	92.1	0.85	115.1	0.57	120	5.4

Note 1. In the Standard, ASCE Table 29.2-1 is erroneously named Table 29.3-1.
Note 2. For a square chimney, wind along diagonal will give higher load than for wind normal to face, as A_f will be higher.

Step 1. ***Risk category***: II (ASCE Table 1.5-1)

Step 2. ***Basic wind speed***: $V = 170$ mph (ASCE Fig. 26.5-1A). See Numerical Example 7.1.1, Step 2.

Step 3.
- ***Wind directionality factor***: $K_d = 0.95$ (ASCE Table 26.6-1, for round chimney)
- ***Exposure category***: The building has Exposure Category D for all directions.
- ***Topographic factor***: $K_{zt} = 1.0$ (ASCE Sect. 26.8). This value corresponds to flat terrain surrounding the building. For an example of how K_{zt} is determined, see Numerical Example 3.4.1.
- ***Gust effect factor***: It is assumed that the chimney is rigid, so $G = 0.85$ (ASCE Sect. 26.9).

Step 4. ***Velocity pressure exposure coefficient***: K_z (ASCE Table 29.3-1, for Exposure D): See Table 7.3.1.

Step 5. ***Velocity pressure***: $q_z = 0.00256 \times K_z \times K_{zt} \times K_d \times V^2$ (ASCE Sect. 29.3.2): See Table 7.1.6.

Step 6. ***Force coefficient***: C_f (ASCE Fig. 29.5-1, for $h/D = 5.0$, moderately smooth surface, $D \times q_z{}^2 > 2.5$ for all z): See Table 7.1.6.

Step 7. ***Wind force***: $F = q_z \times G \times C_f \times A_f$ (ASCE Sect. 29.5, Eq. 29.5-1. For each 10-ft segment, the projected area normal to the wind $A_f = 10 \times 12 = 120$ ft^2): See Table 7.3.1.

7.4 SOLID ATTACHED SIGNS

Pressures on solid signs attached to a wall of a building (ASCE Sect. 29.4.2) are specified as equal to the external pressures on walls considered as C&C. If the sign is in contact with the surface of the wall, this procedure is applicable if the sign does not extend beyond the side or top edges of the wall. If the sign is attached to the wall but not in direct contact with its surface, the procedure is applicable if (1) the gap between sign and wall does not exceed 3 ft and (2) the edges of the sign are at least 3 ft in from the side and top edges of the wall and, if the wall is elevated, from the bottom edge as well. The design is based upon Exposure Requirement 1 (ASCE Sects. 26.7.1, 26.7.4.3).

Numerical Example 7.4.1. *Solid attached sign.* A 7-ft-wide and 6-ft-high sign, with openings whose total area is 25% of the gross area, is attached to one of the longer walls of a low-rise enclosed office building with $h = 60$ ft, horizontal dimensions of $B \times L = 80 \times 100$ ft, and a flat roof. The building is located in Minnesota on a flat terrain with Exposure B. The sign is located at mid-height of the wall, with one of its side edges at 10 ft in from the corner of the wall, and is separated from the wall by a 3-ft gap.

The sign is considered *solid*, as the area of its openings is less than 30% of the gross area. As the gap between the sign and the wall is no more than 3 ft, and as the edge of the sign is at least 3 ft in from the free edges of the wall, we can consider the sign as a *solid attached sign* and use ASCE Sect. 29.4.2. Thus, we determine the wind pressures on the sign using wind pressures on walls in accordance with ASCE Chapter 30, and setting the internal pressure coefficient (GC_{pi}) equal to 0. We follow the steps of ASCE Table 30.4.1 (Steps to Determine C&C Wind Loads for Enclosed and Partially Enclosed Low-rise Buildings).

Step 1. ***Risk category***: II (ASCE Table 1.5-1)

Step 2. ***Basic wind speed***: $V = 115$ mph (ASCE Fig. 26.5-1A).

Step 3.
- ***Wind directionality factor***: $K_d = 0.85$ (ASCE Table 26.6-1, for solid freestanding sign)
- ***Exposure category***: The building has Exposure Category B for all directions.
- ***Topographic factor***: $K_{zt} = 1.0$ (ASCE Sect. 26.8). This value corresponds to flat terrain surrounding the building. For an example of how K_{zt} is determined, see Numerical Example 3.4.1.
- ***Gust effect factor***: $G = 0.85$ (ASCE Sect. 26.9).

Step 4. ***Velocity pressure exposure coefficient***: $K_h = 0.85$ (ASCE Table 30.3-1, for $h = 60$ ft and Exposure B).

Step 5. ***Velocity pressure***: $q_h = 0.00256 \times 0.85 \times 1.0 \times 0.85 \times 115^2 = 24.5$ psf (ASCE Sect. 30.3.2).

Step 6. ***External pressure coefficient*** (ASCE Fig. 30.4-1, $A_{eff} = 42$ ft^2):

As per Note 6 of ASCE Fig. 30.4-1, $a = 8$ ft. Thus the entire sign is in Zone 4 of the wall (see ASCE Fig. 30.4-1). ASCE Fig. 30.4-1, Note 5, allows the use of a 0.9 reduction factor (i.e., a 10% reduction) for the values of (GC_p) for walls, since the roof slope is $\theta \leq 10^\circ$.

1. Zone 4 positive pressures:

$g = [(0.9 \times 0.7) - (0.9 \times 1.0)]/\log_{10}(500/10) = -0.159$, see Eq. 9.2.2d.

For $A_{eff} = 42\ \text{ft}^2$, $(GC_p) = 0.9 \times 1.0 + (-0.159) \times \log_{10}(42/10) = 0.8$, see Eq. 9.2.2c.

2. Zone 4 negative pressures:

$g = [(0.9 \times (-0.8)) - (0.9 \times (-1.1))]/\log_{10}(500/10) = 0.159$, see Eq. 9.2.2d.

For $A_{eff} = 42\ \text{ft}^2$, $(GC_p) = 0.9 \times (-1.1) + 0.159 \times \log_{10}(42/10) = -0.9$, see Eq. 9.2.2c.

Step 7. Wind pressure (ASCE Sect. 30.4.2, Eq. 30.4-1, $GC_{pi} = 0$):

1. Zone 4 positive pressures:

$$p = (0.8 + 0) \times 24.5 = 19.6 \text{ psf (ASCE Eq. 30.4-1).}$$

2. Zone 4 negative pressures:

$$p = (-0.9 + 0) \times 24.5 = -22.1 \text{ psf (ASCE Eq. 30.4-1).}$$

7.5 ROOFTOP STRUCTURES AND EQUIPMENT ON BUILDINGS

The lateral force F_h on rooftop structures and equipment is specified in ASCE Sect. 29.5 (ASCE Fig. 29.5-1) for buildings with heights $h > 60$ ft, and in ASCE Sect. 29.5.1 for buildings with $h \leq 60$ ft. *The two sets of specifications are mutually inconsistent*.

For buildings with heights $h > 60$ ft, the horizontal force is specified as

$$F = q_z G C_f A_f \tag{7.5.1}$$

(see ASCE Sect. 29.5), where the velocity pressure q_z is defined at the centroid of the area A_f, $C_f = 1.3$ (Fig. 29.5-1), and the gust factor is permitted to be taken as $G = 0.85$, so $GC_f = 1.1$. On the other hand, for rooftop structures on buildings with $h \leq 60$ ft, the horizontal force is specified in ASCE Sect. 29.5.1 as

$$F_h = q_h (GC_r) A_f \tag{7.5.2}$$

where q_h is specified at the mean roof height h, $(GC_r) = 1.9$ if the vertical projected area A_f is less than $0.1Bh$ (B is the horizontal dimension of the building normal to the wind direction), $(GC_r) = 1.0$ if the projected area $A_f = Bh$, and linear interpolation between these values is permitted. The designer who uses ASCE Fig. 29.5-1 will obtain an estimate of the horizontal wind force that is approximately 1.9/1.1 = 1.7 times smaller than the estimate based on ASCE Sect. 29.5.1.

For buildings with heights $h \approx 60$ ft, there is a sharp discontinuity between the values yielded by ASCE Sects. 29.5 and 29.5.1.

7.5.1 Recommendation

To correct the inconsistency in the Standard just pointed out, it would be appropriate that (1) in the title of ASCE Fig. 29.5-1 the words "rooftop structures" be omitted, and (2) the provisions in ASCE Sect. 29.5.1, instead of being restricted to buildings with $h \leq 60$ ft, be applied to buildings of all heights. This is also appropriate for the following reason. ASCE Sect. 29.5.1 specifies an uplift force F_v on rooftop structures and equipment

$$F_v = q_h(GC_r)A_r \tag{7.5.3}$$

where A_r is the horizontal projected area of the rooftop structure or equipment, and $(GC_r) = 1.5$ if the projected area A_f is less than $0.1BL$, where B and L are the horizontal dimensions, while $(GC_r) = 1.0$ if the projected area $A_f = BL$. (Linear interpolation between these values is permitted.) On the other hand, for $h > 60$ ft no provision is made for the uplift force.

While this recommendation does not ensure that the pressures on rooftop equipment will be "exact," it will yield more realistic pressures than those specified by ASCE Fig. 29.5-1.

Numerical Example 7.5.1. *Rooftop equipment for building with $h > 60$ ft.* A rectangular office building has dimensions in plan 45 ft × 45 ft, eave height 62.5 ft, and a flat roof. The building is located in Iowa in flat terrain, with Exposure B in all directions. The rooftop equipment has a dimension of $H \times L \times D = 2.4 \times 2.4 \times 2.4$ ft. We follow the flowchart of ASCE Table 29.1-1.

Step 1. Risk category: II (ASCE Table 1.5-1)

Step 2. Basic wind speed: $V = 115$ mph (ASCE Fig. 26.5-1A).

Step 3.

- ***Wind directionality factor***: $K_d = 0.9$ (ASCE Table 26.6-1, equipment with square section)
- ***Exposure category***: The building has Exposure Category B for all directions.
- ***Topographic factor***: $K_{zt} = 1.0$ (ASCE Sect. 26.8). This value corresponds to flat terrain surrounding the building. For an example of how K_{zt} is determined, see Numerical Example 3.4.1.
- ***Gust effect factor***: $G = 0.85$ (ASCE Sect. 26.9).

Step 4. Velocity pressure exposure coefficient: $K_z = 0.86$ (ASCE Table 29.3-1, for $z = h + H/2 = 63.7$ ft and Exposure B).

Step 5. Velocity pressure: $q_z = 0.00256 \times 0.86 \times 1.0 \times 0.9 \times 115^2 = 26.2$ psf (ASCE Sect. 29.3.2).

Step 6. Force coefficient: $C_f = 1.3$ and 1.0, for wind normal to face and wind along diagonal, respectively (ASCE Fig. 29.5-2, for square cross section with $H/D = 1.0$).

Step 7. Wind force:
Wind normal to face:

$$F = 26.2 \times 0.85 \times 1.3 \times (2.4 \times 2.4) = 166.8 \text{ lb}$$

(ASCE Sect. 29.5, Eq. 29.5-1, for equipment projected area normal to the wind $A_f = 2.4 \times 2.4$ sq ft).
Wind along diagonal:

$$F = 26.2 \times 0.85 \times 1.0 \times (2.4 \times 2.4 \times 1.41) = 181.4 \text{ lb}$$

(ASCE Sect. 29.5, Eq. 29.5-1, for equipment projected area normal to the wind $A_f = 2.4 \times 2.4 \times 1.41$ sq ft). The design wind load (lateral force) is 181.4 lb.

Note. As was indicated earlier, if the height of a building is over 60 ft, ASCE Sect. 29.5, Eq. 29.5-1, is applicable. For buildings with height less than or equal to 60 ft, ASCE Sect. 29.5.1, Eq. 29.5-2, is applicable. This will be the source of differences in wind loads on rooftop equipment with $h = 60$ ft and with h slightly exceeding 60 ft. In addition, the vertical uplift force on rooftop equipment is considered for buildings with height less than or equal to 60 ft, but not for buildings with height greater than 60 ft. For example, in Numerical Example 7.5.1, for $h = 62.5$ ft, $F_h = 181.4$ lb, $F_v = 0$, while, as is shown in Numerical Example 7.5.2, for the same rooftop equipment, $h = 60$ ft, $F_h = 400.9$ lb, $F_v = 223.8$ lb.

Note also that: ASCE Table 29.1-1 Step 7 erroneously refers to Eq. 29.6-1 and Eq. 29.6-2, instead of Eq. 29.5-1 and Eq. 29.5-2, respectively, and in ASCE Fig. 29.5-1, the height of the structure (equipment in this case) is denoted by *h,* the same notation as for the mean roof height. The current example designates equipment height as H for clarity.

Numerical Example 7.5.2. *Rooftop equipment for building with $h \leq 60$ ft.* A rectangular office building has dimensions in plan 45 ft × 45 ft, eave height 60 ft, and a flat roof. The building is located in Iowa in flat terrain, with Exposure B in all directions. The rooftop equipment has dimensions $H \times L \times D = 2.4 \times 2.4 \times 2.4$ ft. We follow the steps of ASCE Table 29.1-1.

Step 1. **Risk category**: II (ASCE Table 1.5-1)

Step 2. ***Basic wind speed***: $V = 115$ mph (ASCE Fig. 26.5-1A).

Step 3.
- ***Wind directionality factor***: $K_d = 0.9$ (ASCE Table 26.6-1, equipment with square section)
- ***Exposure category***: The building has Exposure Category B for all directions.
- ***Topographic factor***: $K_{zt} = 1.0$ (ASCE Sect. 26.8). This value corresponds to flat terrain surrounding the building. For an example of how K_{zt} is determined, see Numerical Example 3.4.1.
- ***Gust effect factor***: G value is not applicable, as (GC_r) values are given in ASCE Sect. 29.5.1.

Step 4. ***Velocity pressure exposure coefficient***: $K_z = 0.85$ (ASCE Table 29.3-1, for $z = h = 60$ ft and Exposure B).

Step 5. ***Velocity pressure***: $q_z = 0.00256 \times 0.85 \times 1.0 \times 0.9 \times 115^2 =$ 25.9 psf (ASCE Sect. 29.3.2).

Step 6. ***Force coefficient***: C_f (not applicable).

Step 7.
- ***Lateral force***:

$$F_h = 25.9 \times 1.9 \times (2.4 \times 2.4 \times 1.41) = 400.9 \text{ lb}$$

(ASCE Sect. 29.5.1, Eq. 29.5-2; for wind along diagonal, the equipment's vertical projected area on a plane normal to the direction of wind is $A_f = 2.4 \times 2.4 \times 1.414 = 8.15$ ft sq, which is less than $0.1Bh = 0.1 \times 45 \times 60 = 270$ ft sq).

- ***Uplift force***:

$$F_v = 25.9 \times 1.5 \times (2.4 \times 2.4) = 223.8 \text{ lb}$$

(ASCE Sect. 29.5.1, Eq. 29.5-3; the equipment's horizontal projected area $A_r = 2.4 \times 2.4 = 5.76$ ft sq is less than $0.1BL = 0.1 \times 45 \times 45 = 202.5$ ft sq).

CHAPTER 8

SIMPLIFIED APPROACH: ENCLOSED SIMPLE DIAPHRAGM BUILDINGS, PARAPETS, OVERHANGS (MWFRS)

This chapter presents material on the application of the simplified approach to the determination of wind loads on MWFRS of enclosed simple diaphragm buildings. The Standard contains simplified approach provisions in ASCE Chapter 27 Part 2 (ASCE Sects. 27.5 and 27.6) for buildings with $h \leq 60$ ft and buildings with 60 ft $< h \leq 160$ ft, and in ASCE Chapter 28 Part 2 (ASCE Sect. 28.6) for low-rise buildings (for which the mean roof height is also $h \leq 60$ ft; i.e., ASCE Chapter 27 Part 2 and ASCE Chapter 28 Part 2 overlap). The following Numerical Examples are included in this chapter:

Numerical Example 8.1.1 (Sect. 8.1), *Class 1 buildings* (i.e., buildings with mean roof height $h \leq 60$ ft and ratio of along-wind to across-wind building dimensions in plan $0.2 \leq L/B \leq 5.0$, ASCE Sect. 27.5.2).

Numerical Example 8.2.1 (Sect. 8.2), *Parapets* (ASCE Sect. 27.6.2).

Numerical Example 8.3.1 (Sect. 8.3), *Roof overhangs* (ASCE Sect. 27.6.3).

Numerical Example 8.4.1 (Sect. 8.4), *Class 2 buildings* (i.e., buildings with mean roof height 60 ft $< h \leq 160$ ft, ratio $0.5 \leq L/B \leq 2.0$, and natural frequency $n \geq 75/h$ Hz, where h is in ft; ASCE Sect. 27.5.2).

Numerical Example 8.5.1 (Sect. 8.5), *Simple diaphragm low-rise buildings* (ASCE Sect. 28.6).

8.1 SIMPLIFIED APPROACH: CLASS 1 BUILDINGS, WALLS AND ROOF, MWFRS

Section 8.1.1 presents a Numerical Example on the application of the simplified approach to pressures on a Class 1 building. Section 8.1.2 compares those pressures to those obtained by two methods based on the regular approach applied to buildings of all heights and to low-rise buildings.

8.1.1 Commercial Building with 15-ft Eave Height

Numerical Example 8.1.1. *Commercial building with 15-ft eave height.* The building being considered is an enclosed building with diaphragms (ASCE Sect. 26.2), rectangular shape in plan (45 ft $\times$ 40 ft), 15-ft eave height, gable roof ($\theta = 26.6^\circ$ slope), and mean roof height $h = 20$ ft. The building is located in Arkansas in flat terrain with Exposure B in all directions.

We first ascertain that the building is a Class 1 building. Since $h < 60$ ft, and the ratio between the dimensions of the building in plan is not less than 0.2 nor more than 5.0 (ASCE Sect. 27.5.2), the building is classified as a Class 1 building.

Risk category: II (ASCE Table 1.5-1).

Basic wind speed: $V = 115$ mph (ASCE Fig. 26.5-1a).

Enclosure classification: Enclosed.

Exposure category. Exposure Category B for all directions. For an example of the application of the Standard provisions on exposure see Chapter 3, Numerical Example 3.3.1. As indicated in Table 2.2.3, if the exposure category depends on direction, Exposure Requirement 1 applies; that is, the wind loads are based on the appropriate exposure for each wind direction (ASCE Sect. 26.7.4.1).

Combined internal pressure coefficient for enclosed building: $(GC_{pi}) = \pm 0.18$ (ASCE Sect. 26.11). (For this item, as well as the following five items, required for determining internal pressures, see ASCE Table 27.6-1, Notes 2 and 4 concerning integral pressure effects.)

Elevation for which internal velocity pressures are evaluated: $h = 20$ ft (ASCE Sect. 27.4.1).

Velocity pressure exposure coefficient: For internal pressures, at $z = h = 20$ ft, $K_z = 0.62$ (ASCE Table 27.3.1).

Wind directionality factor $K_d = 0.85$ (ASCE Sect. 26.8).

Velocity pressure (ASCE Sect. 27.3.2):

$$q_h = 0.00256 \times 0.62 \times 0.85 \times 1.0 \times 115^2 = 17.8 \text{ psf.}$$

Internal pressures on walls: The internal pressures are

$$p_i = \pm 0.18 \times 17.8 = \pm 3.2 \text{ psf}$$

Topographic factor K_{zt} (ASCE Sect. 26.11): All pressures are multiplied by the factor K_{zt}, which is evaluated at each height z, but may be evaluated instead at height $h/3$ (ASCE Sect. 27.5.2). The value corresponding to flat terrain is $K_{zt} = 1.0$.

External pressures on walls (ASCE Table 27.6-1, Exposure B): The pressures are calculated for $h = 20$ ft, $V = 115$ mph.

For wind parallel to short building dimension ($L = 40$ ft): $L/B = 40/45 = 0.89$, tabulated pressures: $p_h = 19.2$ psf, $p_0 = 18.8$ psf.

Leeward wall: $p = -0.38 \times 19.2 = -7.3$ psf (ASCE Table 27.6-1, Note 4).

Windward wall: $p_h = 19.2 - 7.3 = 11.9$ psf, $p_0 = 18.8 - 7.3 = 11.5$ psf.

Side walls: $p = -0.54 \times 19.2 = -10.4$ psf (ASCE Table 27.6-1, Note 2).

For wind parallel to long building dimension (L = 45 ft): L/B = 45/40 = 1.125

For $L/B = 1$, $p_h = 19.2$ psf, $p_0 = 18.8$ psf.

For $L/B = 2$, $p_h = 16.6$ psf, $p_0 = 16.2$ psf.

For $L/B = 1.125$, $p_h = 18.9$ psf, $p_0 = 18.5$ psf.

Leeward wall: $p = -0.37 \times 18.9 = -7.0$ psf.

The factor 0.37 is obtained by interpolation between the values 0.38 and 0.27, i.e., $0.38 - (1.125 - 1.00)(0.38 - 0.27)/(2.00 - 1.00) = 0.37$ (see ASCE Table 27.6-1, Note 4).

Windward wall : $p_h = 18.9 - 7.0 = 11.9$ psf, $p_0 = 18.5 - 7.0 = 11.5$ psf.

Side walls : $p = -0.55 \times 18.9 \times 10.4$ psf (ASCE Table 27.6-1, Note 2).

Note. As indicated in Notes 2 and 4 of ASCE Table 27.6-1, the pressures on walls calculated by using ASCE Table 27.6-1 do not include internal pressures. Although internal pressures on opposite exterior walls cancel each other, their effects may need to be taken into account, depending upon the nature of the MWFRS.

Net pressures on roof (ASCE Table 27.6-2): The design is governed by the negative pressures acting on Zones 3, 4, and 5. For Exposure C:

Zone 3	Zone 4	Zone 5
−27.5	−24.5	−20.1

(in psf). (The zones are defined in the schematic of ASCE Table 27.6-2, gable roof.)

Since the building has Exposure B, the given values must be multiplied by the Exposure Adjustment Factor 0.692 (Note 1, ASCE Table 27.6-2). Therefore, the pressures in psf are:

Zone 3	Zone 4	Zone 5
−19.0	−17.0	−13.9

The pressures on all roof zones for Cases 1 and 2 of Table 27.6-2 are:

Net Pressures on Roof (ASCE Table 27.6-2; $V = 115$ mph, Exp. B, $h = 20$ ft, Roof Slope = 26.6°)

Zone	Load Case	Wind Pressure for Exp. C (psf)	Adjustment Factor	Wind Pressure for Exp. B (psf)	K_{zt}	Pressure (psf)
1	1	−14.3	0.692	−9.9	1.0	−9.9
	2	11.3	0.692	7.8	1.0	7.8
2	1	−17.9	0.692	−12.4	1.0	−12.4
	2	−8.6	0.692	−6.0	1.0	−6.0
3	1	−27.5	0.692	−19.0	1.0	−19.0
	2	0.0	0.692	0.0	1.0	0.0
4	1	−24.5	0.692	−17.0	1.0	−17.0
	2	0.0	0.692	0.0	1.0	0.0
5	1	−20.1	0.692	−13.9	1.0	−13.9
	2	0.0	0.692	0.0	1.0	0.0

The wall and roof pressures are multiplied by the topographic factor K_{zt}, which in this example is 1.0.

8.1.2 Comparison of Pressures Based on Regular Approaches (ASCE Sects. 27.4.1 and 28.4.1) and Simplified Approach (ASCE Sect. 27.6)

Zone 3 of the gable roof (ASCE Table 27.6-2, gable roof, simplified approach, wind parallel to the ridge) may be considered to correspond to Zone 2E, Load Case B of ASCE Fig. 28.4-1 (low-rise buildings, regular approach, ASCE Sect. 28.4.1). For the building of Numerical Example 8.1.1, the calculated pressures on the roof near the end wall for flow parallel to the ridge are as follows:

Regular approach, low-rise buildings (Load Case B, Num. Ex. 6.1.1): *−25.1 psf*.

Regular approach, buildings of all heights (Num. Ex. 5.2.2): *−16.8 psf*.

Simplified approach (ASCE Sect. 27.6, Num. Ex. 8.1.1): *−19.0 psf*.

8.2 SIMPLIFIED APPROACH: PARAPETS, MWFRS

Numerical Example 8.2.1. *Parapets for commercial building with flat roof* (ASCE Sect. 27.6.2, ASCE Fig. 27.6-2). The building being considered is an enclosed simple diaphragm building (ASCE Sect. 26.2) with rectangular shape in plan (45 ft × 40 ft), flat roof, and eave height $h = 15$ ft. The building is located in Arkansas ($V = 115$ mph) in flat terrain with Exposure B in all directions.

Note that there is no difference in the calculation of loads on parapets between Class 1 and Class 2 buildings. Assume that the height of the parapet is 3 ft. ASCE Sect. 27.6.2 requires the application to the parapet of a uniform net horizontal pressure equal to 2.25 times the wall pressure evaluated from Table 27.6-1 for $L/B = 1.0$ at the elevation h_p of the top of the parapet (see ASCE Fig. 27.6-2). The height h_p of the top of the parapet is 15 ft + 3 ft = 18 ft. For $V = 115$ mph and $h = 18$ ft, the horizontal pressure on the parapet is $p_p = 2.25 \times [18.2 + (18 - 15)(19.2 - 18.2)/(20 - 15)] =$ 42.3 psf, where 18.2 psf and 19.2 psf are the pressures listed in Table 27.6.1 for Exposure B for $h = 15$ ft and 20 ft, respectively, for $L/B = 1.0$. The 42.3 psf pressure applied uniformly over the height of the parapet is the total additional wind load contributed by the parapets to the building's MWFRS. Since the topographic factor K_{zt} in this example is 1.0, no multiplication by K_{zt} is required for the pressures calculated, as shown in this example.

8.3 SIMPLIFIED APPROACH: ROOF OVERHANGS, MWFRS

Numerical Example 8.3.1. *Roof overhangs for building of Numerical Example* 8.1.1 (ASCE Sect. 27.6.3, ASCE Fig. 27.6-3). In addition to the loading calculated for the building roof and walls, the MWFRS is subjected to an upward wind load acting on the underside of the roof overhangs equal to 75% of the roof pressure for Zone 3 or Zone 1 (see ASCE Fig. 27.6-2 for definition of zones), as applicable. For wind direction parallel to the ridge, the windward roof overhangs (ASCE Sect. 27.6.3 specifies that the pressure shall to applied to the windward overhang only) adjacent to Zone 3 will have an upward wind loading of $0.75 \times (27.5) \times 0.692 = 14.3$ psf[1], where 27.5 psf is the magnitude of the net pressure on Zone 3 for buildings with $h = 20$ ft, $V = 115$ mph, and Exposure C, and 0.692 is the Exposure Adjustment Factor (see Numerical Example 8.1.1).

[1]According to Commentary ASCE Sect. C27.6.3, the multiplier 0.75 is derived from pressures on Zone 3, and applies to "the tabulated pressure for Zone 3" in ASCE Table 27.6-2. There is no explanation in ASCE Sect. C27.6-3 on why that multiplier also applies to Zone 1, as is indicated in ASCE Sect. 27.6.3 and ASCE Fig. 27.6-3.

Similarly, for wind direction normal to the ridge, the windward roof overhangs adjacent to Zone 1 will have an upward wind loading of $0.75 \times (14.3) \times 0.692 = 7.4$ psf, where 14.3 psf is the magnitude of the net pressure on Zone 1 for buildings with $h = 20$ ft, $V = 115$ mph, and Exposure C, and 0.692 is the Exposure Adjustment Factor. Since in this example the topographic factor K_{zt} is 1.0, no multiplication by K_{zt} is required for the pressures calculated, as shown in this example.

Note. There is no difference in the calculation of loads on roof overhangs between Class 1 and Class 2 buildings.

8.4 SIMPLIFIED APPROACH: CLASS 2 BUILDINGS, WALLS AND ROOF, MWFRS

Section 8.4.1 presents Numerical Example 8.4.1 on the application of the simplified approach to pressures on a Class 2 building.

8.4.1 Office Building with *h* = 95 ft

Numerical Example 8.4.1. *Rigid enclosed simple diaphragm office building, $h = 95$ ft* (ASCE Sects. 27.5.2, 27.6.1). The building to be designed has rectangular shape in plan (60 ft × 120 ft), eave height 95 ft, and flat roof; it is located at the southern tip of the Florida peninsula in flat, suburban terrain (Exposure B) in all directions. (For the Florida Building Code requirement on exposure for High Velocity Hurricane Zones, see footnote, Example 5.2.1; in this example, we consider only ASCE 7-10 Standard requirements.) The building complies with the definition of enclosed simple diaphragm buildings (ASCE Sect. 26.2). It is assumed that for this building the fundamental natural frequency in Hz is not less than $75/h$ (h is in feet) (ASCE Sect. 27.5.2).

We first ascertain whether the building is a Class 1 or Class 2 building. Since $h > 60$ ft, the building is not Class 1 (ASCE Sect. 27.5.2). Since 60 ft $< h \leq 160$ ft, and the ratio between the dimensions of the building in plan is not less than 0.5 nor more than 2.0 (ASCE Sect. 27.5.2), the building is a Class 2 building.

Design wind pressures: The following steps are required to determine the design wind pressures:

Risk category (ASCE Table 1.5-1): In accordance with ASCE Table 1.5-1, office buildings are assigned *Risk Category II*.

Basic wind speed: For Risk Category II, use Fig. 26.5-1a. For the southern tip of Florida, the basic wind speed $V = 170$ *mph*.

Enclosure classification (ASCE Sect. 26.10): As stated earlier, the building is assumed to be enclosed. (Given the location of the building, this requires that the glazing be protected or that the conditions listed in Sect. 3.2 for wind-borne debris regions be satisfied.)

Building exposure (ASCE Sect. 26.7): In this example, it is assumed that the site has *Exposure B in all directions.* For an example of the application of the Standard provisions on exposure, see Chapter 3, Numerical Example 3.3.1. If the exposure category depends on direction, Exposure Requirement 1 applies; that is, the wind loads are based on the appropriate exposure for each wind direction (ASCE Sect. 26.7.4.1).

Is the building a low-rise building? (ASCE Sects. 26.2, ASCE Chapter 28). Since $h > 60$ ft, the building is *not a low-rise building.*

Wind directionality factor K_d (ASCE Table 26.6-1): For buildings $K_d = 0.85$.

Topographic factor K_{zt} (ASCE Sect. 26.8): The factor K_{zt} is evaluated at all heights z or may instead be evaluated at height $h/3$ (ASCE Sect. 27.5.2). In this example, it is assumed that the terrain is flat, so $K_{zt} = 1.0$. (For an application to the case where topographic features are present, see Sect. 3.4, Numerical Example 3.4.1).

Combined internal pressure coefficient (GC_{pi}) (ASCE Sect. 26.11): As indicated in Notes 2 and 4 of ASCE Table 27.6-1, the pressures on walls calculated by using ASCE Table 27.6-1 do not include internal pressures. Although internal pressures on opposite exterior walls cancel each other, if walls are supported by members of the MWFRS, the design of such members may need to take internal pressures into account. Since the building is assumed to be enclosed, $(GC_{pi}) = \pm 0.18$.

Internal pressures on walls (ASCE Sect. 27.4): The internal pressures on walls are determined in accordance with ASCE Sect. 27.4.1, so $p_i = \pm 11$ psf (see Numerical Example 5.2.1).

Fundamental natural frequency (ASCE Sect. 27.5.2): It is assumed in the Example that the building's main wind force resisting system consists of sufficiently stiff shear walls that ASCE Eq. 26.9-5 yields $n_a \geq 75/95 = 0.8$ Hz, thus satisfying the requirement of ASCE Sect. 27.5.2.

Wall pressures (ASCE Table 27.6-1): Determine p_h and p_0 for $L/B = 0.5$ (L = along-wind dimension of building, i.e., *wind parallel to short building dimension*).

For $h = 90\text{ft}$:

$V = 160$ mph, $p_h = 65.9$ psf, $p_0 = 49.2$ psf.

$V = 180$ mph, $p_h = 86.0$ psf, $p_0 = 64.2$ psf.

By linear interpolation, For $h = 90$ ft:

$V = 170$ mph, $p_h = 76.0$ psf, $p_0 = 56.7$ psf.

For $h = 100$ ft:

$V = 160$ mph, $p_h = 69.6$ psf, $p_0 = 50.9$ psf.

$V = 180$ mph, $p_h = 91.2$ psf, $p_0 = 66.7$ psf.

By linear interpolation, For $h = 100\text{ft}$:

$V = 170$ mph, $p_h = 80.4$ psf, $p_0 = 58.8$ psf.

For $V = 170$ mph and $h = 95$ ft, by linear interpolation:

$p_h = 78.2$ psf, $p_0 = 57.8$ psf.

The sum of the external pressures on the windward and leeward walls is assumed to vary linearly from $p_0 = 57.8$ psf to $p_h = 78.2$ psf (ASCE Table 27.6-1).

External pressures on leeward wall: Since $L/B = 0.5$, the pressures on the leeward wall are uniformly distributed, acting outward, and are equal to 38% of p_h, that is,

$$p_{leeward} = 0.38 \times (-78.2) = -29.7 \text{ psf.}$$

External pressures on windward wall: The pressures on the windward wall are obtained by subtracting the magnitude of the pressures $p_{leeward}$ from the net sum of the external pressures on the windward and leeward walls, that is, they vary linearly from

$$p_{0\ windward} = 57.8 - 29.7 = 28.1 \text{ psf to}$$

$$p_{h\ windward} = 78.2 - 29.7 = 48.5 \text{ psf at height h.}$$

Side wall external pressures (i.e., pressures on walls parallel to the short building dimension) are negative (i.e., act outward) and are uniform over the wall surfaces. Since $L/B = 0.5$, the side wall external pressures are 54% of the pressure p_h, that is, $0.54 \times (-78.2) = -42.2$ psf.

Similar calculations for wall pressures are now performed for $L/B = 2.0$ ($L =$ along-wind dimension of building, i.e., *wind parallel to the long building dimension*):

The sum of the external pressures on the windward and leeward walls varies linearly from $p_0 = 49.0$ psf to $p_h = 70.8$ psf (ASCE Fig. 27.6-1).

External pressures on leeward wall: Since $L/B = 2.0$, the pressures on the leeward wall are uniformly distributed, acting outward, and are equal to 27% of p_h, that is,

$$p_{leeward} = 0.27 \times (-70.8) = -19.1 \text{ psf.}$$

External pressures on windward wall: $p_{0\ windward} = 49.0 - 19.1 = 29.9$ psf, $p_{h\ windward} = 70.8 - 19.1 = 51.7$ psf.

Side wall external pressures: Since $L/B = 2.0$, the side wall external pressures are 64% of the pressure p_h, that is, $0.64 \times (-70.8) = -45.3$ psf.

Roof pressures (ASCE Sect. 27.6, Table 27.6-2): The pressure zones are defined in ASCE Table 27.6-2 for flat, gable, hip, monoslope, and mansard roofs. For Exposure C, the net pressures induced by *wind* *parallel to the long building dimension*, to be applied simultaneously with the wall pressures, are (in psf):

h (ft)	V (mph)	Zone 3	Zone 4	Zone 5
100	160	−74.7	−66.6	−54.6
90	160	−73.1	−65.2	−53.4
100	180	−94.6	−84.3	−69.2
90	180	−92.5	−82.5	−67.6

The following pressures are obtained for $h = 95$ ft:

V (mph)	Zone 3	Zone 4	Zone 5
160	−73.9	−65.9	−54.0
180	−93.8	−83.4	−68.4
170	−83.7	−74.7	−61.2

The last row is obtained from interpolation between the first two rows. These pressures are applicable for Exposure C. To obtain the pressures corresponding to Exposure B, it is necessary to multiply the pressures applicable to Exposure C by the Exposure Adjustment Factor provided by ASCE 7 Table 27.6-2, Note 1. For $h = 95$ ft, that factor is 0.778. Therefore, the pressures for Exp. B, $V = 170$ mph, $h = 95$ ft, in psf are:

Zone 3	Zone 4	Zone 5
−65.1	−58.1	−47.6

All the pressures calculated as shown in this Numerical Example must be multiplied by the topographical factor K_{zt}. (In this Numerical Example $K_{zt} = 1.0$.)

Design wind load cases (ASCE Sect. 27.6.1; ASCE Appendix D): The MWFRS shall be designed for all wind load cases of ASCE Sect. 27.4-6 (ASCE Fig. 27.4-8), except that no cases involving torsional moments need to be considered for buildings meeting the requirements of ASCE Appendix D. However, the designer may wish to ascertain whether checking those requirements will be more time-consuming than taking the torsional moments into account.

Parapets and roof overhangs: The provisions for the design of parapets and roof overhangs are the same for Class 1 and Class 2 buildings. For Numerical Examples, see Sects. 8.2 and 8.3.

8.4.2 Comparison of Pressures Based on Simplified Approach (ASCE Sect. 27.6.2) and Regular Approach (ASCE Sect. 27.4.1)

Roof pressures. We compare pressures on the office building with $h = 95$ ft obtained by the simplified approach used in Numerical Example 8.4.1 and by the regular approach used in Numerical Example 5.2.1. For roof Zone 3, the simplified approach (ASCE Fig. 27.6-2) yields a pressure of *−65.1 psf* (wind parallel to long building dimension, see Numerical Example 8.4.1). Zone 3 in Numerical Example 8.4.1 corresponds for wind parallel to the long building dimension to the roof zone nearest the windward wall in ASCE Fig. 27.4-1, for which the pressure is *−61.8 psf* (wind parallel to long building dimension, see Numerical Example 5.2.1). The two results are in this case quite close.

Wall pressures: For *wind parallel to long building dimension*, the external pressures (internal pressures are not included, see ASCE Table 27.6-1, notes 2 and 4) at the top of the windward, leeward, and side walls determined by the simplified approach in Numerical Example 8.4.1 are, respectively,

$$51.7, -19.1, \text{ and } -45.3 \text{ psf.}$$

The counterparts of these pressures in Numerical Example 5.2.1 are:

$$41.9, -15.7, \text{ and } -36.7 \text{ psf.}$$

The simplified approach results in overestimation of the pressures by 20–25%.

For *wind parallel to the short building dimension*, the external pressures at the top of the windward, leeward, and side walls calculated in Numerical Example 8.4.1 are:

$$48.5, -29.7, \text{ and } -42.2 \text{ psf}$$

respectively. The counterparts of these pressures in Numerical Example 5.2.1 are:

$$41.9, -26.2, \text{ and } -36.7 \text{ psf.}$$

The simplified approach results in overestimation of the pressures by approximately 15%.

8.5 SIMPLIFIED APPROACH: SIMPLE DIAPHRAGM LOW-RISE BUILDINGS, MWFRS

Section 8.5.1 presents Numerical Example 8.5.1 on the application of the simplified approach to pressures on a low-rise building (ASCE Sect. 28.6.3, ASCE Fig. 28.6-1; Fig. 8.5.1). Section 8.5.2 compares those pressures to pressures on the same building by the regular approach applied to low-rise buildings (ASCE Sect. 28.4.1).

8.5.1 Commercial Building with 15-ft Eave Height

Numerical Example 8.5.1. *Commercial building with 15-ft eave height.* The building being considered is an enclosed simple diaphragm building rectangular in plan (45 ft × 40 ft), eave height 15 ft, gable roof with slope

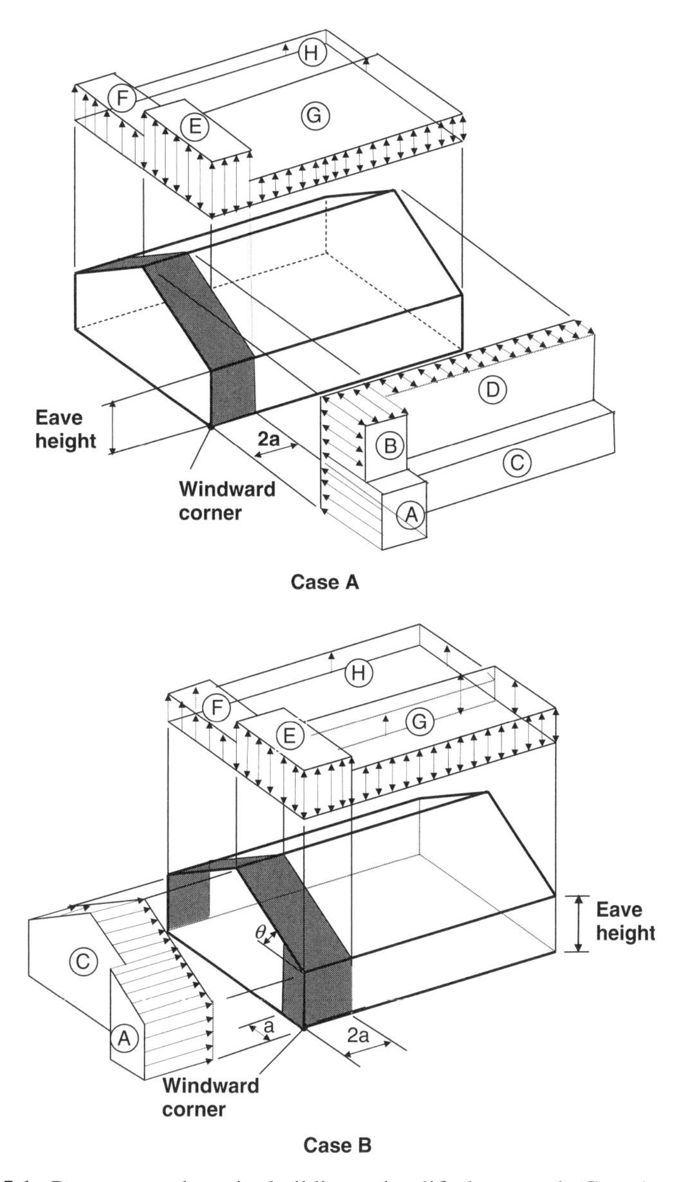

Figure 8.5.1. Pressures on low-rise buildings, simplified approach, Case A and Case B.

$\theta = 26.6^\circ$, and mean roof height $h = 20$ ft. The building is located in Arkansas ($V = 115$ mph) in flat terrain with Exposure B in all directions. The building meets the requirements defining low-rise buildings (ASCE Sect. 26.2), that is, $h \leq 60$ ft, 20 ft/40 ft ≤ 1. This building is the same as the building considered in Numerical Example 8.1.1, and is assumed to satisfy all conditions of ASCE Sect. 28.6.2. The wind loading may be determined in accordance with either ASCE Chapter 27 or ASCE Chapter 28, using in either case the regular or the simplified approach. In this Numerical Example, the determination of the wind loading is made in accordance with Chapter 28, Part 2 (simplified approach).

Risk category (ASCE Table 1.4.-1): In accordance with ASCE Table 1.5-1, office buildings are assigned *Risk Category II*.

Basic wind speed: For Risk Category II, use Fig. 26.5-1a. For Arkansas, the basic wind speed is $V = 115$ *mph*.

Building exposure (ASCE Sect. 26.7). In this example, it is assumed that the site has *Exposure B in all directions.* For an example of the application of the Standard provisions on exposure, see Chapter 3, Numerical Example 3.3.1. If the exposure category depends on direction, Exposure Requirement 2 applies; that is, the wind loads are based on the exposure category resulting in the highest wind load for any wind direction at the site (ASCE Sect. 26.7.4.1).

Topographic factor K_{zt} (ASCE Sect. 26.8): The factor K_{zt} is evaluated at all heights z or may instead be evaluated at height $h/3$ (ASCE Sect. 27.5.2). In this example, it is assumed that the terrain is flat, so $K_{zt} = 1.0$. (For an application to the case where topographic features are present, see Sect. 3.4, Numerical Example 3.4.1.)

Wind pressures for $h = 30$ ft, p_{S30} (ASCE Fig. 28.6-1; Fig. 8.5.1): For $V = 115$ mph and a 26.6° roof slope, the pressures in psf at $h = 30$ ft for Exposure B are as follows:

Pressures p_{S30}, Case A (ASCE Fig. 28.6-1; Fig. 8.5.1)

	Horizontal Pressures				Vertical Pressures				Overhangs*	
Zone	A	B	C	D	E	F	G	H	E_{OH}	G_{OH}
Load Case 1	25.4	8.0	19.0	7.1	−7.4	−15.4	−5.6	−12.6	−17.5	−15.6
Load Case 2	25.4	8.0	19.0	7.1	−0.1	−8.2	1.7	−5.3	−17.5	−15.6

*For windward overhangs only. For leeward and side overhangs, pressures are same as on, respectively, Zones F and H and Zones E and F (ASCE Fig. 28.6-1, Note 8).

Pressures p_{S30}, Case B* (ASCE Fig. 28.6-1; Fig. 8.5.1)

	Horizontal Pressures		Vertical Pressures				Overhangs#	
Zone	A	C	E	F	G	H	E_{OH}	G_{OH}
Case 1	21.0	13.9	−25.2	−14.3	−17.5	−11.1	−35.3	−27.6

*Pressures correspond to the assumption $\theta = 0°$ (ASCE Fig. 28.6-1, Note 3).
#Pressures on overhangs parallel to ridge are the same as pressures on adjacent roof zones (ASCE Fig. 28.6-1, Note 8).

The *design pressures* are

$$p_s = \lambda K_{zt}\, p_{S30}$$

(ASCE Eq. 28.6-1). For the building being considered (Exposure B, $h = 20$ ft), $\lambda = 1.0$, $K_{zt} = 1.0$, so $p_s = p_{S30}$.

Minimum design pressures for Zones A and C are +16 psf. For Zones B and D, they are +8 psf, while assuming the roof pressures (zones E, F, G, H) to be zero (ASCE Sect. 28.6.4).

Load patterns and reference corners: The load patterns of ASCE Fig. 28.6-1 are associated with the reference corner indicated therein. Those patterns must be applied by considering in turn each corner of the building as a reference corner.

8.5.2 Comparison of Pressures Based on Simplified Approach (ASCE Sect. 28.6) and on Regular Approach (ASCE Sect. 28.4.1)

Load Case A. Pressures on Zones A and C (ASCE Fig. 28.6-1) correspond, respectively, to pressures on Zone 1E added to pressures on Zone 4E and pressures on Zone 1 added to pressures on Zone 4 (Fig. ASCE 28.4-1). For Zones A and C, the pressures are *25.4* and *19.0 psf*, respectively (Numerical Example 8.5.1). The ASCE Sect. 27.4.1 pressures for Zone 1E added to pressures for Zone 4E and the pressures for Zone 1 added to pressures for Zone 4 are, respectively, *24.7* and *18.9 psf* (Numerical Example 6.1.1). The pressures based on the simplified and regular approach are comparable.

CHAPTER 9

REGULAR AND SIMPLIFIED APPROACHES: C&C

9.1 INTRODUCTION

Components and cladding (C&C) are designed for the calculated maximum positive and negative net wind pressures. The external pressure coefficients in the expressions for the net pressures depend upon the effective area of the component or cladding element, as explained in Sect. 4.2.5.1: the larger the area tributary to a component or the area of a cladding element, the smaller the extent to which the actual pressures over that area are mutually coherent (that is, the smaller the specified design wind pressures are). If C&C have tributary areas in excess of 700 sq ft, provisions for MWFRS may be used in lieu of the provisions for C&C (ASCE Sect. 30.2.3).

Section 9.2 considers the *regular approach* for determining wind loads. Section 9.3 considers the *simplified approaches* specified in the Standard.

Exposure Requirement 2 is used for all C&C (ASCE Sect. 26.7.3; Sect. 3.4.3).

9.2 REGULAR APPROACH

The Standard provisions based on the regular approach cover:

- *Enclosed and partially enclosed buildings with mean roof height* $h \leq$ 60 ft[1] and gable, hip, stepped, multispan gable, monoslope, and sawtooth

[1]Including both low-rise buildings (i.e., buildings with mean roof height $h \leq 60$ ft and ratio $h/D \leq 1$, where D is the building's least horizontal dimension) and buildings with $h \leq 60$ ft for $h/D > 1$. The title of Part 1 of ASCE Chapter 30 (i.e., "low-rise buildings") is not correct, since ASCE Sect. 30.4 covers (a) all buildings with $h \leq 60$ ft, not just low-rise buildings, as well as (b) domed and arched roofs of all heights, which are inadvertently omitted from the title.

roofs, as well as *domed and arched roofs of all heights* (ASCE Sect. 30.4; Sect. 9.2.1, Numerical Example 9.2.1).

- *Enclosed and partially enclosed buildings with* $h > 60$ ft and flat, pitched, gable, hip, mansard, arched, or domed roof (ASCE Sect. 30.6; Sect. 9.2.2).
- *Open buildings of all heights* with pitched, monoslope, or trough free roof (ASCE Sect. 30.8; Sect. 9.2.3, Numerical Example 9.2.2).
- *Parapets and roof overhangs* (ASCE Sects. 30.9, 30.10; Sect. 9.2.4, Numerical Example 9.2.3).
- *Rooftop structures and equipment for buildings with* $h \leq 60$ ft (ASCE Sect. 30.11; Sect. 9.2.4).

9.2.1 Regular Approach, Enclosed and Partially Enclosed (a) Buildings with $h \leq 60$ ft, and (b) Domed and Arched Roofs of All Heights, C&C (ASCE Sect. 30.4)

This section covers buildings with $h \leq 60$ ft and flat, gable, hip, multispan gable, monoslope, sawtooth roofs (Sect. 9.2.1.1, Numerical Example 9.2.1), and stepped roofs (Sect. 9.2.1.2), as well as domed roofs (Sect. 9.2.1.3) and arched roofs (Sect. 9.2.1.4).

9.2.1.1 Buildings with h ≤ 60 ft and Flat, Gable, Hip, Multispan Gable, Monoslope, or Sawtooth Roofs. The expression for the design pressure is:

$$p = q_h[(GC_p) - (GC_{pi})], \tag{9.2.1}$$

where q_h = velocity pressure at height h (ASCE Eq. 30.4-1), (GC_{pi}) = combined internal pressure coefficient (ASCE Table 26.11-1), and the combined external pressure coefficient is obtained from ASCE Figs. 30.4-1 to 30.4-6.

If desired, the information on the combined external pressure coefficients (GC_p) provided in the figures just listed can be computerized, thereby rendering the determination of the loads more convenient. That information is expressed, exactly, as a function of effective area A_{eff} of the C&C by Eqs. 9.2.2 and Table 9.2.1:

$$A_{eff} < A_{eff\,1}\colon (GC_p) = (GC_p)_1. \tag{9.2.2a}$$

$$A_{eff} > A_{eff\,2}\colon (GC_p) = (GC_p)_2. \tag{9.2.2b}$$

$$A_{eff\,1} \leq A_{eff} < A_{eff\,2}\colon (GC_p) = (GC_p)_1 + g\ \log_{10}\left(A_{eff}/A_{eff\,1}\right). \tag{9.2.2c}$$

$$g = [(GC_p)_2 - (GC_p)_1]/\log_{10}(A_{eff\,2}/A_{eff\,1}). \tag{9.2.2d}$$

The areas $A_{eff\,1}$, $A_{eff\,2}$, and the values $(GC_p)_1$, $(GC_p)_2$, and g are listed in Table 9.2.1. Because in Eqs. 9.2.2c, d and in similar subsequent expressions the logarithmic function depends on nondimensional ratios, the areas may be expressed in either English or SI units.

TABLE 9.2.1. Effective Areas $A_{eff\,1}$, $A_{eff\,2}$ (sq ft), Coefficients $(GC_p)_1$, $(GC_p)_2$, g; Buildings with $h \leq 60$ ft[1]

	Zone(s)	Pressure	$A_{eff\,1}$	$A_{eff\,2}$	$(GC_p)_1$	$(GC_p)_2$	g
Walls[1] (ASCE Fig. 30.4-1)	4, 5	Positive	10	500	1.0	0.7	−0.177
	4	Negative			−1.1	−0.8	0.177
	5	Negative			−1.4	−0.8	0.353
Gable roofs $\theta \leq 7^{\circ}$[2] (ASCE Fig. 30.4-2A)	1, 2, 3	Positive	10	100	0.3	0.2	−0.1
	1	Negative			−1.0	−0.9	0.1
	2	Negative			−1.8	−1.1	0.7
	3	Negative			−2.8	−1.1	1.7
	Overhang, 1, 2	Negative			−1.7	−1.6	0.1
	Overhang, 3	Negative			−2.8	−0.8	2.0
Gable and hip roofs $7^{\circ} < \theta \leq 27^{\circ}$ (ASCE Fig. 30.4-2B)	1, 2, 3	Positive	10	100	0.5	0.3	−0.2
	1	Negative			−0.9	−0.8	0.1
	2	Negative			−1.7	−1.2	0.5
	3	Negative			−2.6	−2.0	0.6
	Overhang, 2	Negative			−2.2	−2.2	0.0
	Overhang, 3	Negative			−3.7	−2.5	1.2
Gable roof $27^{\circ} < \theta \leq 45^{\circ}$ (ASCE Fig. 30.4-2C)	1, 2, 3	Positive	10	100	0.9	0.8	−0.1
	1	Negative			−1.0	−0.8	0.2
	2, 3	Negative			−1.2	−1.0	0.2
	Overhang, 2, 3	Negative			−2.0	−1.8	0.2

(continued overleaf)

TABLE 9.2.1. (*Continued*)

	Zone(s)	Pressure	A_{eff1}	A_{eff2}	$(GC_p)_1$	$(GC_p)_2$	g
Multispan gable roof[3] $10^\circ < \theta \leq 30^\circ$ (ASCE Fig. 30.4-4)	1, 2, 3	Positive	10	100	0.6	0.4	−0.2
	1	Negative			−1.6	−1.4	0.2
	2	Negative			−2.2	−1.7	0.5
	3	Negative			−2.7	−1.7	1.0
Multispan gable roof[3] $30^\circ < \theta \leq 45^\circ$ (ASCE Fig. 30.4-4)	1, 2, 3	Positive	10	100	1.0	0.8	−0.2
	1	Negative			−2.0	−1.1	0.9
	2	Negative			−2.5	−1.7	0.8
	3	Negative			−2.6	−1.7	0.9
Monoslope roofs $3^\circ < \theta \leq 10^\circ$ (ASCE Fig. 30.4-5A)	All zones	Positive	10	100	0.3	0.2	−0.1
	1	Negative			−1.1	−1.1	0.0
	2	Negative			−1.3	−1.2	0.1
	2’	Negative			−1.6	−1.5	0.1
	3	Negative			−1.8	−1.2	0.6
	3’	Negative			−2.6	−1.6	1.0
Monoslope roofs $10^\circ < \theta \leq 30^\circ$ (ASCE Fig. 30.4.5B)	All zones	Positive	10	100	0.4	0.3	−0.1
	1	Negative			−1.3	−1.1	0.2
	2	Negative			−1.6	−1.2	0.4
	3	Negative			−2.9	−2.0	0.9

	Zone(s)	Pressure	$A_{eff\,1}$	$A_{eff\,2}$	$(GC_p)_1$	$(GC_p)_2$	g
Sawtooth roofs (ASCE Fig. 30.4-3)	1	Positive	10	500	0.7	0.4	−0.177
	2	Positive	10	100	1.1	0.8	−0.3
	3	Positive	10	100	0.8	0.7	−0.1
	1	Negative	10	500	−2.2	−1.1	0.674
	2	Negative	10	500	−3.2	−1.6	0.942
	3, Span A[4]	Negative	10	100	−4.1	−3.7	0.4
	3, Spans B, C, D	Negative	10	500	−2.6	−1.9	1.0

[1]For *walls of buildings with roof slopes* $\theta \leq 10^\circ$, values of (GC_p) shall be reduced by 10% (ASCE Fig. 30.4-1, Note 5).
[2]For *gable roof overhangs*, $\theta \leq 7^\circ$ (ASCE Fig. 30.4-2):
(a) Zones 1 and 2:

$$A_{eff\,2} = 100 \text{ sq ft} < A_{eff} \leq 500 \text{ sq ft}: (GC_p) = -1.6 + 0.715 \log_{10}(A_{eff}/A_{eff\,2}). \quad (9.2.2e)$$

$$A_{eff} > 500 \text{ sq ft}: (GC_p) = -1.1. \quad (9.2.2f)$$

(b) If *parapet* of height 3 ft or higher is provided around the perimeter of a gable roof with $\theta \leq 3^\circ$, the negative values of (GC_p) in Zone 3 shall be equal to those of Zone 2, and the positive values of (GC_p) in Zones 2 and 3 shall be equal to those in Zones 4 and 5 for walls (Fig. 30.4-1), respectively (Note 5, ASCE Fig. 30.4-2A).

[3]For *multispan gable roofs*, pressure zones shown in Fig. ASCE 30.4-4 apply to all spans. For roof slopes $\theta \leq 10^\circ$, the values listed in ASCE Fig. 30.4-2A (i.e., in Table 9.2.1 for gable roofs with $\theta \leq 7^\circ$) shall be used, instead of those listed for multispan gable roofs (Note 5, ASCE Fig. 30.4-4).
[4]For *sawtooth roofs*, Zone 3, span A (ASCE Fig. 30.4-6):

$$A_{eff\,2} = 100 \text{ sq ft} < A_{eff} \leq 500 \text{ sq ft}: (GC_p) = -3.7 + 2.29 \log_{10}(A_{eff}/A_{eff\,2}). \quad (9.2.2g)$$

$$A_{eff} > 500 \text{ sq ft}, (GC_p) = -2.1. \quad (9.2.2h)$$

Numerical Example 9.2.1. *C&C, regular approach, enclosed office building with* $h = 60$ ft, *walls* (ASCE Fig. 30.4-1 and ASCE Eq. 30.4-1; Table 9.2.1 and Eqs. 9.2.2a, b, c). Assume effective area of cladding 4 sq ft, flat roof, Exposure B, flat terrain ($K_{zt} = 1.0$). The basic wind speed is $V = 115$ mph. ASCE Fig. 30.4-1, Note 5 (Table 9.2.1, Note 1) allows the use of a 0.9 reduction factor for buildings with roof slope $\theta \leq 10^\circ$.

The combined internal pressure coefficient is $(GC_{pi}) = \pm 0.18$. The velocity pressure exposure coefficient is $K_h = 0.85$ (ASCE Table 30.3-1), the topographic factor is $K_{zt} = 1.0$ (flat terrain), and the directionality factor is $K_d = 0.85$ (ASCE Sect. 26.6). The velocity pressure is $q_h = 0.00256 \times 0.85 \times 1.0 \times 0.85\ V^2 = 24.5$ psf (ASCE Eq. 30.3-1).

Zones 4, 5, positive pressures (ASCE Fig. 30.4-1):

$$A_{eff} \leq 10 \text{ sq ft: } (GC_p) = 0.9 \times 1.0 = 0.9.$$

$$A_{eff} = 100 \text{ sq ft: } (GC_p) = 0.9[1.0 - 0.177 \log_{10}(100/10)] = 0.74.$$

$$A_{eff} \geq 500 \text{ ft: } 0.9 \times (GC_p) = 0.63.$$

For $A_{eff} = 4$ sq ft, $p = (0.9 + 0.18)24.5 = 26.5$ psf (Eq. 9.2.1).

Zone 4 (negative pressures):

$$A_{eff} \leq 10 \text{ sq ft: } (GC_p) = -0.9 \times 1.1 = -0.99 \text{ (Eq. 9.2.1a).}$$

$$A_{eff} = 100 \text{ sq ft: } (GC_p) = 0.9[-1.1 + 0.177 \log_{10}(100/10)] = -0.83 \text{ (Eq. 9.2.1c).}$$

$$A_{eff} \geq 500 \text{ ft: } (GC_p) = 0.9 \times -0.8 = -0.72 \quad \text{(Eq. 9.2.1b).}$$

For $A_{eff} = 4$ sq ft, $p = (-0.99 - 0.18)24.5 = -28.7$ psf (Eq. 9.2.1).

Zone 5 (negative pressures):
For $A_{eff} = 4$ sq ft, $p = (-0.9 \times 1.4 - 0.18)24.5 = -35.3$ psf (Eq. 9.2.1).

9.2.1.2 Buildings with Stepped Flat Roofs (ASCE Fig. 30.4-3). Provisions are restricted to buildings with dimensions $h_1 \geq 10$ ft, $0.3 \leq h_i/h \leq 0.7$, and $0.25 \leq W_i/W \leq 0.75$ ($i = 1, 2$); see notations in ASCE Fig. 30.4-3. For the lower-level roof, zone designations and coefficients (GC_p) for gable roofs with slopes $\theta \leq 7^\circ$ apply (see Table 9.2.1), except that at the lower-roof/upper-wall intersections Zone 3 is treated as Zone 2, and Zone 2 is treated as Zone 1. The width of the crosshatched areas of ASCE Fig. 30.4-3 is $b = 1.5h_1$ or 100 ft, whichever is smaller. Since the positive pressures (GC_p) on those areas are the same as for gable roofs with slopes $\theta \leq 7^\circ$,

$$A_{eff} < 10 \text{ sq ft, } (GC_p) = 0.9 \times 1.0.$$

$$10 \leq A_{eff} < 500 \text{ sq ft, } (GC_p) = 0.9[1.0 - 0.177 \log_{10}(A_{eff}/10)].$$

$$A_{eff} > 500 \text{ sq ft, } (GC_p) = 0.9 \times 0.7.$$

No pressures are specified for walls and upper level of roof, but Eqs. 9.2.2a–d and coefficients from Table 9.2.1 for walls and gable roofs with $\theta \leq 7^\circ$ may be used.

9.2.1.3 Domed Roofs (ASCE Fig. 30.4-7). Using the notations of ASCE Fig. 30.4.7,

$$p = q_{(h_D+f)}[(GC_p) - (GC_{pi})] \tag{9.2.3}$$

where $q_{(h_D+f)}$ = velocity pressure at elevation $h_D + f$ (i.e., at top of dome); $(GC_p) = -0.9$ for $0^\circ < \theta \leq 90^\circ$, $(GC_p) = 0.9$ for $0^\circ < \theta \leq 60^\circ$, and $(GC_p) = 0.5$ for $60^\circ < \theta \leq 180^\circ$. For internal pressure coefficient (GC_{pi}), see ASCE Sect. 26.11. The use of these values is restricted to domes for which $0 < h_D/D \leq 0.5$, and $0.2 \leq f/D \leq 0.5$.

9.2.1.4 Arched Roofs (ASCE Fig. 27.4-3, Note 4). At roof perimeter, use the external pressure coefficients of ASCE Fig. 30.4-2A, B, C (or Table 9.2.1) for gable roofs with slope equal to the spring-line slope of the arched roof. For remaining roof areas, use external pressure coefficients of ASCE Table 27.4-3 divided by 0.87 (*Note:* ASCE Fig. 27.4-2, Note 4, incorrectly states "*multiplied*" instead of "*divided.*" In fact, since wind loads on C&C have smaller effective areas than loads on MWFRS, the external pressures for C&C are larger, not smaller, than the loads used for MWFRS.)

9.2.2 Regular Approach: Enclosed and Partially Enclosed Buildings with h > 60 ft (ASCE Sect. 30.6)

The expression for the pressures is

$$p = q(GC_p) - q_i(GC_{pi}). \tag{9.2.4}$$

For windward walls, $q = q_z$ (ASCE Eq. 30.3-1); for leeward and side walls and for roofs, $q = q_h$; for internal pressures, $q_i = q_h$; for positive internal pressures of *partially enclosed* buildings, the value $q_i = q_{ho}$ may be used in lieu of $q_i = q_h$, where h_o is the elevation of the highest opening that may affect positive internal pressures (Table 2.2.3).

For (GC_{pi}) see ASCE Table 26.11-1.

For walls and for roofs with slopes $\theta \leq 10^\circ$, (GC_p) in Eq. 9.2.1 is obtained from ASCE Fig. 30.6-1 or from Eqs. 9.2.2a–d and Table 9.2.2.

For roofs with slopes $\theta > 10^\circ$ and other geometries, (GC_p) in Eq. 9.2.1 is based on ASCE Figs. 30.4-2A, B, C or, where applicable, Eqs. 9.2.2a, b, c and Table 9.2.1 (ASCE Fig. 30.6-1).

For domes and arched roofs, the procedures for determining external pressure coefficients are the same as for buildings with $h \leq 60$ ft (Sects. 9.2.1.3 and 9.2.1.4).

TABLE 9.2.2. Effective Areas $A_{eff\,1}$, $A_{eff\,2}$(sq ft), External Pressure Coeff. $(GC_p)_1$, $(GC_p)_2$, and Coeff. g; Buildings with $h > 60$ ft; Flat Roofs and Gable Roofs with $\theta \leq 10°$

	Zone[1]		$A_{eff\,1}$	$A_{eff\,2}$	$(GC_p)_1$	$(GC_p)_2$	g
Walls; flat and gable roofs, $\theta \leq 10°$ (ASCE Fig. 30.1-6)	4, 5	Positive	20	500	0.9	0.6	−0.215
	4	Negative	20	500	−0.9	−0.7	0.143
	5	Negative	20	50	−1.8	−1.0	0.572
	1	Negative	10	500	−1.4	−0.9	0.294
	2	Negative	10	500	−2.3	−1.6	0.412
	3	Negative	10	500	−3.2	−2.3	0.53

[1]If a 3 ft high or higher parapet is provided around perimeter of roofs with slope $\theta \leq 10°$, Zone 3 becomes Zone 2 (ASCE Fig. 30.6-1).

9.2.3 Regular Approach: Open Buildings of All Heights with Pitched, Monoslope, or Trough Free Roof, C&C (ASCE Sect. 30.8, Figs. 30.8-1 to 30.8-3)

The net design wind pressures, which include contributions from the top and bottom surfaces of the roofs, have the expression

$$p = q_h G\, C_N \qquad (9.2.5)$$

(ASCE Eq. 30.8-1), where q_h is the velocity pressure evaluated at the mean roof height, and G is the gust effect factor (ASCE Sect. 26.9). The Standard permits the use of the value $G = 0.85$. The net pressure coefficient C_N is given in ASCE Figures 30.8-1, 30.8-2, and 30.8-3 for monoslope, pitched, and troughed roof, respectively. Pressure coefficients are given for *clear* wind flow (50% or less obstructed vertical area under the roof), and for *obstructed* wind flow (more than 50% obstructed vertical area). Both positive and negative pressures shall be used in design.

Numerical Example 9.2.2. *Monoslope roof of open building.* Assume that for a monoslope roof (ASCE Fig. 30.8-1) the mean roof height is $h = 15$ ft, $L = 24$ ft, the length (dimension transverse to the horizontal eave) is 21 ft, and $\theta = 15°$; the basic wind speed is $V = 115$ mph; the site has *Exposure B for all directions*; and the wind flow is *clear*.

The effective area for the components is assumed to be 9 sq ft. The width a is 0.1 × least horizontal dimension = 0.1 × 21 = 2.1 ft or 0.4 × h = 0.4 × 15 = 6 ft, whichever is smaller, but not less than 4% of the least horizontal dimension = 0.04 × 21 = 0.84 ft or 3 ft, so $a = 3$ ft (ASCE Figs. 30.8-1, Note 6). The effective area $A_{eff} = a^2 = 9$ sq ft, so the pressure coefficients are:

Central zone (Zone 1): $C_N = 1.8$ or -1.9

Intermediate zone (Zone 2): $C_N = 2.7$ or -2.9.

Periphery zone (Zone 3): $C_N = 3.6$ or -3.8.

The velocity pressure exposure coefficient is $K_h = 0.7$ (ASCE Table 30.3-1), the topographic factor is $K_{zt} = 1.0$ (flat terrain), and the directionality factor is $K_d = 0.85$ (ASCE Sect. 26.6). The velocity pressure is $q_h = 0.00256 \times 0.7 \times 1.0 \times 0.85\ V^2 = 20.1$ psf (ASCE Sect. 30.3.2, ASCE Eq. 30.3-1). The net design wind pressures $p = q_h GC_N$ are tabulated here:

Zone	Case	q_h (psf)	G	C_N	p (psf)
1	1	20.1	0.85	1.8	30.8
	2	20.1	0.85	−1.9	−32.5
2	1	20.1	0.85	2.7	46.1
	2	20.1	0.85	−2.9	−49.5
3	1	20.1	0.85	3.6	61.5
	2	20.1	0.85	−3.8	−64.9

9.2.4 Regular Approach: Parapets and Roof Overhangs, C&C

9.2.4.1 Parapets (ASCE Sects. 30.9, ASCE Fig. 30.9-1). Except for enclosed buildings with $h \leq 160$ ft, pressures on C&C of parapets are:

$$p = q_p[(GC_p) - (GC_{pi})] \qquad (9.2.6)$$

(ASCE Eq. 30.9-1), where q_p = velocity pressure (e.g., ASCE Eq. 30.3-1) evaluated at the top of the parapet, and (GC_{pi}) is taken from ASCE Sect. 26.11. If internal pressures are present, they shall be based on the porosity of the parapet envelope, and depend on information supplied by the manufacturer.

For enclosed buildings with $h \leq 160$ ft, ASCE Sect. 30.9 requires that parapet loads be determined in accordance with ASCE Sect. 30.7.1.2 (simplified appoach).

ASCE Sect. 30.9 requires that two load cases be considered:

Load Case A pertains to parapets on the building's windward side. The positive external wall pressure coefficients for Zone 4 or 5 (ASCE Fig. 30.4-1 or 30.6-1, depending on height h) are applied to the parapet's windward (outer) face; the negative (edge or corner zone) external roof pressure coefficients are applied to the parapet's leeward (inner) face (ASCE Figs. 30.4-1 to 30.4-7, ASCE Fig. 27.4-3, Note 4, or Fig. 30.6-1, as applicable).

Load Case B pertains to parapets on the building's leeward side. The positive external wall pressure coefficients for Zones 4 and 5 (ASCE Fig. 30.4-1 or 30.6-1) are applied to the parapet's windward (inner) face, and the negative external wall pressure coefficients for Zones 4 and 5 (ASCE Fig. 30.4-1 or 30.6-1) are applied to the parapet's leeward (outer) face.[2]

[2]ASCE Sect. 30.9 excludes the application of its provisions to enclosed buildings with $h \leq 160$ ft, "for which the provisions of Part 4 [i.e., ASCE Sect. 30.7.1.2, author's note] are used." On the other hand, it specifies pressure coefficients for all building types and heights, including therefore enclosed buildings with $h \leq 160$ ft. A more explicit statement of the Standard's intent might be desirable in future versions of the Standard.

Numerical Example 9.2.3. *Parapets, C&C*. An enclosed building with $h >$ 160 ft and flat roof has a 3-ft-high parapet around the perimeter, which is assumed to be nonporous (i.e., there are no internal pressures within the parapets). The components are 3 ft high and 2 ft wide, so $A_{eff} = 3$ ft $\times$ 2 ft $=$ 6 sq ft. The external pressure coefficients are taken from ASCE Figs. 30.7-1.

Case A (windward parapet)

The positive external coefficient for walls, Zones 4 and 5, is $(GC_p) = 0.9$, and is applied to the windward (outside) face of the parapet.

According to Note 7 of ASCE Figure 30.6-1, if a parapet equal to or higher than 3 ft is provided around the perimeter of the roof, with $\theta \leq 10^\circ$, Zone 3 shall be treated as Zone 2. The roof pressure coefficient for the negative Zone 2 (edge) is $(GC_p) = -2.3$ and is applied to the inside (leeward) face of the windward parapet. Therefore, the total external pressure coefficient for determining wind loads on the components of the parapet is $0.9 - (-2.3) = 3.2$.

In the absence of the parapet, for the negative Zone 3 (corner) of the roof, $(GC_p) = -3.2$. However, according to ASCE Fig. 30.6-1 (Note 7), if the building has a 3-ft or higher parapet around the roof perimeter, for negative pressures Zone 3 becomes Zone 2. Therefore, the roof pressure coefficient applied to the inside face of the windward parapet is $(GC_p) = -2.3$ throughout.

Case B (leeward parapet)

The total external pressure coefficient on the leeward parapet is $2.3 - (-0.9) = 3.2$ for Zone 4 of the wall, and $2.3 - (-1.8) = 4.1$ for Zone 5 of the wall (ASCE Fig. 30.6-1). The pressures are directed outward in both cases.

9.2.4.2 Roof Overhangs (ASCE 30.10, ASCE Fig. 30.10-1). Pressures on C&C of parapets are:

$$p = q_h[(GC_p) - (GC_{pi})] \tag{9.2.7}$$

(ASCE Eq. 30.10-1), where q_p = velocity pressure (e.g., ASCE Eq. 30.3-1) evaluated at the mean roof height h, (GC_{pi}) is taken from ASCE Sect. 26.11 and depends on the porosity of the overhangs, and (GC_p) are external pressure coefficients for overhangs of flat, gable, and hip roofs (ASCE Figs. 30.4-2A, B, C; Table 9.2.1).

For enclosed buildings with $h \leq 160$ ft, ASCE Sect. 30.9 requires that roof overhang loads be determined in accordance with ASCE 30.7.1.3 (simplified approach).

Equation 9.2.7 accounts for pressures on both the top side and the underside of the overhangs. Pressures on C&C on the underside of the overhangs are based on pressure coefficients for walls (ASCE Figs. 30.4-1 and 30.6-1, as applicable) consistent with the effective areas of the overhang's underside C&C.

9.2.5 Regular Approach: Rooftop Structures and Equipment for Buildings with $h \leq 60$ ft, C&C (ASCE Sect. 30.11)

The design wind pressures acting inward and outward on each wall of the rooftop structure are equal to the lateral force determined by ASCE Sect. 29.6, divided by the area of the wall. The design wind pressures acting upward on the roof of the rooftop structure are equal to the vertical uplift force determined by ASCE Sect. 29.6, divided by the horizontal projected area of the roof.

The Standard has no provisions for C&C of rooftop structures and equipment for buildings with $h > 60$ ft. In our opinion, to a first approximation, the procedure specified in the Standard for buildings with $h \leq 60$ ft may also be applied for buildings of all heights.

9.3 SIMPLIFIED APPROACHES

9.3.1 Numerical Examples

This section covers two types of building: buildings with $h \leq 60$ ft (ASCE Sect. 30.5; ASCE Fig. 30.5-1)[3] (Numerical Example 9.3.1) and buildings with $h \leq 160$ ft (ASCE Sect. 30.7; Table ASCE 30.7-2)[4] (Numerical Example 9.3.2). Comparisons among results obtained by the regular procedure and the simplified procedures of ASCE Sects. 30.5 and 30.7 are presented in Sect. 9.3.2.

Numerical Example 9.3.1. *C&C, walls of enclosed office building with* $h =$ 60 ft, *simplified approach* (ASCE Sect. 30.5; ASCE Fig. 30.5-1). Assume effective area of cladding 4 sq ft, flat roof, Exposure B, flat terrain ($K_{zt} = 1.0$). The basic wind speed is $V = 115$ mph. The net pressure is $p_{net} = \lambda K_{zt} p_{net30}$. The adjustment factor for building height and exposure is $\lambda = 1.22$ (see last page of ASCE Fig. 30.5-1).

For wall pressures, Zone 4, $p_{net30} = 23.8$ or -25.8 psf, and $p_{net} = 1.22 \times 23.8 = 29.0$ or $1.22 \times (-25.8) = -31.5$ psf; for Zone 5, $p_{net} = 1.22 \times 23.8$ or $1.22 \times (-31.9)$, that is, $p_{net} = 29.0 \text{or} - 38.9$ psf; that is,

$$\text{Zone 4:} \, p_{net} = 29.0 \text{ or } - 31.5 \text{ psf.}$$

$$\text{Zone 5:} \, p_{net} = 29.0 \text{ or } - 38.9 \text{ psf.}$$

Numerical Example 9.3.2. *C&C, walls of enclosed office building with* $h =$ 60 ft, *simplified approach* (ASCE Sect. 30.7; Fig. 30.7-2). Assume effective

[3]Not just low-rise buildings, as is incorrectly indicated in the title of Part 2 of ASCE Chapter 30.
[4]Note the overlap between these two types of building: buildings with $h \leq 60$ ft are a subset of buildings with $h \leq 160$ ft. Since Sect. 30.7 does not specify pressures for buildings with 60 ft $< h \leq$ 160 ft, but rather with $h \leq 160$ ft, it is applicable to buildings with $h \leq 60$ ft.

area of cladding 4 sq ft, flat roof, Exposure B, flat terrain ($K_{zt} = 1.0$). The basic wind speed is $V = 115$ mph. Instead of using the provisions of ASCE Sect. 30.5, we use ASCE Sect. 30.7. The design wind pressures are

$$p = p_{table}(\text{EAF})(\text{RF})K_{zt}$$

(ASCE Eq. 30.7-1), where (EAF) = Exposure Adjustment Factor and (RF) = effective area reduction factor are specified in ASCE Table 30.7-2. RF is 1.0 since the effective area is less than 10 sq ft. Since $p_{table} = 35.3$ or -35.3 psf (Zone 4), and 35.3 or −64.8 psf (Zone 5), and EAF = 0.751, the pressures are:

$$\text{Zone 4: } p = 26.5 \text{ or } -26.5 \text{ psf.}$$

$$\text{Zone 5: } p = 26.5 \text{ or } -48.7 \text{ psf.}$$

9.3.2 Comparison among Pressures Determined by Alternative Procedures

The negative pressures on Zones 4 and 5 yielded by the regular procedure (ASCE Sect. 30.4), the simplified procedure of ASCE Sect. 30.5, and the simplified procedure of Sect. 30.7 are shown here (all values are in psf):

	ASCE Sect. 30.4	ASCE Sect. 30.5	ASCE Sect. 30.7
Zone 4	−28.7	**−31.5**	**−26.5**
Zone 5	**−35.3**	−38.9	**−48.7**

The largest and smallest Zone 4 and Zone 5 pressures are shown in bold. The differences between those pressures are 19% and 38%, respectively.

PART III

WIND ENGINEERING FUNDAMENTALS

CHAPTER 10

ATMOSPHERIC CIRCULATIONS

Wind is fundamentally caused by heat radiated by the sun. Radiation, thermodynamic, and mechanical phenomena transform the thermal energy imparted to the atmosphere into mechanical energy associated with air motion, giving rise to various types of wind. In this chapter, we present elements of atmospheric hydrodynamics (Sect. 10.1) and describe windstorms of interest from a structural engineering viewpoint (Sect. 10.2).

10.1 ATMOSPHERIC HYDRODYNAMICS

The motion of an elementary air mass is determined by forces that include a vertical *buoyancy force*. Depending upon the temperature difference between the air mass and the ambient air, the buoyancy force acts upwards (causing an updraft), downwards, or is zero. These three cases correspond to unstable, stable, or neutral atmospheric stratification, respectively.

The horizontal motion of air is determined by the following forces:

1. The *horizontal pressure gradient force* per unit of mass, which is due to the spatial variation of the horizontal pressures. This force is normal to the lines of constant pressure, called *isobars*, and is directed from high-pressure to low-pressure regions (Fig. 10.1.1). The net force per unit mass exerted by the horizontal pressure gradient is

$$P = (1/\rho)dp/dn \tag{10.1.1}$$

where p denotes the pressure, and ρ is the air density.

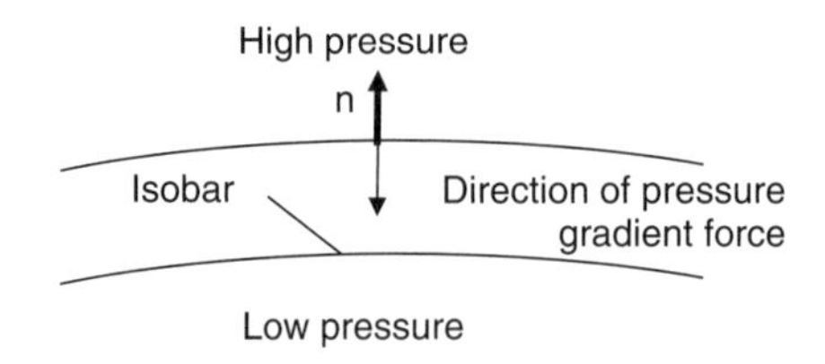

Figure 10.1.1. Direction of pressure gradient force.

2. The *deviating force due to the Earth's rotation.* If defined with respect to an absolute frame of reference, the motion of a particle not subjected to the action of an external force will follow a straight line. To an observer on the rotating earth, however, the path described by the particle will appear curved. The deviation of the particle with respect to a straight line fixed with respect to the rotating earth may be attributed to an apparent force, the *Coriolis force*

$$F_c = mfv \tag{10.1.2}$$

where m is the mass of the particle, $f = 2\omega \sin \varphi$ is the Coriolis parameter, $\omega = 0.7272 \times 10^{-4}\text{s}^{-1}$ is the angular velocity vector of the earth, φ is the angle of latitude, and v is the velocity vector of the particle referenced to a coordinate system fixed with respect to the earth. The force F_c is normal to the direction of the particle's motion, and is directed according to the vector multiplication rule.

3. The *friction force.* The surface of the earth exerts upon the moving air a horizontal drag force that retards the flow. This force decreases with height and becomes negligible above a height δ known as *gradient height.* The atmospheric layer between the earth's surface and the gradient height is called *the atmospheric boundary layer* (see Chapter 11). The wind velocity speed at height δ is called the *gradient velocity*,[1] and the atmosphere above this height is called the *free atmosphere* (Fig. 10.1.2)

In the free atmosphere, an elementary mass of air will initially move in the direction of the pressure gradient force—the driving force for the air motion—in a direction normal to the isobar. The Coriolis force will be normal to that incipient motion, that is, it will be tangent to the isobar. The resultant of these two forces, and the consequent motion of the particle, will no longer be normal to the isobar, so the Coriolis force, which is perpendicular to the particle motion, will change direction, and will therefore no longer be directed along the isobar. The change in the direction of motion will continue until

[1]For "straight winds" (i.e., winds whose isobars are approximately straight), the term "geostrophic" is substituted in the meteorological literature for "gradient." This distinction is not made in the ASCE 7-10 Standard.

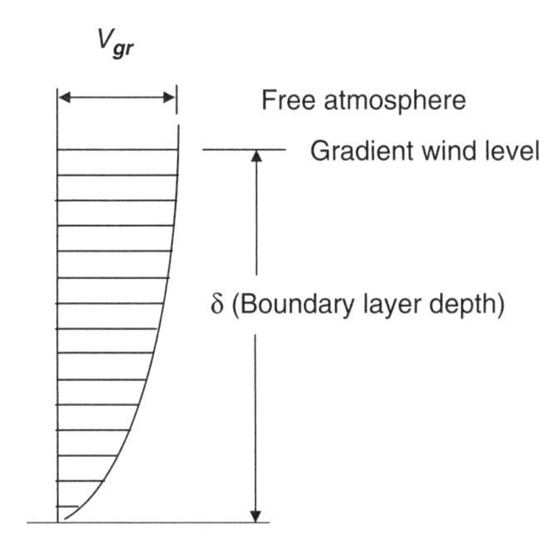

Figure 10.1.2. Schematic of the atmospheric boundary layer.

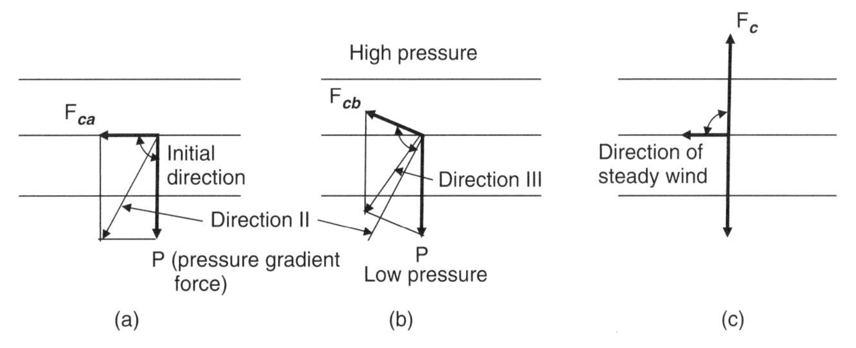

Figure 10.1.3. Frictionless wind balance in geostrophic flow.

the particle is moving steadily *along* the isobar, at which point the Coriolis force will be in equilibrium with the pressure gradient force, as shown in Fig. 10.1.3.

Within the atmospheric boundary layer, the direction of the friction force, denoted by S, coincides with the direction of motion of the particle. During the particle's steady motion, the resultant of the mutually orthogonal Coriolis and friction forces will balance the pressure gradient force—that is, will be normal to the isobars—meaning that the friction force, and therefore the motion of the particle, will cross the isobars (Fig. 10.1.4). The friction force, which retards the wind flow and vanishes at the gradient height, decreases as the height above the surface increases.

Therefore the velocity *increases* with height (Fig. 10.1.2). The Coriolis force, which is proportional to the velocity, also increases with height. The combined effect of the Coriolis force and the friction force is such that the angle between the direction of motion and the isobars increases from zero at the gradient height to its largest value at the Earth's surface. The wind velocity in the boundary layer can therefore be represented by a spiral, as

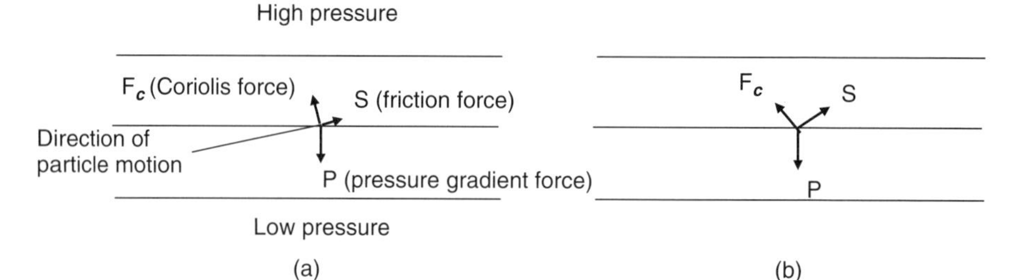

Figure 10.1.4. Balance of forces in the atmospheric boundary layer: (a) motion at higher elevations (low friction); (b) motion at lower elevations (high friction).

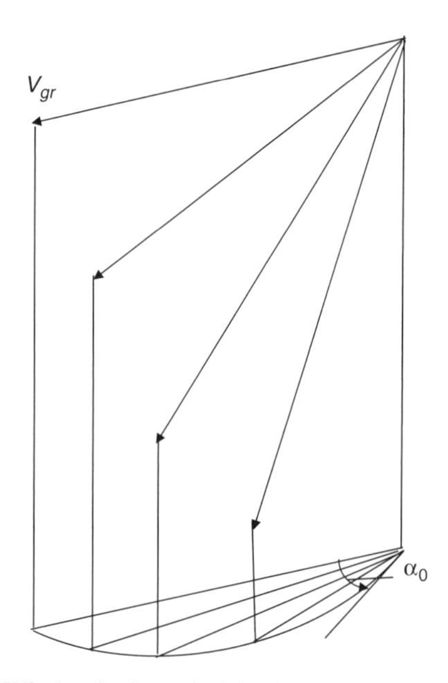

Figure 10.1.5. Wind velocity spiral in the atmospheric boundary layer.

in Fig. 10.1.5. Under certain simplifying assumptions regarding the effective flow viscosity, the spiral is called the *Ekman spiral*.

If the isobars are curved, the horizontal pressure gradient force, as well as the centrifugal force associated with the motion on a curved path, will act on the elementary mass of air in the direction normal to the isobars, and the resultant steady wind will again flow along the isobars. Its velocity results from the relations

$$V_{gr} f \pm \frac{V_{gr}}{r} = \frac{dp/dn}{\rho} \tag{10.1.3}$$

where, if the mass of air is in the northern hemisphere, the positive sign is used when the circulation is cyclonic (around a center of low pressure), and the negative sign is used when the circulation is anticyclonic (around a center of high pressure); r is the radius of curvature of the air trajectory.

10.2 WINDSTORMS

10.2.1 Large-Scale Storms

Large-scale (synoptic) wind flow fields of interest in structural engineering may be divided into two main types of storm: extratropical storms and tropical cyclones.

Extratropical storms occur at and above mid-latitudes. Because their vortex structure is less well defined than in tropical storms, their winds are loosely called "straight winds."

Tropical cyclones, known as *typhoons* in the Far East, and *cyclones* in Australia and the Indian Ocean, generally originate between 5° and 20° latitudes. *Hurricanes* are defined as tropical cyclones with sustained wind speeds at 10 m above water of 74 mph or larger. Tropical cyclones are translating vortices with diameters of hundreds of miles and counterclockwise (clockwise) rotation in the northern (southern) hemisphere. Their *translation speeds* vary from about 3 to 30 mph. As in a stirred coffee cup, the column of fluid is lower at the center than at the edges. The difference between edge and center atmospheric pressures is called *pressure defect*. Rotational speeds increase as the pressure defect increases, and as the *radius of maximum wind speeds*, which varies from about 5 to 60 miles, decreases.

The structure and flow pattern of a typical tropical cyclone is shown in Fig. 10.2.1. The *eye of the storm* (Region I) is a roughly circular, relatively dry core of calm or light winds surrounded by the *eye wall*. Region II contains the storm's most powerful winds. Far enough from the eye, winds in Region V,

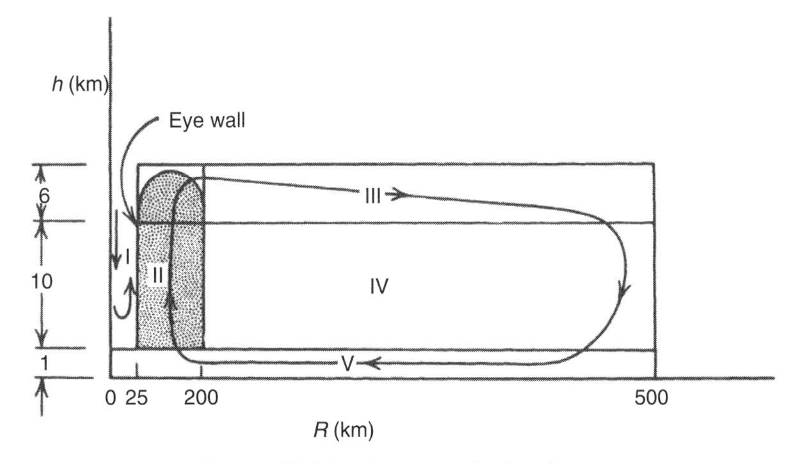

Figure 10.2.1. Structure of a hurricane.

TABLE 10.2.1. Saffir-Simpson Scale and Corresponding Wind Speeds[1]

Hurr. Categ.	Damage Potential	1-min speed at 10 m above open water[2] (mph)[3,4]	3-s gust speed over open terrain (mph)[4]	N. Atlantic examples
1	Minimal	74–95	81–105	Agnes 1972
2	Moderate	96–110	106–121	Cleo 1974
3	Extensive	111–130	122–143	Betsy 1965
4	Extreme	131–155	144–171	David 1979
5	Catastrophic	>155	>171	Andrew 1992

[1]For the definition of 1-min wind speeds, see Sect. 11.1.1.
[2]See Commentary, ASCE 7-10 Standard, Table C26.5-2.
[3]Official speeds are in mph.
[4]1-min and 3-s wind speeds at 10 m above open terrain can be shown to be, respectively, lower by approximately 15% than 1-min speeds at 10 m over open water, and higher by approximately 22% than 1-min speeds at 10 m above open terrain (see Sects. 11.2.4.1, 11.2.4.2).

which decrease in intensity as the distance from the center increases, are parallel to the surface. Where Regions V and II intersect, the wind speed has a strong updraft component that alters the mean wind speed profile and is currently not accounted for in structural engineering practice. The source of energy that drives the storm winds is the warm water at the ocean surface. As the storm makes landfall and continues its path over land, its energy is depleted and its wind speeds gradually decrease.

In the United States, hurricanes are classified in accordance with the Saffir-Simpson scale (Table 10.2.1).[2]

10.2.2 Local Storms

Foehn winds (called chinook winds in the Rocky Mountains area) develop downwind of mountain ridges. Cooling of air as it is pushed upwards on the windward side of a mountain ridge causes condensation and precipitation. The dry air flowing past the crest warms as it is forced to descend, and is highly turbulent. A similar type of wind is the *bora*, which occurs downwind of a plateau separated by a steep slope from a warm plain.

[2]It has been argued that the destructive potential of tropical cyclones is inadequately indicated by the Saffir-Simpson scale—which is based primarily on the largest local wind velocity within the storm—and that more realistic scales, global in nature, would be warranted. A scale based on the tropical cyclone integrated kinetic energy was recently proposed, which would reflect the potential for damage due to wind and storm surge development throughout the area covered by the storm [10-1]. However: "The National Hurricane Center does not believe that such scales would be helpful and effective in conveying the storm surge threat. For example, if 2008's Hurricane Ike had made landfall in Palm Beach Florida, the resulting storm surge would have been only 8 ft, rather than the 20 ft that occurred where Ike actually made landfall on the upper Texas Coast. These greatly differing storm surge impacts arise from differences in the local bathymetry (the shallow Gulf waters off Texas enhance storm surge while the deep ocean depths off southeastern Florida inhibit surge). The proposed storm surge scales that consider storm size do not consider these local factors that play a crucial role in determining actual surge impacts" (www.nhc.noaa.gov/sshws_statement.shtml).

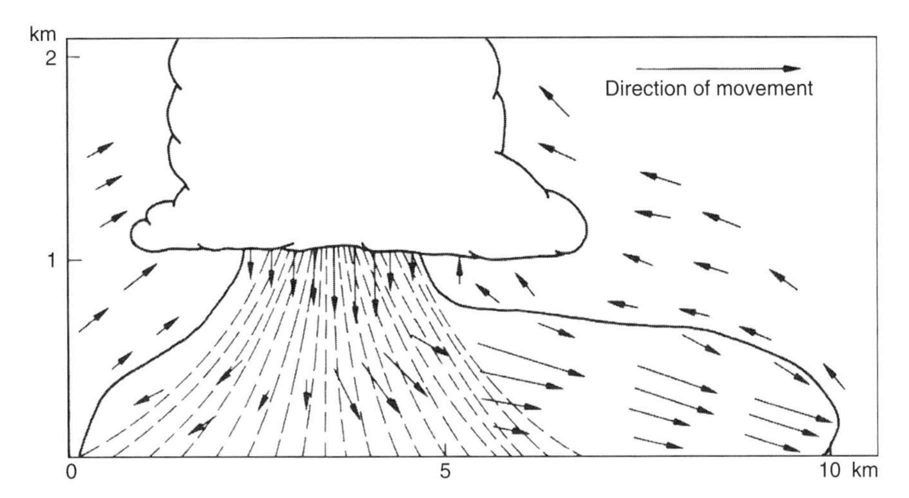

Figure 10.2.2. Section through a thunderstorm in the mature stage.

Jet effect winds are produced by features such as gorges.

Thunderstorms occur as heavy rain drops, due to condensation of water vapor contained in ascending warm, moist air, drag down the air through which they fall, causing a *downdraft* that spreads on the earth's surface (Fig. 10.2.2). The edge of the spreading cool air is the *gust front*. If the wind behind the gust front is strong, it is called a *downburst*.

Tornadoes are small vortex-like storms that can contain winds in excess of 250 mph. A U.S. map of tornado wind speeds with probability of exceedance of 10^{-5} per year, excerpted from the 1983 ANSI/ANS-2.3 Standard [10-2], is included in the Commentary to the ASCE 7-10 Standard (Sect. C26.5-2). For the regions for which the 1983 ANSI/ANS-2.3 Standard specifies 200 mph and 150 mph speeds, the speeds specified in the draft 2010 ANSI/ANS-2.3 Standard are approximately 165 mph and 140 mph, respectively.

CHAPTER 11

THE ATMOSPHERIC BOUNDARY LAYER

Wind flows that affect buildings and other structures are characterized by two fundamental features: (1) the increase of the wind speeds with height (Fig. 10.1.2), and (2) the atmospheric turbulence. Details on atmospheric boundary layer features of interest to structural engineers include:

- The dependence of wind speeds on averaging time (Sect. 11.1), which results from the fluctuating nature of the turbulent wind speeds.
- The variation of wind speed with height (Sect. 11.2), which depends on surface roughness, fetch (i.e., distance over which that surface roughness extends upwind of the structure), topography, and storm type and intensity.
- The atmospheric turbulence (Sect. 11.3), which not only affects the definitions of wind speeds as functions of averaging time (e.g., 3-s peak gusts, 1-min speeds, 10-min speeds, mean hourly speeds), but can also strongly influence the aerodynamic loading, as well as causing dynamic motions in flexible structures.

The models described in Sects. 11.1, 11.2, and 11.3 apply to strong straight winds, but are commonly also used for other types of wind, including hurricanes. Moderate wind speeds have features, including lower turbulence intensities, that differ from those of strong winds in ways that can affect the design of structures susceptible to vortex-shedding effects (Sect. 11.3.4).

11.1 WIND SPEEDS AND AVERAGING TIMES

Definitions of wind speeds as functions of averaging time are presented in Sect. 11.1.1. The dependence of wind speeds on averaging time for the reference case of speeds at 10 m elevation in open exposure is considered in Sect. 11.1.2. The case of wind speeds above surfaces other than open terrain is considered in Sect. 11.2.4.2.

11.1.1 Definitions of Wind Speeds as Functions of Averaging Time

If the flow were laminar, wind speeds would be the same for all averaging times. However, owing to turbulent fluctuations, such as those recorded in Fig. 11.1.1, the definition of wind speeds depends on averaging time.

- *The peak 3-s gust speed* is a storm's largest speed averaged over 3 s. In 1995 it was adopted in the ASCE Standard as a measure of wind speeds. Similarly, the peak 5-s gust speed is the largest speed averaged over 5 s. The 5-s speed is reported by the National Weather Service ASOS (Automated Service Observations System), and is about 2% less than the 3-s speed. The 28 mph peak of Fig. 11.1.1 is, approximately, a 3-s speed.

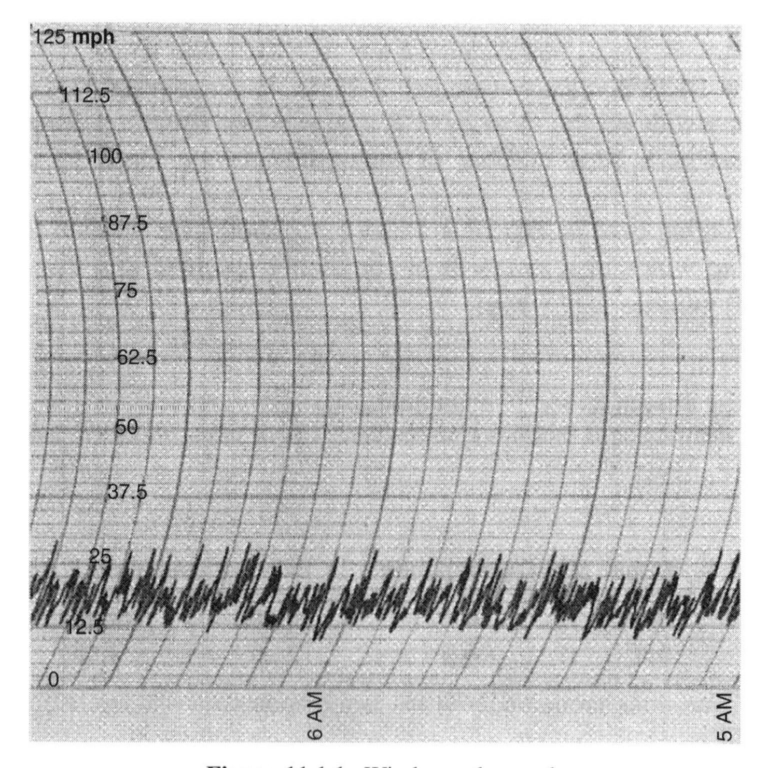

Figure 11.1.1. Wind speed record.

- *The hourly wind speed* is the speed averaged over one hour. It is commonly used as a reference wind speed in wind tunnel simulations. Hence the need to relate the hourly speed corresponding to a 3-s (or 1-min, or 10-min) speed specified for design purposes or recorded at weather stations. The hourly wind speed is used to define extreme speeds by the National Building Code of Canada [11-1; see also 19-20]. In Fig. 11.1.1, the statistical features of the record do not vary significantly (i.e., the record may be viewed as *statistically stationary*) over an interval of several hours; the hourly wind speed is about 18.5 mph, or about 1/1.52 times the peak 3-s gust. The hurricane winds of Fig. 11.1.2 can be viewed as statistically stationary for about 20 min (10:20 to 10:40 hrs), and their largest *20-min speed* (i.e., wind speed averaged over 20 min) is about 64 mph. The thunderstorm wind record of Fig. 11.1.3 can be viewed as statistically stationary for about 4 min (around 11:20 hrs).

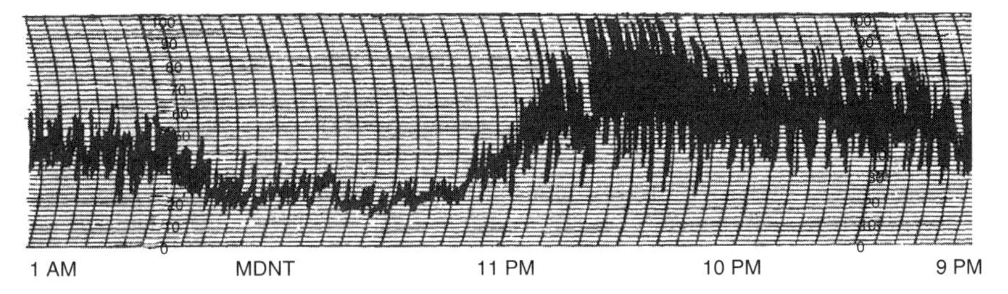

Figure 11.1.2. Hurricane wind speed record.

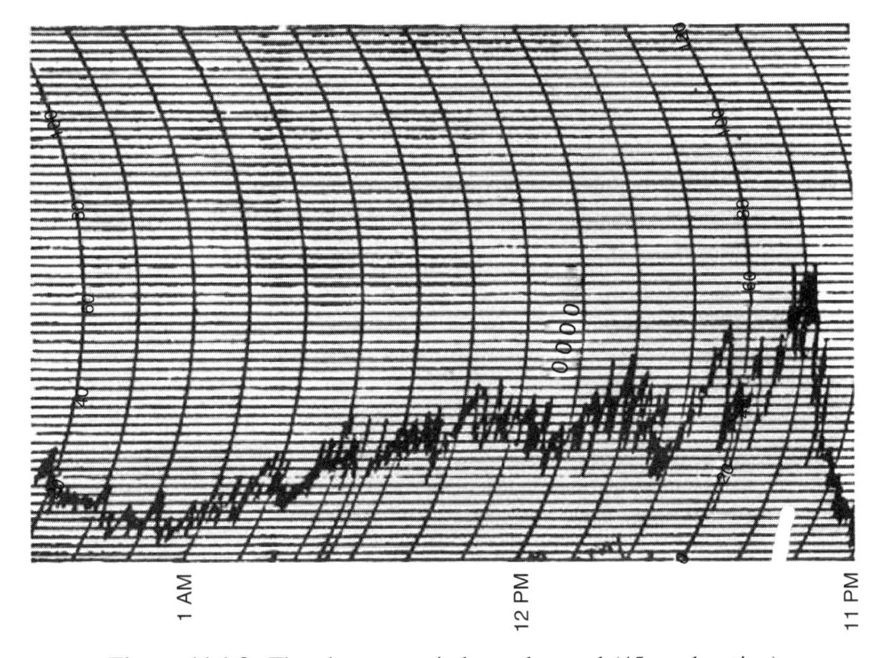

Figure 11.1.3. Thunderstorm wind speed record (45 m elevation).

- *Sustained wind speeds*, defined as wind speeds averaged over intervals of the order of one minute, are used in both engineering and meteorological practice. The *fastest 1-minute wind speed* or, for short, the 1-min speed, is the storm's largest 1-min average wind speed. Before 1995, the ASCE 7 Standard used the *fastest-mile wind speed*, defined as a storm's largest wind speed averaged over the time required for the passage at a point of a volume of air with a horizontal length of one mile. (This definition is based on a wind speed recording device used until the 1990s.) For fastest-mile speeds V_f, in mph, the averaging time in seconds is $t = 3600/V_f$.
- *10-min wind speeds* are wind speeds averaged over 10 minutes, and are used for design purposes by the Eurocode [11-2], in conformity with World Meteorological Organization (WMO) practice.

11.1.2 Relation between Wind Speeds with Different Averaging Times, Open Terrain

The approximate *mean ratio* r of the t-s speed to the hourly (3,600-s) speed, at a height of 10 m over open terrain is listed for selected values of t as follows:

t (s)	3	5	40	60	600	3,600
r	1.52	1.49	1.29	1.25	1.1	1.0

(see ASCE 7-10 Commentary Fig. C26.5-1). For winds over surfaces other than open terrain, see Sect. 11.2.4.2.

It is commonly assumed in wind engineering practice that wind speed records are statistically stationary over a period of 10 min to one hour. This assumption is currently made even for *thunderstorm winds*, meaning that, over open terrain, thunderstorm winds with a peak 3-s gust speed V_{3s} are commonly represented for aerodynamic and structural purposes by winds in a large-scale storm with mean hourly speed $\overline{V} = V_{3s}/1.52$. This representation is typically conservative from a structural engineering viewpoint.

Numerical Example 11.1.1. *Conversion of fastest-mile wind speed to mean hourly speed and to peak 3-s gust for open terrain.* For a fastest-mile wind speed at 10 m over open terrain of 90 mph, the averaging time is $3{,}600/90 = 40$ s, and the corresponding hourly speed and peak 3-s gust are $90/1.29 = 69.8$ mph and $69.8 \times 1.52 = 106$ mph, respectively.

Numerical Example 11.1.2. *Conversion of peak 3-s gust speed to mean hourly speed for open terrain.* Let the peak 3-s gust speed at 10 m above ground in open terrain be 64 mph. For wind tunnel testing and structural purposes, winds characterized by that gust speed are modeled by winds with a $64 \text{ mph}/1.52 = 42$ mph mean hourly speed at 10 m above ground in open terrain.

11.2 WIND SPEED PROFILES

It was pointed out in Chapter 10 that wind speeds are lower near the surface owing to flow retardation by friction between the moving air and the surface, and that the retardation becomes increasingly weaker at higher elevations. For high speeds averaged over 10 minutes or longer, above horizontal surfaces with uniform roughness over a sufficiently long fetch, the *logarithmic law* (Sect. 11.2.1) describes mean wind profiles up to elevations that define the atmospheric boundary layer's *surface layer*. The depth of the surface layer increases as the wind speed increases, and can reach several hundred meters. Wind profiles can alternatively be described by the *power law* (Sect. 11.2.2).

The logarithmic law has long superseded the power law in meteorological practice. It is used in the Eurocode [11-2] and in the Commentary to the ASCE 7-10 Standard [2-1]. However, the power law remains in use in the Canadian Building Code (for mean hourly wind speeds) and in the ASCE 7 Standard (for fastest-mile speeds before 1995 and peak 3-s gust speeds as of 1995). For tentative information on *veering angles*, that is, on the shape of the spiral of Fig. 10.1.5, see Sect. 11.2.3. Wind speeds over surfaces other than open terrain can be calculated from wind speeds over open terrain by using *relations between wind speeds in different roughness regimes* (Sect. 11.2.4). The wind profile is affected by the length of the *fetch* (Sect. 11.2.5), and by *topographical features* such as hills (Sect. 11.2.6). Measurements of wind profiles in *hurricanes* are discussed in Sect. 11.2.7.

11.2.1 Wind Profiles over Horizontal Terrain with Long Fetch: The Logarithmic Law

11.2.1.1 The Logarithmic Law: Roughness Length, Drag Coefficient, and Zero Plane Displacement. The logarithmic (or log) law describes the variation with height and surface roughness of strong mean speeds with averaging times of 10-min to 1-hr in straight winds. Its expression may be written as

$$\overline{V}(z) = \overline{V}(z_{ref}) \frac{\ln \dfrac{z}{z_0}}{\ln \dfrac{z_{ref}}{z_0}}, \tag{11.2.1}$$

in which $\overline{V}(z)$ and $\overline{V}(z_{ref})$ are the mean wind speeds at elevation z and z_{ref}, respectively, z_{ref} is a reference elevation, and z_0 is an empirical measure of the surface roughness called *roughness length*.

For a derivation of Eq. 11.2.1, see Appendix A2.1. Table 11.2.1 shows approximate ranges for z_0 and, in the footnotes, values of z_0 which, according to Sect. C26.7-1 of the ASCE 7-10 Commentary, are approximately consistent

TABLE 11.2.1. Roughness Lengths z_0 Proposed in the ASCE 7-10 Commentary

Type of Surface	Roughness Length, ft (m)
Water*	0.016–0.03 (0.005–0.01)
Open terrain**	0.05–0.5 (0.015–0.15)
Urban and suburban terrain, wooded areas***	0.5–2.3 (0.15–0.7)

*The larger values apply over shallow waters (e.g., near shorelines). Approximate typical value corresponding to ASCE 7-10 Exposure D: 0.016 ft (0.005 m) (ASCE Commentary Sect. C26.7). According to [11-4], for strong hurricanes $z_0 \approx 0.001-0.003$ m. See also [11-27].
**Approximate typical value corresponding to ASCE 7-10 Exposure C: 0.066 ft (0.02 m) (ASCE Commentary Sect. C26.7).
***Value corresponding approximately to ASCE 7-10 Exposure B: 0.5 ft (0.15 m); this value is smaller than the typical value for ASCE 7-10 Exposure B [1 ft (0.3 m)] and accounts, conservatively, for the possible presence of open spaces such as parking lots (ASCE Commentary Sect. C26.7); see also [11-5].

The values of Table 11.2.1 differ from those specified in the Eurocode, in which:

$z_0 = 0.003$ m for sea or coastal areas exposed to the open sea.

$z_0 = 0.01$ m for lakes or flat and horizontal area with negligible vegetation and no obstacles.

$z_0 = 0.05$ m for areas with low vegetation and isolated obstacles such as trees or buildings with separations of minimum 20 obstacle heights (e.g., villages, suburban terrain, permanent forest).

$z_0 = 0.3$ m for areas with regular cover of vegetation or buildings or with isolated obstacles with separations of maximum 20 obstacle heights.

$z_0 = 1$ m for areas in which at least 15% of the surface is covered with buildings whose average height exceeds 15 m.

with ASCE 7-10 Standard exposure categories. Meteorologists sometimes use *the drag coefficient* C_d as a measure of surface roughness, where

$$C_d = \left\{ \frac{k}{\ln\left(\dfrac{10}{z_0}\right)} \right\}^2 \tag{11.2.2}$$

$k = 0.4$ is the von Kármán constant, and z_0 is expressed in meters. For example, for $z_0 = 0.3$ m, $C_d = 0.0013$.

The following relation was proposed by Lettau [11-3]:

$$z_0 = 0.5\, H_{ob} \frac{S_{ob}}{A_{ob}} \tag{11.2.3}$$

where H_{ob} is the average height of the roughness elements in the upwind terrain, S_{ob} is the average vertical frontal area presented by the obstacle to the wind, and A_{ob} is the average area of ground occupied by each obstruction, including the open area surrounding it.

Numerical Example 11.2.1. *Application of the Lettau formula.* Check Eq. 11.2.3 against the Eurocode value $z_0 = 1$ m indicated earlier, assuming the average building height is $H_{ob} = 15$ m, the average dimensions in plan of the buildings are 16 m × 16 m, and $A_{ob} = 1600$ m^2. We have $S_{ob} = 15 \times 16 = 240$ m^2, so the average area occupied by buildings is $16 \times 16/1600 = 16\%$. Equation 10.2.3 yields $z_0 = 1.125$ m.

On account of the finite height of the roughness elements, the following empirical modification of Eq. 11.2.1 is required. The quantity z, rather than denoting height above ground, is defined as

$$z = z_{ground} - z_d \tag{11.2.4}$$

where z_{ground} denotes height above ground, z is the effective height, and z_d is called the *zero plane displacement*. It has been suggested [11-6] that, denoting the general roof-top level by $\overline{H}$, it is reasonable to assume

$$z_d = \overline{H} - 2.5z_0 \tag{11.2.5}$$

This correction to the wind profile was originally suggested for flow over centers of large cities. However, it is also applicable to heavily built-up suburbs, and should be considered when simulating flows affecting individual homes, especially in wall-of-wind facilities (Sect. 13.3).

11.2.1.2 Depth of the Surface Layer. The logarithmic law is not valid throughout the boundary layer, but only within its lowest part, called the *surface layer*. According to results of micrometeorological research, the depth z_s of the surface layer *increases* with wind speed and terrain roughness, and depends on angle of latitude φ, as follows:

$$z_s \approx 100 \frac{\overline{V}(z_{ref})}{\sin\varphi \, \ln \dfrac{z_{ref}}{z_0}}, \text{ in feet, } [\overline{V}(z_{ref}) \text{ in mph}], \tag{11.2.6}$$

and is about one-tenth of the depth of the atmospheric boundary layer (see Appendix A2.1 and references quoted therein).

Numerical Example 11.2.2. *Application of the log law. Surface layer depth.* For a mean hourly speed $\overline{V}(32.8 \text{ ft}) = 56$ mph, and a roughness length $z_0 = 0.1$ ft, Eq. 11.2.1 yields a mean hourly speed at 65.6 ft above ground $\overline{V}(65.6 \text{ ft}) = 62.7$ mph. For an angle of latitude $\varphi = 30°$, Eq. 11.2.6 yields a theoretical surface layer depth $z_s \approx 1900$ ft (590 m).

Before the development of atmospheric boundary layer theory, it was widely believed—incorrectly—that the surface layer depth is $z_s \approx$ 50–100 m, regardless of wind speed [11-7].

11.2.2 Wind Profiles over Horizontal Terrain with Long Fetch: The Power Law

The variation of wind speed with height can be expressed approximately by the power law:

$$V(z) = V(z_{\text{ref}}) \left(\frac{z}{z_{ref}} \right)^{1/\alpha} \tag{11.2.7}$$

In Eq. 11.2.7, the exponent $1/\alpha$ depends upon surface roughness and upon the averaging time, the profiles being flatter as the averaging time decreases. The power law applied to 3-s peak gust wind profiles has the same form as Eq. 11.2.7, and in the ASCE 7 Standard its power law exponent is denoted by $\hat{\alpha}$ rather than by α. Table 11.2.2 shows power law exponents and gradient heights specified by the ASCE 7-93 Standard for sustained wind speeds (including fastest-mile wind speeds), by the National Building Code of Canada (NBC) for mean hourly speeds, and by the 1995 and subsequent versions of the ASCE 7 Standard for peak 3-s gusts [2-1]. Differences between exponents corresponding to 3-s and 5-s peak gusts may be neglected for practical purposes. Equation 11.2.7 is valid up to a height z_g purported to represent the gradient height.

The ASCE Commentary Sect. C26.7 states: "The ground surface roughness is best measured by a roughness length parameter called z_0." A similar statement was advanced in [11-8]. However, at the time, the slide rule was still being extensively used, and it was opined that the use of the logarithmic law would be unwieldy. In addition, as noted earlier, it was believed that the logarithmic law is valid only in the lowest 50 m or so of the atmospheric boundary layer. The power law is still in use in the ASCE Standard 7-10.

TABLE 11.2.2. Power Law Exponents and Gradient Heights Used in 1993 to 2010 Versions of ASCE 7 Standard, and in the National Building Code of Canada [11-1]

	Exposure	A[a]	B[b]	C[c]	D[d]
ASCE 7-93[e]	$1/\alpha$	1/3	1/4.5	1/7	1/10
	z_g ft (m)	1500 (457)	1200 (366)	900 (274)	700 (213)
NBC[f]	$1/\alpha$	0.4	0.28	0.16	—
	z_g ft (m)	1700 (520)	1300 (400)	900 (274)	—
ASCE 7[g]	$1/\hat{\alpha}$	—	1/7	1/9.5	1/11.5
(1995–2010)	z_g ft (m)	—	1200 (366)	900 (274)	700 (213)

[a] Centers of large cities;
[b] Suburban terrain, towns;
[c] Open terrain (e.g., airports);
[d] Water surfaces;
[e] Sustained speeds;
[f] Mean hourly speeds;
[g] Peak 3-s gust speeds.

Numerical Example 11.2.3. *Application of the power law.* Let $z_{\text{ref}} = 32.8$ ft (10 m), $V_{3s}(z_{\text{ref}}) = 55$ mph, and $\hat{\alpha} = 1/9.5$ (open terrain). From Eq. 11.2.5, at 100 ft above ground, $V_{3s}(100 \text{ ft}) = 55(100/32.8)^{1/9.5} = 62$ mph.

11.2.3 Veering Angles

It was pointed out earlier that the wind speed describes a spiral (Fig. 10.1.5). The veering angle can be defined as the angle between the gradient wind direction and the wind direction at a lower elevation, and is counterclockwise in the northern hemisphere (Fig. 11.2.1) and clockwise in the southern hemisphere. Assuming that the wind direction is nearly the same at the gradient height and at 500 m elevation, preliminary measurements suggest that: At 10 m above ground, the veering angle is about 10° for open terrain exposure and 15° for suburban terrain exposure; above open terrain, the veering angle decreases by about 3° at 100 m elevation and 7° at 300 m elevation; and above suburban terrain, the veering angle decreases by about 5° at 100 m elevation and 10° at 300 m elevation [11-9]. Actually, veering angles decrease as the wind speeds increase (this follows from Eqs. A214 and A2.16). Also, according to theory, veering angles do not change within the surface layer, that is, within the lower portion of the atmospheric boundary layer in which the logarithmic law holds.

In practical design terms, a consequence of veering is that, as the height above ground increases, the wind direction changes with respect to the direction of the reference winds at 10 m above ground. Consider, for example, winds in the northern hemisphere. Winds blowing from the north at 10 m elevation over terrain with open exposure will blow at 100 m elevation from a direction 3° clockwise from the north, and at 300 m elevation from a direction 7° clockwise from the north.

Winds from the north at 10 m above ground over terrain with open exposure will blow in suburban exposure at 10 m above ground from a direction (15° − 10°) = 5° counterclockwise from the north, since winds near the surface in suburban terrain make a 15° angle counterclockwise from the direction of the gradient speed, while winds near the surface in open terrain make a 10° angle counterclockwise from that direction; at 100 m above suburban terrain, the winds will blow from the north, and at 300 m above suburban ground, they

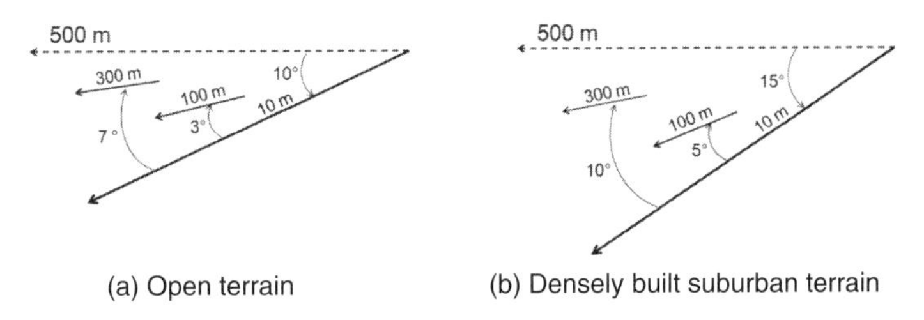

Figure 11.2.1. Tentative values of veering angles [11-9].

will blow from a direction 5° clockwise from the north (Fig. 11.2.1). It is emphasized that these suggestions are tentative and approximate.

11.2.4 Relation between Wind Speeds in Different Roughness Regimes

Information on design wind speeds is usually based on wind speed measurements over open terrain (airports). Owing to the stronger effects of ground friction in rougher terrain, wind speeds at any given elevation within the atmospheric boundary layer are lower over built-up than over open terrain. At the top of the boundary layers, where ground friction effects are negligible, wind speeds are the same over surfaces with any type of roughness. The relation between wind speeds at various elevations in different roughness regimes is based on this fact. Strictly speaking, this relation is valid only for straight winds, but it is also used in current practice for hurricane and thunderstorm winds.

11.2.4.1 Logarithmic Law Description. For a storm with mean hourly or 10-min speed $\overline{V}(z_{open})$ at elevation z_{open} over open terrain with roughness z_{0open}, the corresponding speed at elevation z over a site with roughness z_0 for a sufficiently long fetch may be written as

$$\overline{V}(z) = \overline{V}(z_{0open}) \left(\frac{z_0}{z_{0open}} \right)^{0.07} \frac{\ln \dfrac{z}{z_0}}{\ln \dfrac{z_{open}}{z_{0open}}} \tag{11.2.8}$$

Equation 11.2.8, first proposed by Biétry in 1976, is used in the Eurocode.

Numerical Example 11.2.4. *Relation between wind speeds in different roughness regimes: log law description.* The mean hourly speed is $\overline{V}(z_{open}) =$ 56 mph (25 m/s) at $z_{open} = 32.8$ ft (10 m) above open terrain with $z_{0open} =$ 0.066 ft (0.02 m). Calculate the mean hourly wind speed $\overline{V}(z)$ at 150 ft (45.7 m) above *suburban terrain* with $z_0 = 2.3$ ft (0.7 m).

Equation 11.2.8 yields $\overline{V}$ (150 ft) = 48.3 mph (21.6 m/s) above suburban terrain. Above *open terrain*, $\overline{V}$ (150 ft) = 69.7 mph (31 m/s) (Eq. 11.2.1). Note the significant retardation of the mean wind speed at 150 ft elevation over suburban terrain, where the wind speed is 48.3 mph, while at the same elevation above open terrain the speed is 69.7 mph.

11.2.4.2 Relation between Wind Speeds Averaged over Different Time Intervals for Winds above Any Type of Surface. In Sect. 11.1.2, we considered the relation between wind speeds averaged over various time intervals for the case of open terrain. In this section, we consider this relation for the case of winds over any type of surface.

The following approximate relation may be used [7-1, p. 69]:

$$V_t(z) = \overline{V}(z)\left\{1 + \frac{\eta(z_0)c(t)}{2.5\ln\dfrac{z}{z_0}}\right\} \tag{11.2.9}$$

where $V_t(z)$ is the peak speed averaged over t s within a record of approximately one hour, and $\overline{V}(z)$ is the mean wind speed for that record, over terrain with surface roughness z_0. The function $\eta(z_0)$ (Table 11.2.3a) is the ratio of the r.m.s of the longitudinal velocity fluctuations to the friction velocity (see Eq. 11.3.7b). The coefficient $c(t)$ (Table 11.2.3b) is an empirical peak factor which increases as t decreases. Therefore, the lower the averaging time t, the larger the peak of the longitudinal velocity fluctuations (i.e., the second term within brackets in Eq. 11.2.9) and, hence, the larger the peak speed $V_t(z_0)$.[1]

For an application, see Numerical Example 11.2.6.

11.2.4.3 Power Law Description. For strong winds, given the mean hourly speed $\overline{V}(z_{open})$ at the reference height z_{open} above open terrain with power law exponent $1/\alpha_{open}$, the mean hourly wind speed at height z above built-up terrain with power law exponent $1/\alpha$ is

$$\overline{V}(z) = \overline{V}(z_{open})\left(\frac{z_{g,open}}{z_{open}}\right)^{1/\alpha_{open}}\left(\frac{z}{z_g}\right)^{1/\alpha} \tag{11.2.10}$$

where the product of the first two terms in the right-hand side is the gradient speed above open terrain, $\overline{V}(z_{g,open})$. Since gradient speeds are not affected by surface roughness, the gradient speed over built-up terrain, $\overline{V}(z_g) = \overline{V}(z_{g,open})$. The last term in Eq. 11.2.6 transforms $\overline{V}(z_g)$ into $\overline{V}(z)$ at height z above built-up terrain. A relation similar to Eq. 11.2.10 is also used

TABLE 11.2.3a. Factor $\eta(z_0)$

z_0 (m)	0.005	0.03	0.30	1.00
$\eta(z_0)$	2.55	2.45	2.30	2.20

TABLE 11.2.3b. Factor $c(t)$

t	1	10	20	30	50	100	200	300	600	1000	3600
$c(t)$	3.00	2.32	2.00	1.73	1.35	1.02	0.70	0.54	0.36	0.16	0

[1]The following alternative equation relating 3-s and 1-min speeds was proposed in [11-10]: $V_{3\,s} = V_{60\,s}[1 + 2/\ln(z/z_0)]$. It can be verified, for example, that for $z_0 = 0.03$ m (open terrain) and $z = 10$ m, this equation yields $V_{3\,s} = 1.34\,V_{60\,s}$. In contrast, for extratropical storms Eq. 11.2.10 and ASCE Fig. C26.5-1 yield $V_{3\,s} = 1.39\,V_{60\,s}$ and $V_{3\,s} = 1.42\,V_{60\,s}$, respectively. Ratios $V_{3\,s}/V_{60\,s}$ tend to be higher for hurricanes than for extratropical storms [11-11]; see Sect. 11.3.1.

(with the appropriate values of the parameters z_g and $\hat{\alpha}$ from Table 11.2.2), for 3-s peak gust speeds, denoted by V or V_{3s}, instead of $\overline{V}$, and for sustained wind speeds such as fastest-mile speeds or 1-min speeds. In the ASCE 7 Standard, $V(z_{open} = 32.8\text{ ft})$, denoted in the Standard by V, is the 3-s basic wind speed, and the product of the last two terms in Eq. 11.2.10 is denoted by $\sqrt{K_z}$.

Numerical Example 11.2.5. *Relation between wind speeds in different roughness regimes, power law description.* Denote the 3-s peak gust speed by V. Let $V(32.8\text{ ft}) = 86$ mph above open terrain ($\hat{\alpha} = 9.5$, $z_g = 900$ ft, Table 11.2.2). Equation 11.2.10 yields $V(148\text{ ft}) = 101$ mph (*open terrain*). Using Table 11.2.2 and Eq. 11.2.10, above *suburban terrain* ($\hat{\alpha} = 7.0$ and $z_g = 1200$ ft), $V(32.8\text{ ft}) = 73$ mph and $V(148\text{ ft}) = 90$ mph.

Numerical Example 11.2.6. *Conversion of Saffir-Simpson scale winds to wind speeds above open terrain.* In this example, the 1-min wind speeds at 10 m above open terrain are calculated for Category 4 hurricanes as defined in the Saffir-Simpson scale. From Table 10.2.1, the 1-min speeds at 10 m above open water that define weakest and strongest Category 4 hurricanes are 131 mph and 155 mph, respectively. The conversion is performed in this example by using the log law description.

The conversion depends on the assumed values of the surface roughness lengths z_0 for open water and open terrain. Relative large values of z_0 are applicable to wind flow over water near shorelines where the water is shallow, as opposed to flow over open water. Assuming that for hurricane winds over open water $z_0 = 0.003$ m, Eq. 11.2.9 yields, with $\eta(z_0 \approx 0.003\text{ m}) \approx 2.55$ and $c(60\text{ s}) \approx 1.29$ (Tables 11.2.3a and 11.2.3b):

$$V^w_{60\text{ s}}(10\text{ m}) = \overline{V}^w(10\text{ m})\left[1 + \frac{2.55 \times 1.29}{2.5\ln\dfrac{10}{0.003}}\right],$$

where $\overline{V}^w(10\text{ m})$ is the mean hourly wind speed at 10 m above water, that is,

$$\overline{V}^w(10\text{ m}) = 0.86\,V^w_{60\text{ s}}(10\text{ m}).$$

Assuming conservatively that over open terrain $z_0 = 0.04$ m, Eq. 11.2.8 yields

$$\overline{V}^w(10\text{ m}) = \overline{V}(10\text{ m})\left[\frac{0.003}{0.04}\right]^{0.07}\frac{\ln\dfrac{10}{0.003}}{\ln\dfrac{10}{0.04}},$$

where $\overline{V}(10\text{ m})$ is the mean hourly wind speed at 10 m above open terrain, that is,

$$\overline{V}(10\text{ m}) = 0.816\overline{V}^w(10\text{ m}).$$

It follows that

$$\overline{V}(10\text{ m}) = 0.86 \times 0.816 V_{60\text{ s}}^{w}(10\text{ m})$$
$$= 0.7 V_{60\text{ s}}^{w}(10\text{ m}).$$

Therefore, the peak 3-s gust over open terrain is

$$V_{3\text{ s}}(10\text{ m}) = 1.52 \times 0.7 \times V_{60\text{ s}}^{w}\ (10\text{ m})$$
$$= 1.06 V_{60\text{ s}}^{w}(10\text{ m}).$$

To the speed $V_{60\text{ s}}^{w}(10\text{ m}) = 155$ mph there corresponds, then, a peak 3-s gust at 10 m over open terrain of about 164 mph. A slightly more conservative value (171 mph) was adopted in the Commentary of the ASCE 7-10 Standard (ASCE Table C26.5-2) to account for the fact that the turbulence intensity is larger in hurricanes than in synoptic (extratropical) storms.

11.2.5 Wind Profiles Near a Change in Terrain Roughness; Fetch and Terrain Exposure

The surface roughness is not the sole factor that determines the wind profile at a site. The profile also depends upon the distance (the *fetch*) over which that surface roughness prevails upwind of the site. The terminology used in the ASCE 7 Standard therefore distinguishes between *surface roughness* and *exposure*. For example, a site is defined as having Exposure B if it has surface roughness B *and* surface roughness B prevails over a sufficiently long fetch. For design purposes, the wind profile at a site with Exposure B may then be described by the power law with parameters corresponding to surface roughness B. Sections 11.2.1 to 11.2.3 consider only the case of long fetch. ASCE 7 Standard criteria on the fetch required to assume a given exposure are reproduced in Sect. 3.4.2; see also ASCE 7-10 Commentary, Sect. C26.7.

Useful information on the flow in transition zones can be obtained by considering the simple case of an abrupt roughness change along a line perpendicular to the direction of the mean flow. Upwind of the discontinuity the flow is horizontally homogeneous and, near the ground, it is governed by the parameter z_{01}. Downwind of the discontinuity the flow will be affected by the surface roughness z_{02} over a height $h(x)$, where x is the downwind distance from the discontinuity. This height, known as the height of the *internal boundary layer*, increases with x until the entire flow adjusts to the roughness length z_{02}. A well-accepted model of the internal boundary layer, which holds for both smooth-to-rough and rough-to-smooth transitions, is

$$h(x) = 0.28\, z_{0r} \left(\frac{x}{z_{0r}} \right)^{0.8} \tag{11.2.11}$$

where z_{0r} is the largest of the roughness lengths z_{01} and z_{02} [11-12] (Fig. 11.2.2).

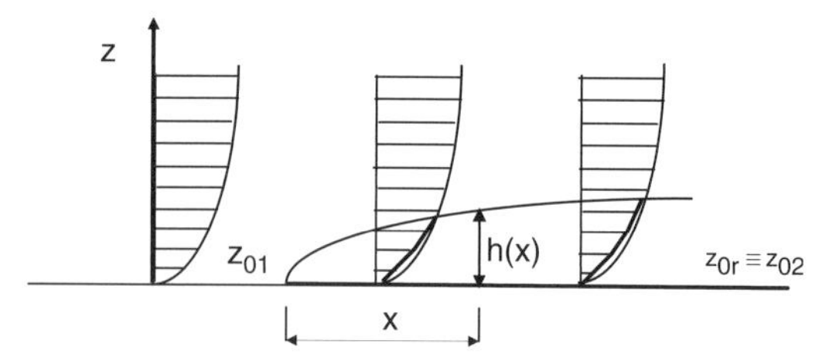

Figure 11.2.2. Internal boundary layer $h(z)$. Mean wind speed profile within the internal boundary layer is adjusted to the terrain roughness $z_{02} > z_{01}$.

11.2.6 Wind Profiles over Nonhorizontal Terrain

Topographic features alter the local wind environment and create wind speed increases (*speed-up* effects), since more air has to flow per unit of time through an area decreased, with respect to the case of flat land, by the presence of the topographical feature. A procedure specified in the ASCE 7 Standard for the calculation of speed-up effects on 2- or 3-dimensional isolated hills and 2-dimensional ridges and escarpments is presented in Sect. 3.5 (ASCE Sect. 26.8).

11.2.7 Hurricane Mean Wind Speed Profile Measurements

Most of the research on wind profiles up to elevations of hundreds of meters has been performed for non-cyclonic storms. More recently, Geophysical Positioning System (GPS) sondes have been used for measurements of hurricane mean wind speed profiles up to 3 km elevations [11-4]. The measurements yielded the following results: (1) On average the mean wind profiles are approximately logarithmic up to elevations of a few hundred meters. (2) The mean wind profile in tropical cyclones is not monotonically increasing, as suggested in Fig. 10.1.2; rather, on average, the mean wind speed profiles are approximately logarithmic up to elevations of about 300–400 m, beyond which they decrease with height. (3) Because the physical features of the ocean surface depend on the wind speed, so does the roughness length z_0 (see references in [11-4]). Mean wind profile measurements in the eye of hurricanes are also reported in [11-13].

11.3 ATMOSPHERIC TURBULENCE

Except for winds with relatively low speeds under special temperature conditions, the wind flow is not laminar (smooth). Rather, it is *turbulent*—it fluctuates in time and space; that is, at any one point in space, the wind speed is a random function of time (Fig. 11.1.1), and at any one moment in time, the wind speed is a random function of position in space.

Atmospheric flow turbulence characterization is of interest in structural engineering applications for four reasons. First, turbulence affects the definition of the wind speed specified in engineering calculations (Sect. 11.1). Second, by transporting particles from flow regions with high speeds into low-speed regions, turbulence can influence significantly the wind flow around a structure and, therefore, the aerodynamic forces (Sect. 13.2.2). Third, flow fluctuations produce dynamic effects in flexible structures such as tall buildings or long-span bridges (Sects. 14.3.6, 15.4.2). Fourth, to replicate aerodynamic and dynamic effects in the laboratory, it is typically attempted in current practice to reproduce the features of atmospheric turbulence (Sect. 13.3).

Flow features in thunderstorms (including associated gust fronts and downbursts), and their effects on buildings have been the object of relatively few full-scale measurements. Useful laboratory and numerical simulations of such transient atmospheric flows and their effects have been reported, for example in [11-14 to 11-17]; however, the results are still tentative, and have not yet been used in engineering practice.

Descriptions of turbulence intensities, integral turbulence lengths, and turbulence spectra and cross-spectra for extratropical and tropical storms are presented in Sects. 11.3.1, 11.3.2, and 11.3.3, respectively.

11.3.1 Turbulence Intensities

The longitudinal turbulence intensity at a point with elevation z is defined as

$$I_u(z) = \frac{\overline{u^2(z,t)}^{1/2}}{\overline{V}(z)}, \tag{11.3.1a}$$

that is, as the ratio of the r.m.s. of the longitudinal wind speed fluctuations $u(z, t)$ to the mean speed $\overline{V}(z)$, $u(z, t)$ being parallel to $\overline{V}(z)$. The following empirical relation holds:

$$\overline{u^2(z,t)}^{1/2} = \eta(z_0)\frac{\overline{V}(z)}{2.5 \ln \dfrac{z}{z_0}}, \tag{11.3.1b}$$

where approximate values of $\eta(z_0)$ are given in Table 11.2.3a. It follows from Eqs. 11.3.1a and b that the expression for the longitudinal turbulence intensity is

$$I_u(z) \approx \frac{\eta(z_0)}{2.5 \ln \dfrac{z}{z_0}}. \tag{11.3.2}$$

Equation 11.3.2 shows that the turbulence intensity decreases as the height above the surface increases. Measurements suggest that the turbulence intensity is typically higher by roughly 10–15% in tropical cyclones than in extratropical storms [11-11]. Definitions similar to Eq. 11.3.1a are applicable to the lateral and vertical turbulence intensities $I_v(z)$ and $I_w(z)$ (see Sect. 11.3.3.1).

Numerical Example 11.3.1. *Calculation of longitudinal turbulence intensity.* For $z_0 = 0.03$ m, $z = 20$ m, Eq. 11.3.2 and Table 11.2.3a yield $I_u(z) \approx 0.16$.

11.3.2 Integral Turbulence Scales

The velocity fluctuations in a flow passing a point may be considered to be caused by an overall (resultant) eddy consisting of a superposition of conceptual component eddies transported by the mean wind. Each component eddy is viewed as causing, at that point, a periodic fluctuation with frequency n. Integral turbulence scales are measures of the size (i.e., of the spatial extent) of the overall turbulent eddy.

The integral turbulence scale L_u^x is an indicator of the extent to which an overall eddy associated with the longitudinal wind speed fluctuation u will engulf a structure in the along-wind direction, and will thus affect at the same time both its windward and leeward sides. If L_u^x is large in relation to the along-wind dimension of the structure, the gust will engulf both sides. The scales L_u^y and L_u^z are measures of the transverse and vertical spatial extent of the fluctuating longitudinal component u of the wind speed. The scale L_w^x is a measure of the longitudinal spatial extent of the vertical component w. If the mean wind is normal to a bridge span and L_w^x is large in relation to the deck width, the vertical wind speed fluctuation w will act at any given time on the whole width of the deck.

Mathematically the integral turbulence length L_u^x is defined as follows:

$$L_u^x = \frac{1}{\overline{u^2}} \int_0^{\infty} R_{u_1 u_2}(x)\, dx \qquad (11.3.3)$$

in which $u_1 = u(x_1, y_1, z_1, t)$, $u_2 = u(x_1 + x, y_1, z_1, t)$, and the denominator is the variance of the longitudinal velocity fluctuations, a statistic that, for given elevation z, is the same throughout the flow. The integrand is the cross-covariance of the signals u_1 and u_2. Equation 11.3.3 may be interpreted as follows (see also Appendix A1): At any time t, the fluctuation u at a point x_1, y_1, z_1 differs from its counterpart at a point $x_1 + x, y_1, z_1, t$. The difference increases as the distance x increases. If the distance x is small in relation to the average eddy size, the two fluctuations will be nearly the same, so in Eq. 11.3.3 the elemental length dx is multiplied by a factor of almost unity. If x is large in relation to the average eddy size, the two fluctuations differ randomly from each other, and on average their contribution is small or nil. The integral length is therefore a measure of average eddy size.

Measurements show that L_u^x increases with height above ground and as the terrain roughness decreases [11-18]. Roughly, at 15 m above ground, in suburban terrain on average $L_u^x \approx 50$ m, and in open terrain $L_u^x \approx 75$ m; in open terrain, at 30 m above ground, 60 m $< L_u^x <$ 450 m (average 200 m), and at 150 m above ground, 120 m $< L_u^x <$ 630 m (average 400 m) [11-19]. Note the very large variability in the reported values of L_u^x.

Measurements reported in [11-20, 11-21] suggest that

$$L_u^y \approx 0.2L_u^x,\ L_u^z \approx 6z^{0.5}(L_u^z \text{ and } z \text{ in meters}),\ L_w^x \approx 0.4z \qquad (11.3.4a, b, c)$$

Turbulence scales L_u^x for flow at 10 m above open terrain, estimated from 20-min tropical cyclone records, are reported in [11-22].

11.3.3 Spectra and Cross-Spectra of Turbulent Wind Speed Fluctuations

As indicated in Sect. 11.3.2, integral turbulence scales are measures of average turbulent eddy sizes. In some applications, it is necessary to refine the description of the turbulent fluctuations. One example of such an application is the excitation of the dynamic resonant response of a flexible structure by velocity fluctuation components with frequencies equal or close to a natural frequency of vibration of a flexible structure (Sect. 14.3.6).

11.3.3.1 Spectral Density Functions. A statistically stationary random signal may be viewed as a superposition of many harmonic components, each of which has a distinct frequency. The *spectral density function* $S_g(n)$ or, for short, the spectrum, provides a measure of the contribution to the signal of each of those components. The area under the spectral density curve is by definition equal to the mean square value of the fluctuation excitation $g(t)$:

$$\int_0^{n_{\max}} S_g(n)dn = \overline{g^2(t)}, \qquad (11.3.5)$$

where n denotes frequency and $n_{\max}$ is the frequency beyond which $S_g(n) \equiv 0$. Mathematically, the ordinates of a spectral density function are counterparts of the squares of the amplitudes of a Fourier series. In a Fourier series the frequencies are discrete, and the contribution of each harmonic component to the signal's variance is finite. In a spectral density plot the frequencies are continuous, and each component $S_g(n)dn$ has an infinitesimal contribution to the variance of the signal. Thus, spectral density plots have a relation to plots of squares of Fourier series harmonic components that is similar to the relation of a probability density function to a histogram.[2]

A useful expression for the spectral density of the longitudinal velocity fluctuations at elevation z is

$$\frac{nS_u(z,n)}{u_*^2} = \frac{200f}{(1+50f)^{5/3}} \qquad (11.3.6)$$

[2]A mathematical introduction to random processes, including their spectral representation, is presented in Appendix A1.

where the friction velocity u_* is given by the expression

$$u_* = \frac{\overline{V}(z)}{2.5 \ln \frac{z}{z_0}} \tag{11.3.7a}$$

$$= \frac{1}{\eta} \overline{u^2(z,t)}^{1/2}, \text{ and} \tag{11.3.7b}$$

$$f = \frac{nz}{\overline{V}(z)} \tag{11.3.8}$$

$\overline{u^2(z,t)}^{1/2}$ is the r.m.s. of the longitudinal velocity fluctuations, $\overline{V}(z)$ is the mean wind speed at z, and it is assumed that $\eta = 2.45$. As follows from Table 11.2.3a, the errors inherent in the application of Eq. 11.3.6 for suburban terrain or over-water exposure are relatively small. Equation 11.3.6 was proposed in [11-23] in a slightly different form, which corresponded to a significantly smaller value of η than the values typically applicable to atmospheric flows with neutral stratification. The coordinate f is called the Monin coordinate.

Other expressions for the spectral density of the longitudinal velocity fluctuations have been proposed in the literature. Among those expressions, some of the most common disregard the significant variation of the spectrum with height z, or are functions of an integral turbulence scale whose dependence on height z is typically disregarded in wind tunnel simulation practice.

In some applications, expressions for the spectra of vertical and lateral turbulent fluctuations are required. According to [11-24], up to an elevation of about 50 m, the expression for the *vertical* velocity fluctuations is

$$\frac{nS_w(z,n)}{u_*^2} = \frac{3.36f}{1+10f^{5/3}} \tag{11.3.9}$$

The expression for the spectrum of the *lateral* turbulent fluctuations proposed in [11-23] is of the form

$$\frac{nS_v(z,n)}{u_*^2} = \frac{15f}{(1+10f)^{5/3}}. \tag{11.3.10}$$

In Eqs. 11.3.9 and 11.3.10, the variable f is defined as in Eq. 11.3.8.

Hurricane wind spectra were found to contain more energy at low frequencies than the spectra listed in this section [11-22]. Theoretical considerations on velocity spectra are presented in Appendix A2.

11.3.3.2 Cross-Spectral Density Functions. The *cross-spectral density function* of turbulent fluctuations occurring at two different points in space indicates the extent to which harmonic fluctuation components with frequencies n at those points are in tune with each other or evolve at cross-purposes (i.e., are or are not mutually *coherent*). For components with high frequencies, the distance in space over which wind speed fluctuations are mutually coherent is small. For low-frequency components, that distance is relatively large—on the order of integral turbulence scales. An eddy corresponding to a component with frequency n is said to envelop a structure if the distance over which the fluctuations with frequency n are relatively coherent is comparable to the relevant dimension of the structure.

The expression for the cross-spectral density of two signals u_1 and u_2 is

$$S^{cr}_{u_1u_2}(r,n) = S^{C}_{u_1u_2}(r,n) + iS^{Q}_{u_1u_2}(r,n) \tag{11.3.11}$$

in which $i = \sqrt{-1}$, r is the distance between the points M_1 and M_2 at which the signals occur, and the first and second term on the right-hand side of Eq. 11.3.11 denote the co-spectrum and the quadrature spectrum of the two signals, respectively. The coherence function is defined as

$$\mathscr{C}(r,n) \equiv [Coh(r,n)]^2 = c^2_{u_1u_2}(r,n) + q^2_{u_1u_2}(r,n) \tag{11.3.12}$$

where

$$c^2_{u_1u_2}(r,n) = \frac{\left[S^{C}_{u_1u_2}(r,n)\right]^2}{S(z_1,n)S(z_2,n)}, \quad q^2_{u_1u_2}(r,n) = \frac{\left[S^{Q}_{u_1u_2}(r,n)\right]^2}{S(z_1,n)S(z_2,n)} \tag{11.3.13a, b}$$

In Eq. 11.3.13a and b, $S(z_1,n)$ and $S(z_2,n)$ are the spectra of the signals at points M_1 and M_2. Larger integral turbulence scales correspond to stronger coherence.

In the atmosphere, it is typically assumed that the quadrature spectrum is negligible in relation to the co-spectrum. The following expression for the co-spectrum has been proposed:

$$S^{C}_{u_1u_2}(r,n) = S^{1/2}(z_1,n)S^{1/2}(z_2,n)\exp(-\hat{f}) \tag{11.3.14}$$

where

$$\hat{f} = \frac{n\left[C_z^2(z_1^2 - z_2^2) + C_y^2(y_1^2 - y_2^2)\right]^{1/2}}{(1/2)[\overline{V}(z_1) + \overline{V}(z_2)]} \tag{11.3.15}$$

y_i, z_i are the coordinates of point $M_i\,(i = 1,2)$, and according to wind tunnel measurements, $C_z = 10$, $C_y = 16$ [11-25]. For *lateral* fluctuations, the expression for the co-spectrum is similar, except that values $C_z = 7$, $C_y = 11$ have been proposed [11-26]. For two points with the same elevation, the expression for the co-spectrum of the *vertical* fluctuations is also assumed to be similar, with $C_y = 8$ [11-26].

11.3.4 Turbulence in Flows with Stable Stratification

The models presented in the preceding sections were developed for strong winds in which, owing to turbulent mixing, the flow stratification is for practical purposes neutral. However, in flows with relatively low velocities, low temperatures of the ground or water surface cause the lower layers of the air flow to be colder than the upper layers (i.e., the flow stratification is stable; Sect. 10.1). In the absence of turbulent mixing brought about by strong winds, colder and heavier air flows near the surface, while warmer and lighter air flows at higher elevations. For structural engineering purposes, such flows may be considered to be approximately laminar (turbulence-free). The vortices shed in the wake of a structure are typically more coherent and stronger in laminar than in turbulent flows, a fact that needs to be accounted for in the design of structures subjected to relatively low wind speeds—for example, bridges experiencing motions due to vortices shed in their wake.

CHAPTER 12

EXTREME WIND SPEEDS AND WIND-INDUCED EFFECTS

Structures are designed so that the nominal probabilities of attaining or exceeding the limit states of interest are acceptably small. For example, as required by the ASCE 7-10 Standard, members of Risk Category II structures are designed so that, on average, the strength design limit state is exceeded at intervals of at least 700 years (Sects. 3.1, 3.2). For a given structure, the probability of exceedance of a wind-induced limit state depends, for any specified member, upon the extreme wind speeds at the structure's site. This chapter is concerned with probabilistic estimation of extreme wind speeds and of wind effects induced by extreme winds.

Section 12.1 discusses *nondirectional and directional wind speed data* in non-hurricane and hurricane-prone regions. Section 12.2 provides simple, intuitive definitions of *exceedance probabilities* and *mean recurrence intervals* (MRIs), and extends those definitions to wind speeds in *mixed wind climates* (e.g., climates with both hurricane and non-hurricane winds, or with synoptic storm and thunderstorm winds). Section 12.3 describes *parametric methods* for estimating extreme wind speeds with specified MRIs. Section 12.4 describes procedures for the *estimation of extreme wind effects and their MRIs*. Such procedures are currently based on either nondirectional or directional wind speeds. The estimation of extreme wind effects based on directional wind speeds typically requires the development of large synthetic directional wind speed data samples from relatively small measured data samples (i.e., samples obtained over periods of, say, 20 to 100 years). Sections 12.5 and 12.6 are concerned with the *development of large directional wind speed data sets* for hurricane and non-hurricane winds, respectively. Section 12.7 presents an example of the *non-parametric estimation of extremes*. Section 12.8 discusses *errors* in the estimation of extreme wind speeds and extreme wind effects.

12.1 WIND SPEED DATA

12.1.1 Micrometeorological Homogeneity of the Data

Extreme wind speed analyses must use micrometeorologically homogeneous data, meaning that all the data in a set must correspond to (1) the same height above surfaces with the same exposure (e.g., ocean surface, open terrain, suburban built-up terrain), and (2) the same averaging time (e.g., 3 s, 1 min, 10 min, or 1 hour). Where data do not satisfy the micrometeorological homogeneity requirement, they must be transformed so that the requirement is satisfied (see, e.g., Sects. 11.2.1.1 and 11.2.4).

12.1.2 Directional and Nondirectional Wind Speeds

Standard provisions for wind loads are based primarily on the use of *nondirectional* extreme wind speeds (i.e., largest wind speeds in any one year or storm event, regardless of wind direction). *Directional* extreme wind speeds (i.e., largest wind speeds in any one year or storm event for each of the directions being considered) are used to estimate wind effects on structures for which aerodynamic data corresponding to a sufficient number of wind directions are available.

Denote the directional wind speeds by v_{ij}, where the subscript i indicates the year or the storm event, and the subscript j indicates the wind direction. For fixed i, the corresponding nondirectional wind speed is $v_i = \max_j(v_{ij})$.

Numerical Example 12.1.1. *Directional and nondirectional wind speeds.* To illustrate the definitions of directional and nondirectional wind speeds, we consider the following largest peak 3-s gusts in mph recorded over two consecutive 1-year periods:[1]

	Directional speeds v_{ij}								Nondirect. speeds $\max_j(v_{ij})$
j	1 (NE)	2 (E)	3 (SE)	4 (S)	5 (SW)	6 (W)	7 (NW)	8 (N)	
$i = 1$	90	**100**	81	96	87	89	93	78	100
$i = 2$	78	94	83	**107**	80	83	72	76	107

The nondirectional speeds are also shown in bold type in the list of directional speeds.

12.1.3 Wind Speed Data Sets

Selected sources of data are listed in the following sections.

12.1.3.1 Data in the Public Domain. Data indicated in Sects. 12.1.3.1.1 to 12.1.3.1.3 are available at the www.nist.gov/wind datasets link.

[1]In the statistical literature, a fixed time period is called an epoch.

12.1.3.1.1 Largest Yearly Directional 3-s Peak Gust Speeds. In m/s at 10 m above ground in open terrain at about 35 U.S. weather stations for up to about 30 years of record, listed for each of eight octants (at 45° intervals).

12.1.3.1.2 Nondirectional Yearly 3-s Peak Gust Speeds. In m/s at 10 m above ground in open terrain for about 15- to 50-year periods of record at about 100 U.S. weather stations.

12.1.3.1.3 Simulated (Synthetic) Directional Tropical Storm/Hurricane 1-min Speeds. In knots at 10 m elevation in open terrain for 55 locations along the Gulf and Atlantic coasts (Fig. 12.1.1). The speeds were obtained by Monte Carlo simulation from roughly 100 years of recorded hurricane

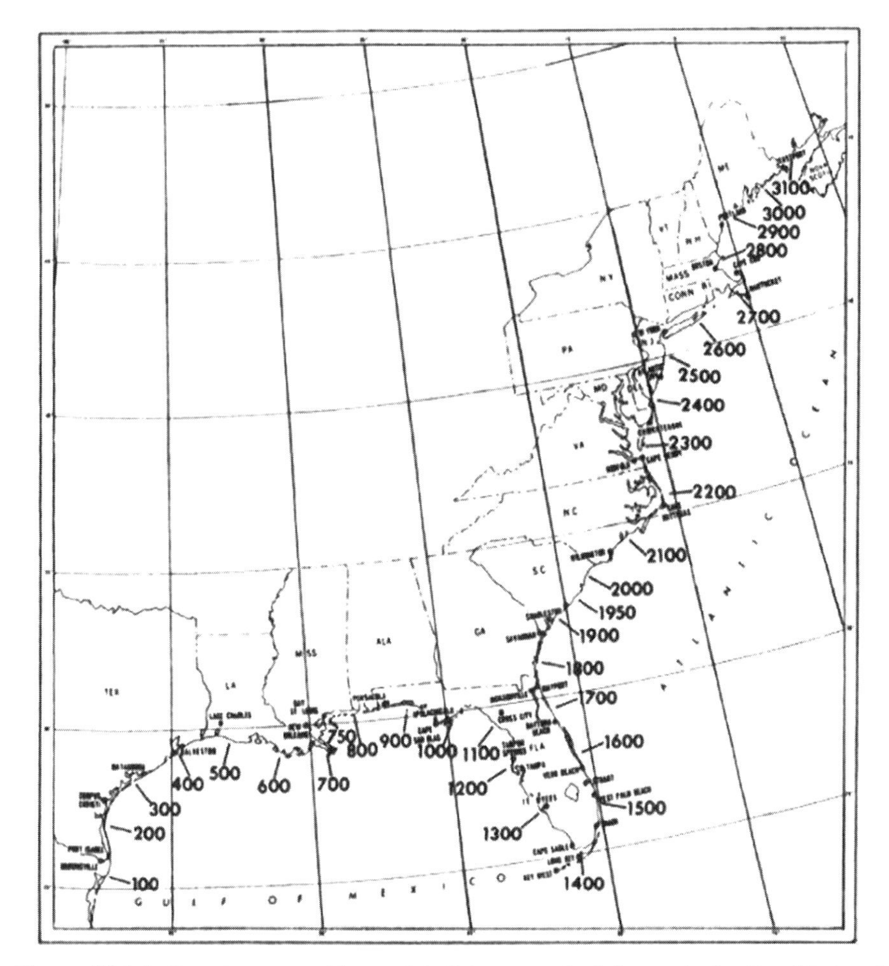

Figure 12.1.1. Locator map with coastal distance marked, in nautical miles (National Oceanic and Atmospheric Administration).

climatological data, such as pressure defects, radii of maximum wind speeds, and translation speed and direction (see Sect. 10.2.1). For each location, the data consist of estimated hurricane arrival rates, and sets of 999 1-min speeds in knots for 16 directions at 22.5° intervals (1 knot ≈ 1.15 mph; ratio between 1-min speeds and 3-s speeds: ≈0.82; ratio between 1-min speeds and 1-hr speeds: ≈1.25; 1 mph = 0.447 m/s). At any given site, as many as 20% to 40% of the total number of simulated hurricane wind speeds is negligibly small. Such small or vanishing speeds occur, for example, where the hurricane's translation velocity counteracts the rotational velocity.

12.1.3.1.4 HURDAT Database. Full details on the database are available at www.aoml.noaa.gov/hrd/data_sub/hurdat.html. The database covers information on hurricanes dating back to 1851.

12.1.3.1.5 University of Florida Hurricane Data (fcmp.ce.ufl.edu/)

12.1.3.1.6 Texas Tech University Data (www.atmo.ttu.edu/TTUHRT/)

12.1.3.2 Data Available Commercially. Such data include both hurricane and non-hurricane wind speeds.

12.1.3.2.1 Hurricane Directional Wind Speed Data Sets. That also cover interior areas, in addition to the coastline, are available commercially (Applied Research Associates, Raleigh, NC).

12.1.3.2.2 Directional Peak Gust Speeds. Above open terrain for each of 36 directions (at 10° intervals) are recorded, during every hour for periods of up to about 20 years, at Automated Surface Observations System (ASOS) stations. Software is available for the separate extraction of non-thunderstorm and thunderstorm wind speeds from the mass of ASOS data at any one station [12-1]. Before being used in statistical analyses, the ASOS data must be reduced to a common elevation by using information on anemometer elevation history (see Sect. 12.1.1). Specialized software for effecting the extraction, and anemometer elevation histories, are available for ASOS stations at www.nist.gov/wind (Extreme Winds, Data Sets, item 6). If peak 5-s gust speeds are listed, they can be transformed to 3-s peak gust speeds through multiplication by the factor 1.02.

A list of ASOS data is available at www.ncdc.noaa.gov/hofnasos/HONAsosStn or www.ncdc.noaa.gov/oa/hofn/asos/asos-home.html. A sample ASOS record is listed at www.nist.gov/wind/asos-wx/asos-wx.htm.

12.1.4 Super-Stations

Extreme wind speed estimates improve as the sizes of the samples on which they are based increase (Sect. 12.8). Some analysts have suggested basing statistical estimates on larger data sets obtained by pooling data from several stations (i.e., by creating a "super-station" from several stations). A uniform extreme wind climate based on those estimates is then assumed to prevail over the entire super-station zone. This approach has pitfalls that can lead to incorrect estimates. A super-station must satisfy the following criteria: (1) its component stations must be comparable in meteorological and micrometeorological terms (e.g., a super-station consisting of two stations, one of which is in a hurricane-prone region while the other is not, would yield incorrect wind speed estimates for both stations); (2) the data of the component stations should be mutually independent (e.g., creating a super-station by pooling the data measured at two immediately neighboring stations would not necessarily increase the total amount of useful information and result in more precise estimates); (3) a station cannot be used as a component station in more than one super-station, since this would artificially create large zones of climatological uniformity. The development of the ASCE 7 peak gust map was shown to violate criteria (1) and (3),[2] thereby compromising the quality of the estimates of wind speeds specified in the map for non-hurricane regions [12-2 to 12-4, 12-30].

12.2 CUMULATIVE DISTRIBUTIONS, EXCEEDANCE PROBABILITIES, MEAN RECURRENCE INTERVALS

Sections 12.2.1 and 12.2.2 introduce probabilistic terms needed for the definition and estimation of extreme values by using the example of a fair die and extending it to extreme wind speeds. Section 12.2.3 considers mixed distributions, which are of interest for regions with, for example, non-hurricane and hurricane winds, or non-thunderstorm and thunderstorm winds.

12.2.1 Estimation of Probability of Exceedance and Mean Recurrence Interval of an Event for a Fair Die

Denote the outcome of throwing a die once by O. The probability, denoted by $P(O \leq n), (n = 1, 2, \ldots, 6)$, that the trial outcome (the event) O is less than or equal to n is called the *cumulative distribution function* (CDF) of the event O. The probabilities $P(O \leq n)$ for $n = 1, 2, \ldots, 5, 6$ are 1/6, 2/6, ..., 5/6, 6/6, respectively. The probability that the outcome is larger than n is $P(O > n) = 1 - P(O \leq n)$, and is called the *probability of exceedance* of the outcome n. The probabilities $P(O > n)$ for $n = 1, 2, \ldots, 5, 6$ are 5/6, 4/6, ..., 1/6, 0, respectively. The *mean recurrence interval* (MRI) of the event $O > n$ is defined as the inverse of the probability of exceedance of that event, and is the average number of trials (throws) between consecutive occurrences of the event $O > n$. The MRI is also called the *mean return period*.

[2]For a full listing of super-stations used in the development of the ASCE 7 Standard map, including sets of two or more super-stations having at least one common station, see [12-4].

Numerical Example 12.2.1. *Mean recurrence interval of the outcome of the throw of a die.* For a fair die, the probability of exceedance $P(O > 5) = 1 - P(O \leq 5) = 1 - 5/6 = 1/6$. The MRI of the event $O > 5$ is $1/(1/6) = 6$ trials, that is, the outcome "six" occurs, on average, once in 6 trials. For $n = 6$, $P(O > 6) = 0$, and the corresponding MRI is infinity, which is another way of stating that the event $O > 6$ will never occur.

Since, for any given n, $P(O \leq n)$ is the same for any one trial (throw of a die), and is independent of the outcomes of other trials, the probability of non-exceedance of the outcome n in m trials is $[P(O \leq n)]^m$ (i.e., it is the probability of non-exceedance in a first trial, *and* in the second trial, ..., *and* in the mth trial.) The probability that the outcome n is exceeded at least once in m trials is $1 - [P(O \leq n)]^m$. For example, the probability of non-exceedance of the outcome "five" in 2 throws of a die is $(5/6)^2 = 0.69$, and the probability of exceedance of that outcome is $1 - 0.69 = 0.31$, versus a probability of exceedance of the outcome "five" in one trial of $1/6 \approx 0.17$.

12.2.2 Extension to Extreme Wind Speeds

Conceptually there is little difference between the statement "the outcome of throwing a die once exceeds n" and the statement "the largest wind speed V occurring in any one year exceeds v," except that the CDF of the largest speed in a year, $P(V \leq v)$, is continuous, whereas $P(O \leq n)$ is discrete. Just as $P(O \leq n)$ is the same for any one trial (throw of a die), and is independent of the outcomes of other trials, $P(V \leq v)$ is the same for any one trial (year or storm event), and is independent of speeds occurring in other years or storm events (provided that such factors as global warming do not come into play).

The speed v with an $\overline{N}$-year MRI is called the $\overline{N}$*-year speed.* The MRI, in years, is

$$\overline{N}(v) = \frac{1}{1 - P(V \leq v)}. \qquad (12.2.1)$$

Numerical Example 12.2.2. *Probability of exceedance of the largest wind speed in a given data sample.* Consider the sample of size 9 of measured largest yearly wind speeds 40, 37, 43, 49, 34, **50**, 44, 39, 31 (in mph; we show in bold type the largest speed in the sample). There are $n = 9$ outcomes for which $V \leq 50$ mph, out of $n + 1 = 10$ possible outcomes (the 10th outcome being $V > 50$ mph). Hence, the estimated probability $P(V \leq 50 \text{ mph}) = 9/10 = 0.9$. The probability of exceedance of a 50 mph largest yearly speed is $1 - 0.9 = 0.1$, and the MRI of the 50-mph wind speed is $1/0.1 = 10$ largest yearly speed wind events, that is, 10 years.

Numerical Example 12.2.3. *Mean recurrence interval of the event that the wind speed exceeds a specified value v within an m-year time interval.* Assume, as in the previous example, that for $v = 50$ mph, $P(V \leq 50 \text{ mph}) = 0.9$ in any one year. The probability of non-exceedance of the speed $V = 50$ mph in $m = 30$ years is $0.9^{30} = 0.04$. The probability of exceedance of the event $V > 50$ mph in 30 years is $1 - 0.04 = 0.96$.

12.2.3 Mixed Wind Climates

In this section, we consider wind speeds in regions exposed to both non-hurricane and hurricane winds. We are interested in the probability that, in any one year, wind speeds regardless of their meteorological nature are less than or equal to a specified speed v. For example, let the random variables V_H and V_{NH} denote, respectively, the largest hurricane wind speed in any one year, and the largest non-hurricane wind speed in any one year. Further, let the probability that $V_H \leq v$ and the probability that $V_{NH} \leq v$ be denoted, respectively, by $P(V_H \leq v)$ and $P(V_{NH} \leq v)$. The random variable of interest is the maximum yearly speed regardless of whether it is a hurricane or a non-hurricane wind speed, and it is denoted by $\max(V_H, V_{NH})$. The statement "$\max(V_H, V_{NH}) \leq v$" and the statement "$V_H \leq v$ and $V_{NH} \leq v$" are equivalent. Therefore, $P[\max(V_H, V_{NH}) \leq v] = P(V_H \leq v \text{ and } V_{NH} \leq v)$. Since it is reasonable to assume that V_H and V_{NH} are independent random variables, it follows that

$$P[\max(v_H, v_{NH}) \leq v] = P(v_H \leq v)P(v_{NH} \leq v). \tag{12.2.2}$$

(see footnote, Sect. A4.2).

The probability distributions $P(v_{NH} \leq v)$ and $P(v_H \leq v)$ can be obtained as shown in Sect. 12.3. With an appropriate change of notation, Eq. 12.2.2 is also applicable, for example, to non-thunderstorm and thunderstorm wind speeds or to any number of storm types.

Numerical Example 12.2.4. *Mean recurrence interval of the event that the wind speed exceeds a specified value v in a mixed wind climate.* Assume that, at a given site, the MRI of the 100-mph peak 3-s gust speed due to non-hurricane winds is $\overline{N}_{NH} = 120$ years, and that the MRI of the 100-mph peak 3-s gust speed due to hurricanes is $\overline{N}_H = 50$ years. The respective CDFs are $P(V_{NH} \leq 100 \text{ mph}) = 1 - 1/\overline{N}_{NH} = 0.99167$, and $P(V_H \leq 100 \text{ mph}) = 1 - 1/\overline{N}_H = 0.98$. By Eq. 12.2.2, the CDF of the 100-mph wind speed due to either non-hurricane or hurricane winds is $P[\max(v_H, v_{NH}) \leq 100 \text{ mph}] = P(v_H \leq 100 \text{ mph})P(v_{NH} \leq 100 \text{ mph}) = 0.99167 \times 0.98 = 0.9718$. The MRI of the 100-mph wind speed at the site is $1/(1 - 0.9718) = 35.5$ years.

12.3 PARAMETRIC ESTIMATES OF $\overline{N}$-YEAR WIND SPEEDS; CLOSED FORM ESTIMATORS; SOFTWARE

Section 12.3.1 discusses two approaches to the use of measured wind speed data for the estimation of extreme wind speeds: the *epochal approach* and the *peaks-over-threshold approach*. Section 12.3.2 presents *parametric methods* for estimating extreme speeds with any specified MRI $\overline{N}$. Estimation methods are presented for the Extreme Value Type I (Gumbel) distribution and the reverse Extreme Value Type III (reverse Weibull) distribution,[3] both of which are applicable to epochal wind speed data, and for the generalized Pareto distribution (GPD), which is applicable to peaks-over-threshold data. Point processes, which include Poisson processes, are discussed in an extreme value context in [12-8, 12-29]. Statistical estimates of extreme speeds based on large simulated data sets can be obtained by convenient *non-parametric* methods (i.e., methods that do not require the estimation of parameters in the cumulative probability distribution of the variate being sought); see Sect. 12.7.

Software for estimates that use the distributions covered in this section is available at www.nist.gov/wind. Under I Extreme Winds, click Software; scroll down to Software for Extreme Wind Speeds, and click 3. Dataplot. Links are available for various estimation methods (e.g., De Haan, see Sect. 12.3.2.2.2), and for various types of associated plots.

All parametric methods discussed in this section pertain to non-directional wind speeds, or to wind speeds blowing from a single direction. No multivariate Extreme Value probability distributions exist to date; therefore, direction-dependent wind speeds cannot be fitted to a multidirectional Extreme Value distribution.

12.3.1 Epochal Approach and Peaks-Over-Threshold Approach to Estimation of Extremes

In the *epochal* approach, a cumulative probability distribution function is fitted to a set of wind speed data consisting of the largest speed recorded at the site of interest in each of a number of consecutive fixed epochs. The epoch being chosen most commonly is one year. The data set then consists of the largest yearly wind speed for each year of the period of record.

In the *peaks-over-threshold* (POT) approach, a cumulative probability distribution function is fitted to all the independent speeds that exceed a specified threshold. Speeds are assumed to be independent if they do not belong to the same storm system. A reasonable way to ensure that this is indeed the case is to use data separated from each other in time, by at least five days, say; other criteria can be considered, however [e.g., 12-6, 12-7]. The advantage of the POT approach is that it allows the use of larger data samples, since speeds other than the largest annual speeds can also be included in the data sample [12-8, 12-29].

Numerical Example 12.3.1. *Sample sizes in epochal and POT approaches.* Assume that in Year 1 the largest speed is 69 mph, while in Year 2 the largest

[3] Statistical studies have shown that the Extreme Value Type II distribution is typically not appropriate for modeling extreme wind speeds [12-5].

speed is 85 mph, and the second and third largest speeds are 70 mph and 67 mph. In the POT approach, if a threshold of 68 mph is chosen, the speeds in Years 1 and 2 included in the sample are 85 mph, 70 mph, and 69 mph (three speeds). In the epochal approach, only two speeds are included: 69 mph (Year 1) and 85 mph (Year 2).

If the threshold is too high, the advantage of a larger sample size is lost. For example, if in Numerical Example 12.3.1 the threshold were 80 mph, only one speed—85 mph—would be included in the two-year sample. If the threshold is too low, the sample will include wind speeds that may not be representative of the extreme wind climate, resulting in biased estimates of the extreme wind speeds.

Figure 12.3.1 shows estimates of 100-year, 1,000-year, and 100,000-year fastest-mile wind speeds at 6.1 m above ground in open terrain at Green Bay, Wisconsin. The estimates are functions of threshold speed (in mph) and sample size (i.e., number of data above the threshold speed). The data consisted of the maximum wind speed for each of the successive 8-day intervals within a 15-year record, and included no wind speeds separated by less than 5 days. For thresholds between about 38 mph and 32 mph (sample sizes of about 35 to 127), the estimated 100-year speeds are stable around 60 mph. This is indicative of a reliable estimate. The reliability of the estimate is poorer as the

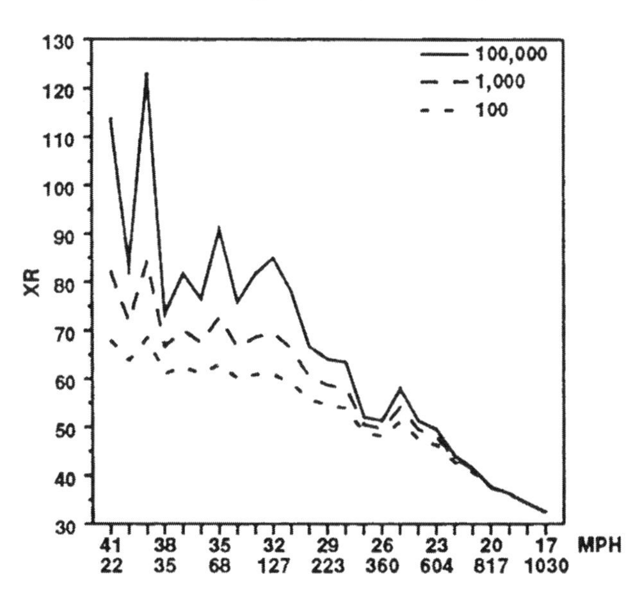

Figure 12.3.1. Estimated wind speeds, with 100-, 1,000-, and 100,000-yr mean recurrence interval at Green Bay, Wisconsin, as functions of threshold (mph). For low wind speed thresholds (below 32 mph, say), the estimates are increasingly biased; that is, the wind speeds are increasingly underestimated.

MRI increases (this is clearly seen for the 100,000-yr estimates). For thresholds higher than 38 mph, the estimates are less stable; that is, they vary fairly strongly as a function of threshold. For thresholds lower than about 32 mph, the estimates are increasingly biased with respect to the 60 mph estimate, owing to the presence in the data sample of low speeds unrepresentative of the extremes. *Including low speeds in a sample used for inferences on extreme speeds can result in biased estimates*, as would be the case if the heights of children were included in a sample used to estimate the height of adults. For example, estimates of extreme wind speeds based on wind speed data recorded every hour, the vast majority of which are low and meteorologically unrelated to the extreme speeds, would be unrealistic. Modern extreme value statistics recognizes that to obtain dependable estimates of extreme values it is necessary to "let the tails speak for themselves," instead of allowing estimates to be biased by data with small values, as is the case in Fig. 12.3.1 for wind speeds below about 32 mph.

12.3.2 Gumbel, Reverse Weibull, and Generalized Pareto Distributions

A reasonably persuasive theoretical and empirical basis exists for the assumption that Extreme Value distributions of the largest values are adequate for describing extreme wind speeds probabilistically. It has been proven that three types of such distributions exist, depending upon the nature of the distribution tail. Of these distributions, two are in practice applicable to extreme wind speeds: the *Gumbel distribution* (also known as the Extreme Value Type I, or EV I, distribution) and the *reverse Weibull (RW) distribution* (also known as the Extreme Value Type III distribution of the largest values, or EV III). Statistical evidence has been adduced in support of the hypothesis that, for some stations, the RW is a more appropriate model of extreme wind speeds than the EV I distribution [12-9, 12-11, 12-12, 12-13]. The evidence appears to be particularly strong for hurricane wind speed data obtained by Monte Carlo simulation (see Sect. 12.4) [12-7]. The RW distribution has been adopted in the Australian/New Zealand Standard [12-10, Commentary Section C3.2].

It has been argued that because the EV III distribution is only valid in an asymptotic sense, it is a less appropriate model of the extreme winds than the EV I distribution [12-13]. In fact, the EV I distribution—like all Extreme Value distributions—is also valid only in an asymptotic sense. Therefore, to the extent that the EV I model is acceptable, it would appear that the EV III model should be acceptable as well, particularly if it does not differ much from the Gumbel distribution. It is widely believed that asymptotic models are acceptable in practice. Nevertheless, the theory of penultimate Extreme Value distributions and its application to extreme wind speeds continues to be the object of careful study (see, e.g., [12-28]).

The *generalized Pareto distribution (GPD)* is an asymptotic expression applicable to the difference $y = v - u$, where v is the variate of interest (e.g., the wind speed), and u is a sufficiently high threshold. The expression for the GPD describes the tails of all Extreme Value distributions. In particular, for negative values of the tail length parameter, the GPD corresponds to upper tails of RW distributions [12-8, 12-14]. See also [12-16, 12-17].

Estimation methods based on the distributions discussed in this section are provided in Sects. 12.3.2.1 and 12.3.2.2. The distributions depend on parameters estimated to achieve a good distributional fit to the available data samples. The estimation methods covered in this section are therefore called *parametric* methods. *Non-parametric* estimation methods do not require parameter estimation and are discussed in Sect. 12.7.

12.3.2.1 Epochal Approach, Method Of Moments. In this section, we describe an approximate method, called the method of moments, for estimating the $\overline{N}$-year speed from a sample of the n maximum yearly wind speeds. The method relies on calculated sample means $\overline{V}$ and standard deviations $s(v)$ of the sample of m wind speeds.[4] We apply the method of moments to both the EV I and the RW distribution.

12.3.2.1.1 Gumbel (EV I) Distribution. The method of moments yields in this case the following estimator of the $\overline{N}$-year speed:[5]

$$v_{EV\,I}(\overline{N}) \approx \overline{V} + 0.78(\ln \overline{N} - 0.577)s(v). \qquad (12.3.1)$$

12.3.2.1.2 Reverse Weibull (RW) Distribution. The method of moments yields the following estimator of the $\overline{N}$-year speed:

$$v_{RW}(\overline{N}) = \overline{V} + s(v)A(c)\left\{B(c) - \left[-\ln\left(1 - \frac{1}{\overline{N}}\right)\right]^{-c}\right\}, \qquad (12.3.2)$$

$$A(c) = \frac{1}{\{\Gamma(1-2c) - [\Gamma(1-c)]^2\}^{1/2}}; \quad B(c) = \Gamma(1-c), \quad (12.3.3a,b)$$

where the tail parameter $c < 0$, and Γ denotes the gamma function. For non-hurricane speeds, if the estimated $c \leq -0.1$, it is prudent to assume $c = -0.1$, and therefore $A(c) = 8.726, B(c) = 0.95135$.

Numerical Example 12.3.2. *EV I and RW extreme wind estimates, epochal approach, method of moments.* Assume that for an $n = 17$-year record at a site, the yearly peak 3-s gust speeds from any direction (in mph) are: 80, 76, 80, 80, 74, 80, 90, 82, 91, 81, 74, 74, 63, 76, 66, 72, 64. The epochal approach

[4] $\overline{V}$ denotes here the mean of a sample of wind speeds, rather than the mean hourly speed during a storm.

[5] Estimated quantities are distinguished in the statistics literature by the circumflex symbol ˆ. For the sake of simplicity, we omit this symbol in the present chapter.

makes use of the mean $\overline{V} = 76.65$ mph and standard deviation $s = 7.83$ mph of the n largest annual speeds.

From Eq. 12.3.2 and Eq. 12.3.3 (in which we set $c = -0.1$), we obtain, respectively, the Extreme Value Type I and the reverse Weibull estimates:

$$\overline{N} = 50 \text{ years: } v_{\text{EV I}} = 97.0 \text{ mph}; \ \overline{N} = 500 \text{ years: } v_{\text{EV I}} = 111.1 \text{ mph}$$

$$\overline{N} = 50 \text{ years: } v_{\text{RW}} = 95.4 \text{ mph}; \ \overline{N} = 500 \text{ years: } v_{\text{RW}} = 105.0 \text{ mph.}$$

For $\overline{N} = 50$ years, the estimated extreme speeds based on the reverse Weibull are only slightly lower than those based on the Gumbel distribution, but the differences increase for higher MRIs.

12.3.2.2 Peaks-Over-Threshold Approach, Generalized Pareto Distribution. The wind speed with an $\overline{N}$-yr mean recurrence interval obtained from data larger than a threshold speed u may be estimated by using the generalized Pareto distribution

$$v_{\text{GP}}(\overline{N}) = u - a[1 - (\lambda\, \overline{N})^c]/c \tag{12.3.4}$$

[12-27, p. 131], where $\lambda = k/n_{yrs}$ is the annual rate of arrival of speeds greater than u, k is the number of wind speed data greater than u, and n_{yrs} is the length of the record in years. The parameters a and c may be estimated as shown in Sect. 12.3.2.2.1 or 12.3.2.2.2.

12.3.2.2.1 Method of Moments. The following estimators for the parameters a and c were proposed in [12-14]:

$$a = \tfrac{1}{2} E(y)\{1 + [E(y)/s(y)]^2\}; \quad c = \tfrac{1}{2}\{1 - [E(y)/s(y)]^2\} \tag{12.3.5a,b}$$

where $y = v - u$ is the excess of the wind speed v over the threshold speed u, and $E(y)$ and $s(y)$ denote the sample mean and the sample standard deviation of y.

12.3.2.2.2 De Haan Method. The following estimators for the parameters c and a were proposed in [12-15]:

$$c = M_n^{(1)} + 1 - \frac{1}{2\{1 - [M_n^{(1)}]^2/[M_n^{(2)}]\}}; \quad a = uM_n^{(1)}/\rho_1 \tag{12.3.6a,b}$$

$$\rho_1 = 1, c \geq 0; \ \rho_1 = 1/(1-c), c \leq 0. \tag{12.3.7}$$

$$M_n^{(r)} = \frac{1}{k}\sum_{i=0}^{k-1}[\log(X_{n-i,n}) - \log(X_{n-k,n})]^r, r = 1, 2 \tag{12.3.8}$$

where k is the number of data above the threshold, $\lambda = k/n_{yrs}$ is the annual rate of arrival of speeds greater than u, and n_{yrs} is the length of the record in years. The highest, second highest, ..., kth, $(k+1)$th highest speeds are denoted by $X_{n,n}, X_{n-1,n}, \ldots, X_{n-(k-1),n}, X_{n-k,n} = u$, respectively, where n is the total number of data points. If c is specified (e.g., $c = -0.1$), the corresponding value of a can be obtained by substituting the specified value of c in Eq. 12.3.7 to obtain ρ_1, and then using Eq. 12.3.6b.

It is useful to estimate the confidence bounds for the parameter c. The 95% confidence bounds correspond to the values $c \pm 2$ s.d. (c), where s.d. denotes standard deviation. The following relations hold:

$$\text{s.d.}(c) = [(1+c^2)/k]^{1/2}, \quad c \geq 0 \tag{12.3.9a}$$

$$\text{s.d.}(c) = \left\{(1-c^2)(1-2c)\left[4 - \frac{8(1-2c)}{(1-3c)} + \frac{(5-11c)(1-2c)}{(1-3c)(1-4c)}\right]\frac{1}{k}\right\}^{1/2}, \quad c < 0. \tag{12.3.9b}$$

12.4 PROBABILISTIC ESTIMATES OF WIND EFFECTS BASED ON NONDIRECTIONAL AND DIRECTIONAL WIND SPEED DATA

This section presents a brief review of the probabilistic estimation of wind effects based on nondirectional wind speed data, as specified, for example, in the ASCE 7 Standard (Sect. 12.4.1). It then presents a physically and probabilistically rigorous approach to the estimation of wind effects based on directional wind speed data (Sect. 12.4.2). An instructive comparison between results obtained in a simple example for the two cases is presented in Sect. 12.4.3.

12.4.1 Probabilistic Estimates of Wind Effects Based on Nondirectional Wind Speed Data

The design wind speeds specified in the ASCE 7 Standard are *nondirectional* (see Sect. 12.2 and Numerical Example 12.1.1). It is implicit in the ASCE 7 Standard provisions that the MRIs of calculated wind effects due to nondirectional wind speeds are the same as the MRIs of those wind speeds (e.g., that a 700-yr nondirectional wind speed induces in a structure wind effects with a 700-yr MRI.) Wind effects corresponding to specified MRIs can then be estimated by applying parametric estimation methods (such as those presented in Sect. 12.3) to the nondirectional wind speeds inducing those effects. To account for directionality, the Standard requires the application to wind effects estimates of a blanket wind directionality reduction coefficient K_d (Sect. 4.2.3). Extreme wind speed and wind effect estimates are based on samples of data measured over periods of 20 to 100 years, say. They are therefore affected by sampling errors that depend upon the size of the data sample (Sect. 12.8).

12.4.2 Probabilistic Estimates of Wind Effects Based on Directional Wind Speed Data

Inherent in the ASCE 7 Standard approach to wind directionality effects (Sect. 12.4.1) are approximations that may be acceptable for the design of ordinary structures but are deemed unacceptable for the design of special structures, including in particular tall buildings. For such structures, alternative approaches have been developed and are currently being used in wind and structural engineering practice.

This section describes a procedure for estimating extreme wind speeds and their MRIs based on directional wind speed data. The procedure yields simple and transparent probabilistic estimates of wind effects, while accounting for directionality in a physically and probabilistically rigorous manner, even though no parametric expressions for the probability distribution of direction-dependent variates are available in the literature. The procedure entails the following steps:

1. Develop a matrix of directional wind speeds from measured data. The number d of columns in the matrix is equal to the number of wind directions being considered (e.g., 8, 16, or 36). The number n of rows is equal to the number of storm events or of years of record, and must be sufficiently large to allow the use of non-parametric estimates of wind effects with MRIs of the order of thousands of years. Techniques for the development of directional wind speed matrices from data measured over periods of, say, 20 to 100 years are discussed in Sect. 12.5 for hurricane wind speeds and Sect. 12.6 for non-hurricane wind speeds.
2. Transform the $n \times d$ matrix of directional wind speeds into an $n \times d$ matrix of wind effects, each element of which is the wind effect induced in the structure by its counterpart in the directional wind speed matrix. Numerical Example 12.4.1 illustrates this transformation in a deliberately simple case.[6]
3. Create a *vector of dimension n* of the wind effect of interest. The vector consists of the largest wind effect in each row of the matrix developed in Step 2. (All the other elements in the rows of the wind effects matrix are discarded, since they are of no interest from a structural design viewpoint.)

[6]For tall buildings, the transformation of the directional wind speeds matrix v_{ij} involves: (1) databases of time-dependent pressure coefficients measured simultaneously at large numbers of points on the building envelope for a sufficient number of mean flow directions, (2) dynamic analyses producing inertial forces and torsional moments induced by the resultant wind forces and torsional moments applied at the center of mass of each floor, (3) influence coefficients converting the resultant forces and moments (applied *and* inertial) into internal forces, inter-story drift, and top floor accelerations, and (4) summations of demand-to-capacity ratios used for strength design purposes—see Chapter 19, where public domain software for matrix transformations and the other requisite calculations is referenced, described, and illustrated.

4. Use *non-parametric estimates* to obtain statistics of the wind effect for which the vector was created in Step 3. An example of the use of non-parametric methods is presented in Sect. 12.7.

Like estimates of nondirectional wind speeds and wind effects, the estimated wind speeds in the matrix developed in Step 1, and therefore the wind effects obtained in Step 4, are affected by sampling errors (Sect. 12.8).

Section 12.4.2.2 presents a deliberately simple Numerical Example that illustrates the procedure just described on the one hand and the ASCE 7 procedure on the other. The Numerical Example illustrates, for every MRI, differences in wind effects based on directional data on the one hand and on nondirectional data on the other.

Numerical Example 12.4.1. *Wind effects based on nondirectional and on directional wind speeds.* For the sake of simplicity and without loss of generality, we consider in this example directional wind speeds in $n = 3$ successive wind storm events, and assume that winds blow from $d = 2$ directions.

Denote the wind speeds matrix by v_{ij} (in mph, say), where $i = 1, 2, 3$ indicates the number of the storm event, and $j = 1, 2$ indicates the wind direction. The speeds v_{ij} and the nondirectional wind speeds $v_i \equiv \max_j(v_{ij})$ are listed in Table 12.4.1. (The values $\max_j(v_{ij})$ are also indicated in bold type in the matrix v_{ij}, as in Numerical Example 12.1.1.)

The second step in the procedure based on directional wind speeds is the transformation of the matrix v_{ij} into a matrix whose elements are the directional wind effects induced by the elements of v_{ij}. In this example, the wind effects being considered are aerodynamic forces. Consider a hypothetical structure with direction-dependent aerodynamic wind effect coefficients (i.e., force coefficients) C_{fj} $(j = 1, 2)$. For the range of directions $j = 1, 2$ the force coefficient regardless of wind direction (i.e., the nondirectional force coefficient) envelops the force coefficients C_{fj} and is denoted by $C_f \equiv \max_j(C_{fj})$ (Table 12.4.2).

The transformation of the v_{ij} matrix is achieved by calculating the directional wind effects induced in the structure by the elements v_{ij} of the matrix. To within a constant dimensional factor, the wind effects induced by the

TABLE 12.4.1. Largest Directional Wind Speeds and Largest Wind Speeds Regardless of Their Direction

	Directional Wind Speed Matrix v_{ij}		Nondirectional Wind Speed Vector v_i
	Dir. $j = 1$	Dir. $j = 2$	
Event $i = 1$	**54**	47	54
Event $i = 2$	41	**46**	46
Event $i = 3$	**47**	39	47

TABLE 12.4.2. Directional and Non-directional Force Coefficients

	Directional Force Coefficients C_{fj}		Non-directional Force Coefficient C_f
	Dir. $j = 1$	Dir. $j = 2$	
Force Coeff.	0.8	1.0	1.0

TABLE 12.4.3. Directional Wind Effects $C_{fj} v_{ij}^2$ and Largest Directional Wind Effects $\max_j (C_{fj} v_{ij}^2)$

	Matrix of Directional Wind Effects $C_{fj} v_{ij}^2$		Vector of Wind Effects $\max_j (C_{fj} v_{ij}^2)$
	Dir. $j = 1$	Dir. $j = 2$	
Event $i = 1$	**2,333**	2,209	2,333
Event $i = 2$	1,345	**2,116**	2,116
Event $i = 3$	**1,747**	1,521	1,747

speeds v_{ij} are $C_{fj} v_{ij}^2$ (e.g., for event $i = 1$, dir. $j = 1$, $C_{fj} v_{ij}^2 = 0.8 \times 54^2 =$ 2,333) (Table 12.4.3).

The third step consists of creating the time series of wind effects, $\max_j (C_{fj} v_{ij}^2)(i = 1, 2, 3)$, since for each storm event i only the *largest* of the effects induced by the directional wind speeds $C_{fj} v_{ij}^2$ is of interest for design purposes. These largest effects are listed in the last column of Table 12.4.3.

It follows from these calculations that, over the time period during which the three events $i = 1, 2, 3$ occur, the largest, second largest, and third largest of the wind effects $\max_j (C_{fj} v_{ij}^2)(i = 1,2,3)$ are, respectively,

2,333, 2,116, and 1,747.

If the calculations were performed on the basis of nondirectional wind speeds in accordance with the ASCE 7 Standard procedure, that is, by using the largest force coefficient $C_f = 1.0$ and the largest nondirectional wind speeds $v_i (i = 1, 2, 3)$, then, before considering directionality effects, the largest, second-largest, and third-largest wind effects $\max_j (C_{p,j}) \max_j (v_{ij})^2$ would be $1.0 \times 54^2 = 2916$, $1.0 \times 47^2 = 2{,}209$, and $1.0 \times 46^2 = 2116$, respectively. After multiplication by the directionality reduction factor $K_d = 0.85$, the largest, second-largest, and third-largest wind effects based on the ASCE 7 procedure would be

2,480, 1,878, and 1,799.

rather than 2,333, 2,116, and 1,747, the values obtained by using the procedure based on directional wind speeds.

In this example, the number n of storm events is 3, and the one-dimensional response vector approach can therefore only yield effects with an MRI of at most $\overline{N} = n + 1 = 4$ epochs (an epoch being the average time interval between successive storm events). Section 12.7 presents an application of the non-parametric approach to estimating the MRIs of a variate for which, unlike in the case presented in this section, a large data sample is available. Note that the procedure illustrated in this example is applicable to any wind effects, including aerodynamic forces, dynamic response, and sums of demand-to-capacity ratios used in design interaction equations. To use structural reliability terminology (Chapter 16): The analyses are not performed in the space of *wind speed variables*, but rather in the space of *wind effect variables*. The reason for doing so is that, unless wind directionality effects are accounted for by using the simple but generally inaccurate ASCE 7 Standard procedure, there are no known ways of estimating probabilities of exceedance, or MRIs of strength, or serviceability limit states other than by considering time series of *wind effects*, rather than time series of *wind speeds*.

For an alternative approach based on outcrossing of limit state boundaries by directional wind effects, see Sect. A1.6 (Appendix A1), and Sect. A4.1 (Appendix A4). For the sector-by-sector approach to the wind directionality problem, see Sect. A4.2 (Appendix A4).

12.4.3 Mean Recurrence Intervals of Wind Effects: Directional versus Nondirectional Approach

In the example of Sect. 12.4.2, the nondirectional approach yielded the ranking 1, 3, and 2 for the wind effects induced by storm events 1, 2, and 3, respectively. In contrast, the directional approach yielded for the wind effects induced by those storm events the ranking 1, 2, and 3, respectively. This illustrates the fact that the mean recurrence interval of the nondirectional wind speed $v_i \equiv \max_j(v_{ij})$. This illustrates the fact that the mean recurrence interval of the nondirectional wind speed $v_i \equiv \max_j(v_{ij})$ is not necessarily the same as the mean recurrence interval of the directional wind effect induced by the storm event i.

12.5 DEVELOPMENT OF DIRECTIONAL DATABASES OF HURRICANE WIND SPEEDS

Sets of measured hurricane wind speed data at any one site are too small to allow the reliable estimation of extreme wind speeds or wind effects. For this reason, synthetic hurricane wind speed data sets are created by Monte Carlo simulation from measured climatological parameters (radius of largest wind speeds, pressure defect, speed and direction of translation) used in conjunction with a physical hurricane flow model dependent on those parameters

[12-18 to 12-22]. Probability distributions of the climatological parameters are estimated from historical records. Sets of climatological parameter values are obtained through Monte Carlo simulations based on those distributions. Given the physical hurricane flow model, to each set of climatological parameters so obtained there corresponds a hurricane, and therefore a set of largest directional speeds at the site of concern. The physical model also depends on a parameter B introduced in [12-23].

The Monte Carlo simulation allows the creation of directional hurricane wind speed data sets of any desired size. Errors in the estimation of the hurricane wind speeds or wind effects are not reduced significantly as the size of the simulated data set increases, since they are determined primarily by the size of the sets of measured climatological data (pressure defect and so forth), which in the U.S. corresponds to a period of the order of 100 years. For an area within which the climatological parameters may be assumed to have approximately the same probability distributions (say, a 100 km × 100 km area), the number of available measured climatological data sets is of the order of 100 (or, at higher latitudes, less than 100). Such data have been used to develop synthetic directional hurricane/tropical cyclone wind speed data samples of size 999, available at www.nist.gov/wind and, more recently, proprietary data sets of larger size, covering not only the coastline but areas adjacent to the coastline as well [12-22].

Given a nondirectional or directional set of hurricane wind speeds, it is of interest to estimate the probability distributions that fit those speeds. Parametric estimation methods have been used to that effect, and indicated that the generalized Pareto distribution with negative values of the tail length parameter typically fits hurricane wind speed data satisfactorily; that is, the distribution tails are of the reverse Weibull type [12-7]. Figure 12.5.1 shows an example of the dependence on wind speed threshold and sample size of the estimate of the tail length parameter c and of nondirectional wind speeds (speeds irrespective of direction) for a coastline location. Non-parametric methods are more convenient for some applications, however, and can be used if sample sizes available are sufficiently large. An example is presented in Sect. 12.7.

To an extent that is perhaps greater than for other types of wind speeds, the estimation of hurricane wind speeds with specified mean recurrence intervals is not an exact science. Estimates by various researchers exhibit differences that are in some instances significant. Four sets of such estimates (hourly mean speeds in m/s; mileposts refer to Fig. 12.1.1) are plotted in Fig. 12.5.2. Some of the differences among the various estimates can be attributed to the use of different data samples in the simulations. In particular, the ASCE 7-10 Standard estimates were performed after the 2004 hurricane season, the most active Atlantic hurricane season in recorded history, which strongly affected the Gulf coast states from eastern Texas to Florida.

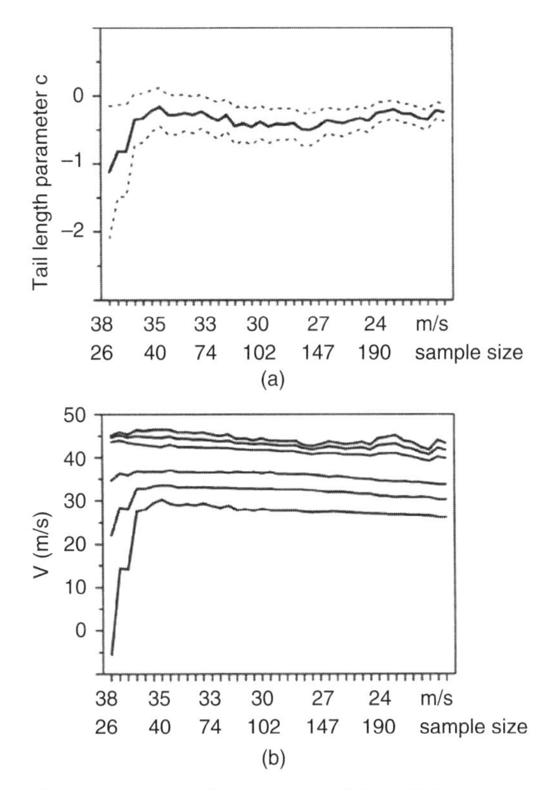

Figure 12.5.1. Peaks-over-threshold estimates of (a) tail length parameter of reverse Weibull distribution, and (b) wind speeds with 25-, 50-, 100-, 500-, 1,000-, and 2,000-yr mean recurrence intervals, milepost 2,550 (1-min speeds at 10 m above ground in open terrain), as functions of wind speed threshold and sample size. Estimates tend to be most stable for thresholds between about 32 m/s and 34 m/s (sample sizes 85 and 56).

12.6 DEVELOPMENT OF DIRECTIONAL DATABASES OF NON-HURRICANE WIND SPEEDS

The probabilistic model used to generate by Monte Carlo simulation large sets of directional wind speeds from measured directional wind speed data is based on nonlinear transformations of a d-dimensional Gaussian vector, where d is the number of directions being considered. The rate of arrival of the extreme wind events is denoted by λ. For example, if the observed number of wind events at a site during a 30-yr period is 66, then the estimated rate of arrival is $\lambda = 2.2$/yr.

Denote by $\boldsymbol{V}$ a wind speed matrix with d columns and n rows whose elements $V_{k,i}(i = 1, \ldots, d; k = 1, 2, \ldots, n)$ are wind speeds from direction i recorded during the kth wind event. It is assumed that the rows k of the matrix

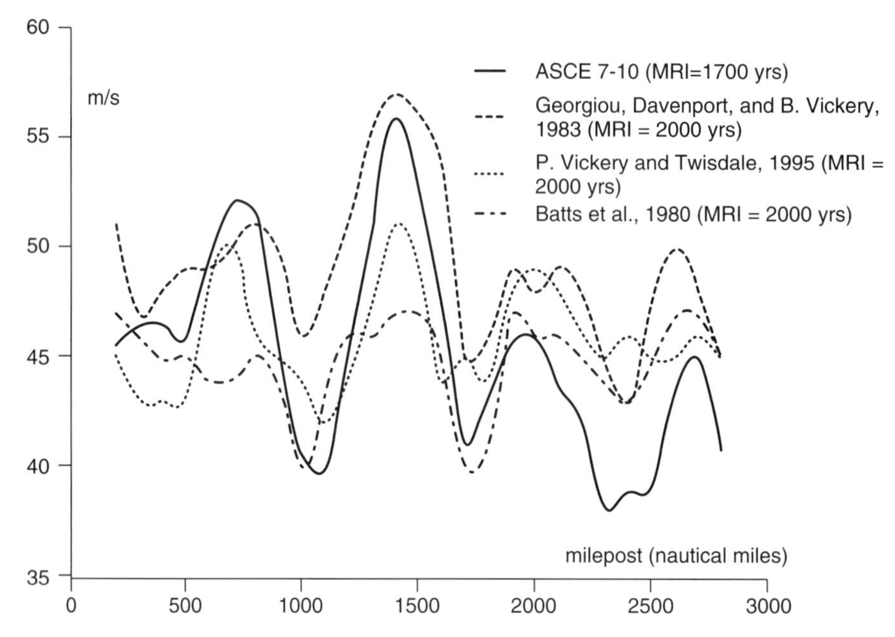

Figure 12.5.2. Approximate estimates of mean hourly hurricane wind speeds at 10 m above ground in open terrain (after [2-1] and [7-1]).

$\boldsymbol{V}\,(k = 1, 2, \ldots\}$ are independent copies of a d-dimensional random vector $\{V_i, i = 1, \ldots, d\}$ with joint distribution F. The model for the multidirectional extreme wind events is completely characterized by the rate of arrival λ and the distribution F. The components $\{V_i\}$ are defined as follows:

$$V_i = F_i^{-1}[\Phi(G_i)], \quad i = 1, \ldots, d, \tag{12.6.1}$$

where $\{F_i, i = 1, \ldots, d\}$ denote arbitrary distributions, Φ denotes the distribution of the standard Gaussian variable with mean 0 and variance 1, $\{G_i, i = 1, \ldots, d\}$ are correlated standard Gaussian variables with covariance matrix $\{\rho_{ij} = E[G_i G_j], i, j = 1, \ldots, d\}$, and E denotes expectation. Equation 12.6.1 establishes a one-to-one correspondence between the wind speed vector $(V_1, \ldots, V_d)$ and the Gaussian vector $(G_1, \ldots, G_d)$ [12-23]. The correlation coefficients between G_i and G_j may be assumed to be the same as the correlation coefficients between V_i and V_j [12-24].

If the number of measured data is too small to allow the reliable estimation of the correlations, the latter may be assumed to be zero. In that case, the probability of exceedance of the wind speed irrespective of direction is larger than its counterpart obtained by accounting for positive correlations $E[V_i, V_j]$; that is, estimates of the extreme nondirectional wind speed based on the assumption that the directional sets of wind speeds are mutually independent are conservative. However, the conservatism inherent in this assumption is weak (i.e., of the order of 1%, unless the correlations are large, e.g., $\rho_{ij} > 0.7$). Therefore, assuming that the directional wind speed vectors are mutually independent (and, therefore, that their correlations are zero) has no significant effect on the estimated extreme nondirectional wind speeds, even for MRIs as large as 10,000 years.[7]

The distributions $\{F_i\}$ of the directional wind speeds $\{V_i\}$ can be modeled by the generalized Pareto distribution (GPD), and the parameters of the distribution can be estimated by the de Haan method (see Sect. 12.3.2.2). To avoid unrealistic or unconservative modeling of non-hurricane wind speeds, it is recommended that if the estimated value of the GPD tail length parameter is $c > 0$, the value used in the calculations be taken as $c = -0.01$ (corresponding, within a close approximation, to a Gumbel distribution tail), and if the estimated value is $c < -0.1$, the value used in the calculations be taken as $c = -0.1$, thereby avoiding distribution tails that may be unconservatively short.

The generation of the requisite data is performed in two steps: (1) Generate samples $(g_1, \ldots, g_d)$ of a d-dimensional standard Gaussian vector with covariance matrix $[E[V_i, V_j]]$, and (2) calculate the image $(v_1, \ldots, v_d)$ of $(g_1, \ldots, g_d)$, defined by the mapping

$$v_i = F_i^{-1}[\Phi(g)], \quad i = 1, \ldots., d. \tag{12.6.2}$$

Steps 1 and 2 complete the generation of one line of the matrix $v_{ki}(k = 1, \ldots, n)$, where n is the requisite number of wind events or epochs. Steps 1 and 2 are repeated n times.

12.7 NON-PARAMETRIC STATISTICS, APPLICATION TO ONE-DIMENSIONAL TIME SERIES

Consider a set of n nondirectional wind speed data at a location (milepost) where the mean storm arrival rate is λ/year. If the rate were $\lambda = 1$ storm/year,

[7]Consider, for example, winds blowing from two directions. Wind speeds from directions 1 and 2 are denoted by V_1 and V_2, respectively. For simplicity, we assume that the probability distributions of V_1 and V_2 are identical. If V_1 and V_2 are perfectly positively correlated, the event $V_1 \leq v$ implies the event $V_2 \leq v$; therefore, $P(V_1 \leq v \text{ and } V_2 \leq v) = P(V_1 \leq v)$. If V_1 and V_2 are independent, $P(V_1 \leq v \text{ and } V_2 \leq v) = P(V_1 \leq v)P(V_2 \leq v) < P(V_1 \leq v)$. The assumption of independence, therefore, results in (a) estimates of the probability of non-exceedance of a speed v that are smaller, (b) estimates of the probability of exceedance of v that are larger, and (c) estimates of the MRI of v that are smaller than is the case under the assumption of positive correlation. The estimates based on the assumption of independence are therefore conservative, meaning that if for a specified MRI $\overline{N}$ this assumption yields a wind speed v, the actual MRI of v inherent in the positively correlated directional wind speed vectors is greater than $\overline{N}$. For example, if the assumption of independence yields a design wind speed $v = 90$ mph corresponding to $\overline{N} = 700$ years, that design wind speed may actually correspond to an MRI $\overline{N} = 800$ years, say.

the estimated probability that the highest speed in the set would be exceeded is $1/(n+1)$, and the corresponding estimated MRI would be $\overline{N} = n + 1$ years (on average $n + 1$ "trials"), that is, $n + 1$ storms would be required for a storm to exceed that highest speed (see Numerical Example 12.2.2). The estimated probability that the qth highest speed in the set is exceeded is $q/(n+1)$, the corresponding estimated MRI in years is $\overline{N} = (n+1)/q$, and the rank of the wind speed with MRI $\overline{N}$ is $q = (n+1)/\overline{N}$.

In general $\lambda \neq 1$, and the estimated MRI is therefore $\overline{N} = (n+1)/(q\lambda)$ years. For example, if $n = 999$ hurricane wind speed data, and $\lambda = 0.5$/year, the estimated MRI of the event that the highest wind speed in the sample will occur is $\overline{N} = (n+1)/(q\lambda) = 1000/0.5 = 2000$ years, the estimated MRI of the second highest speed is 1,000 years, and so forth. The rank of the speed with a specified MRI $\overline{N}$ is $q = (n+1)/(\overline{N}\lambda)$.[8]

Numerical Example 12.7.1. *Non-parametric MRI estimates for hurricane wind speeds from a specified directional sector at a specified coastal location.*[9] The use of non-parametric estimates of MRIs is illustrated for quantities forming a vector v_k ($k = 1, 2, \ldots, n$, where n is the number of trials). The methodology is the same regardless of the nature of the variate, which can represent wind effects or, as in this example, hurricane wind speeds. We consider speeds blowing from the 22.5° sector centered on the SW (i.e., 225°) direction at milestone 2,250 (near New York City), where $\lambda = 0.305$/year. The data being used were obtained from the site www.nist.gov/wind, as indicated in Sect. 12.1. They are rank-ordered in Table 12.7.1. It is sufficient to consider the first 55 rank-ordered data, since higher-rank data are small.

The qth largest speed in the set of 999 speeds corresponds to a mean recurrence interval $\overline{N} = (n+1)/(q\lambda) = 1{,}000/(0.305q)$. For the first-highest and second-highest speeds listed in Table 12.7.1, $\overline{N} = 1{,}000/0.305 = 3{,}279$ years and $\overline{N} = 1{,}000/(0.305 \times 2) = 1{,}639$ years. The peak 3-s gust speed with a 100-year mean recurrence interval has rank $q = 1{,}000/(0.305 \times 100) = 32.78$, that is, 33, and is seen from Table 12.7.1 to be 17 m/s. Note that the precision of the estimates is poorer for higher-ranking speeds, owing to the relatively large differences between successive higher-ranking speeds in Table 12.7.1 (e.g., 54 m/s vs. 39 m/s for the highest vs. the second-highest speed). This problem is less acute for winds blowing from any direction (i.e., for nondirectional wind speeds), and for data samples of size significantly larger than 999.

[8]A formula based on the theory of Poisson processes, which takes into account the possibility that two or more hurricanes may occur at a site in any one year, and is more exact for short MRIs (e.g., 5 years), is: $\overline{N} = 1/\{1 - \exp[-\lambda q/(n+1)]\}$. For example, for $n = 999, \lambda = 0.5$, and $q = 2$, $\overline{N} = 1{,}000.5$ years.

[9]This example was contributed by Dr. W. P. Fritz.

TABLE 12.7.1. Rank-Ordered Peak 3-s Gust Speeds (in m/s) from SW Direction at 10 m above Open Terrain for 22.5° Sector at Milepost 2,550 (1-min Speed in Knots = 0.625 x 3-s Speed in m/s)

Rank, q	SW 225°	Rank, q	SW 225°	Rank, q	SW 225°
1	54	19	19	39	14
2	39	20	19	40	14
3	33	21	18	41	14
4	30	22	18	42	13
5	27	23	18	43	13
6	26	24	17	44	13
7	26	25	17	45	13
8	23	26	17	46	13
9	23	27	17	47	13
10	22	28	17	48	12
11	22	29	17	49	12
12	21	30	17	50	12
13	20	31	17	51	11
14	20	32	17	52	10
15	20	33	17	53	10
16	19	34	16	54	9
17	19	35	16	55	0
18	19	36	16		

12.8 ERROR ESTIMATES

Inherent in estimates of extreme values are errors of five types:

1. *Observation errors*, that is, errors in the measurement of the wind speeds or of the parameters used in wind speed simulation models (e.g., the radius of maximum wind speeds).
2. *Physical modeling errors*, which pertain to the transformation of wind speeds at a given elevation, in a given type of surface exposure, and averaged over a given time interval (e.g., 1 min), into wind speeds at a different elevation, and/or in a different type of exposure, and/or averaged over a different time interval (e.g., 3 s). Modeling errors are also present, and may be significant, for the physical hurricane storm model used to generate simulated wind speeds.
3. *Probabilistic modeling errors*, which pertain to the choice of probability distribution fitted to the wind speed data, or to climatological data such as, for example, the hurricane pressure defect and the radius of maximum wind speeds.

4. *Sampling errors* which, given the probabilistic model being used, are due to the limited size of the data sample on which estimates of extreme wind speed are based.[10] Sampling errors affect inferences on wind speeds based on samples of both measured and simulated wind speeds. It has been argued that because the size of simulated samples of hurricane wind speeds can be arbitrarily large (e.g., $n_{\max} = 1{,}000$ or even 100,000), the corresponding sampling errors are in practice negligible. However, this view disregards the fact that sampling errors in the estimation of hurricane speeds are predominantly due to the relatively small size (of the order of 100 years) of the samples of parameters used in hurricane wind speed simulations (Sect. 12.5).
5. If calculated mean or median values of the peaks are used in the calculations of wind effects, errors arise due to the difference between those values on the one hand and the actual values of the peaks that may occur in a storm event on the other.

Engineering judgment informed by statistical considerations is typically used to assess the first three types of errors. The fourth type of errors is considered in Sect. 12.8.1. For the fifth type of errors, see Sect. 13.3.5.

12.8.1 Sampling Errors in the Estimation of Non-Hurricane Wind Speeds

For the *epochal approach*, the standard deviation of the sampling errors in the estimation of $\overline{N}$-year speeds under the assumption that a Gumbel (EV I) distribution is appropriate is, approximately,

$$SD[v(\overline{N})] = 0.78\{1.64 + 1.46(\ln \overline{N} - 0.577) + 1.1(\ln \overline{N} - 0.577)^2\}^{1/2} \frac{s}{\sqrt{n}}. \tag{12.8.1}$$

[12-25]. The use of Eq. 12.8.1 for speeds assumed to be best fitted by a reverse Weibull distribution is conservative.[11]

[10] Assume a coin is thrown twice, and the outcomes are "heads" after both throws. An inference that the probability of getting "heads" is unity, based on the sample of two outcomes, would be incorrect; owing to the small size of the sample of outcomes, that inference would be affected by sampling errors in the estimation of the frequency of occurrence of "heads."

[11] For the POT approach, sampling errors may be estimated by using Monte Carlo simulations to generate a large number of samples of size n, and calculating the respective $\overline{N}$-year wind speeds, from which the requisite error statistics in the estimation of the $\overline{N}$-year wind speeds can be obtained. Given a data sample, the parameters $\hat{a}$ and $\hat{c}$ of the GPD are estimated as shown in Sect. 12.3.2.2. A normal distribution of the parameter c of the generated samples is constructed, with $\hat{c}$ and s.d.(c) as its mean and standard deviation. The parameters c of the samples generated by Monte Carlo simulation are taken from this distribution. The corresponding parameters a for those samples are calculated by using Eqs. 12.3.7 and 12.3.6b.

Numerical Example 12.8.1. *Sampling errors estimates for non-hurricane wind speeds.* We refer to the data of Numerical Example 12.3.2. For a sample of size $n = 17$ of largest annual peak 3-s gust speeds, $s = 7.83$ mph. Equation 12.8.1 yields $SD[v(\overline{N} = 50 \text{ years})] = 6.41$ mph, and $SD[v(\overline{N} = 700 \text{ years})] = 10.62$ mph.

12.8.2 Sampling Errors in the Estimation of Hurricane Wind Speeds

Sampling errors based on hurricane recording periods of about 100 years have estimated coefficients of variation (COV)[12] of about 8%, 10%, 15%, and 20% for $\overline{N} = 50, 100$, 1,000, and 10,000 years, respectively [12-26]. According to [12-22], however, for $\overline{N} = 100$ years estimated COVs are about 6–15%, while for $\overline{N} = 1{,}000$ years estimated COVs have approximately the same or even smaller values—a result that may be an artifact of the error estimation procedure, as suggested by the results of [12-26] and those of Numerical Example 12.8.1. The size of hurricane wind speed samples created by numerical simulation can be very large (e.g., 10,000 or larger). However, as was pointed out earlier, for such large samples, the sampling errors are due predominantly to the relatively short period of record (i.e., about 100 years) from which the physical information used in the simulations is obtained. Increasing the size of the simulated sample does not decrease these irreducible errors.

[12]The estimated coefficient of variation (COV) of the sampling error is defined as the ratio between the estimated standard deviation of the error and the estimated mean wind speed.

CHAPTER 13

BLUFF BODY AERODYNAMICS BASICS; AERODYNAMIC TESTING

13.1 INTRODUCTION

A body immersed in a turbulent flow experiences flow-induced fluctuating pressures due to (1) flow fluctuations caused by the presence of the body, and (2) turbulent wind speed fluctuations in the oncoming flow. The purpose of this chapter is to present basic bluff body aerodynamics principles that will help structural engineers to understand and interpret procedures for determining wind-induced aerodynamic effects on structures (Section 13.2). Those principles are also used to discuss basic testing requirements, limitations, and techniques for wind tunnels and large-scale testing facilities (Sect. 13.3). Given the large differences among results of aerodynamic tests conducted at various laboratories, a possible testing technique applicable to residential homes is described that lends itself to standardization and is capable of significantly reducing errors associated with the simulation of low-frequency flow fluctuations in the laboratory (Sect. 13.4).

Computational fluid dynamics (CFD) has in recent years made useful strides and may be expected to become increasingly important in the future. However, current techniques for computing time-dependent pressures induced by turbulent flows on buildings and other structures do not yet allow the confident use of CFD as a structural engineering tool.

13.2 BLUFF BODY AERODYNAMICS

This section presents basic elements of fluid dynamics for inviscid and viscous flows, including the Bernoulli equation and the Reynolds number

(Sect. 13.2.1), boundary layers, flow separation and reattachment, negative pressures, and drag (Sect. 13.2.2), vortex shedding, the Strouhal number, and vortex-induced fluctuating lift (Sect. 13.2.3), pressure and force coefficients, internal pressures (Sect. 13.2.4), and the effect of Reynolds number and turbulence on drag (Sect. 13.2.5).

13.2.1 The Bernoulli Equation, Pressures, and Shear Stresses; the Reynolds Number

13.2.1.1 The Bernoulli Equation. For an inviscid (viscosity-free) steady flow, the Bernoulli equation relates the velocity V and the static pressure p along a streamline (i.e., a curve tangent at each instant to the velocity vector of the flow), as follows:

$$p + {}^{1}\!/_{2}\rho V^2 = \text{const} \tag{13.2.1}$$

where the first and second term in the sum denote the static and the dynamic pressure, respectively, and ρ denotes the fluid density. Applying Eq. 13.2.1 along the streamline between two points, one of which is located far upstream in the undisturbed flow field, while the other is the *stagnation point* on the windward face of the building where $V = 0$, yields the *stagnation pressure* p_{st}

$$p_{st} = p_0 + {}^{1}\!/_{2}\rho V_0^2 \tag{13.2.2}$$

In general, at any point along a streamline where $V > V_0$ $(V < V_0)$, the pressure $p - p_0$ is negative (positive).

To derive Eq. 13.2.1, consider a fluid element (Fig. 13.2.1) subjected in the direction of the streamline to the force $p\,dy\,dz$, the force $-(p + dp)\,dy\,dz$, and the *inertial force*

$$\rho\,dx\,dy\,dz\frac{dV}{dt} = \rho\,dy\,dz\,V\,dV, \tag{13.2.3}$$

where $V = dx/dt$. (The inertial force is equal to the mass of the element $\rho\,dx\,dy\,dz$ times the acceleration dV/dt.) The equation of equilibrium among these three forces yields $-dp = \rho V\,dV$ or, upon integration, Eq. 13.2.1. The term ${}^{1}\!/_{2}\rho V^2$ is called the *dynamic pressure*. The derivative $\frac{dp}{dx}$ is called the *pressure gradient* in the x direction.

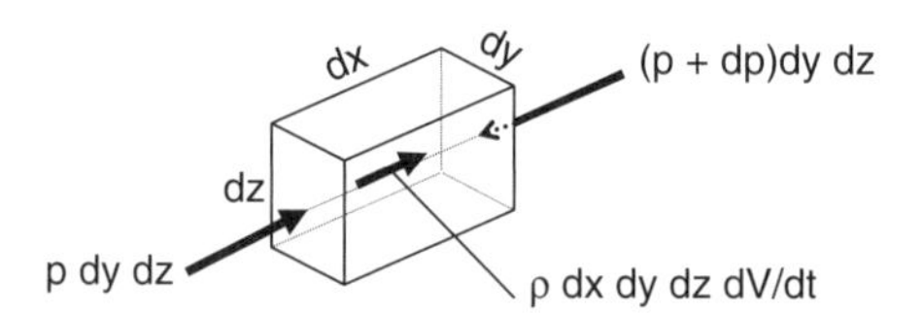

Figure 13.2.1. Flow-induced pressures and inertial force on an elemental volume of a fluid in motion.

13.2.1.2 Shear Stresses and the Reynolds Number. In addition to pressures, real flows experience *shear stresses.* For example, if the flow velocity changes across the coordinate axis normal to a streamline, the shear stress is

$$\tau = \mu \frac{dV}{dz}. \tag{13.2.4}$$

The constant μ is the *fluid viscosity*. Viscosity causes the fluid to adhere to the boundaries, where the flow velocity vanishes (no-slip condition).

Bernoulli's equation shows that, in a steady flow, the velocity induces on the body a dynamic pressure of the order of $^1/_2\rho V^2$, which has the dimensions of force per unit area. The inertial force—the force associated with the flow acceleration—due to this pressure acting on a volume of fluid with linear dimensions L is of the order of $\rho V^2 L^2$ (i.e., it has the dimension of mass times acceleration). The viscous stresses are of the order of $\mu V/L$ and create viscous (friction) forces of the order of $(\mu V/L)L^2 = \mu VL$. The ratio between the inertial force and the friction force acting on the volume of fluid is therefore of the order of $VL/(\mu/\rho)$. This ratio is nondimensional and is called the *Reynolds number* of the flow. The ratio $\nu = \mu/\rho$ is called the fluid's *kinematic viscosity*. The Reynolds number can then be written

$$\mathrm{Re} = \frac{VL}{\nu}. \tag{13.2.5a}$$

For air flow at usual temperature and atmospheric pressure conditions, Eq. 13.2.5a is

$$Re = 67{,}000\, VL \tag{13.2.5b}$$

where V and L are given in m/s and m, respectively.

Numerical Example 13.2.1. *Reynolds numbers for a prototype and its model.* For a circular cylinder with diameter $D = 30$ m in flow with characteristic velocity $V = 40$ m/s, $Re = 67{,}000 \times 40 \times 30 = 8 \times 10^7$ (Eq. 13.2.5b). For a 1:400 model of the cylinder in flow with a characteristic velocity of the order of $V_{\mathrm{model}} = 10$ m/s (typical of commercial wind tunnels), $Re = 5 \times 10^4$, that is, about three orders of magnitude less than in the full-scale case.

The choice of the characteristic velocity V and length L is arbitrary, provided that it is indicated explicitly and that experimental or theoretical results are reported in a manner consistent with that choice. For example, for a cylinder with diameter D immersed in a flow with mean oncoming velocity V, Re may be based on a characteristic velocity V and a characteristic length L equal to the cylinder diameter D (this is the standard choice). Alternatively, Re could be based on a characteristic velocity V and, say, a characteristic length $L = D/2$.

13.2.2 Boundary Layers, Flow Separation and Reattachment, Negative Pressures, Drag

Friction, which causes the velocity to vanish at the flow boundary, retards the flow within a region called the *boundary layer* (Fig. 13.2.2), in which shear stresses are present (Eq. 13.2.4). The retardation effect—and the shears—are weaker as the distance from the boundary increases. Outside the boundary layer, the effects of friction become negligible, and in the absence of turbulence in the oncoming flow an ideal, frictionless flow prevails.

In a flow through a nozzle, the velocity through the nozzle's constricted area is higher than the velocity through the larger area upstream of the nozzle, since the same amount of fluid must flow per unit of time through both areas. A similar explanation applies to *hills*, which, like a nozzle, tend to constrict the area available for the passage of the fluid, and cause the occurrence of speed-up effects (Sect. 3.5). Conversely, if the area through which the flow occurs increases downwind, the flow decelerates. In addition, friction at a boundary further decelerates the flow. *Flow reversal* and *flow separation* can thus occur, and are typically associated with the formation of a turbulent *shear layer* (Figs. 13.2.2, 13.2.3).

A visualization of flow separation for a bluff shape, and of the turbulent flow in the separation zone, is shown in Fig. 13.2.4a. If the shape of the deck is streamlined, as opposed to bluff (Fig. 13.2.4b), the separation zone decreases substantially, and the turbulent flow above the upper face of the deck almost disappears. Figure 13.2.5a shows a visualization of flow around

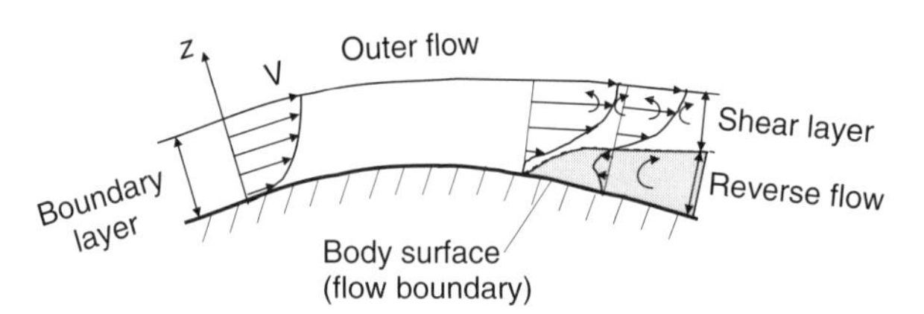

Figure 13.2.2. Velocity profile in the boundary layer and in the separation zone of a flow near a curved body surface (after *Aérodynamique*, Centre Scientifique et Technique du Bâtiment, Nantes, France, 1980).

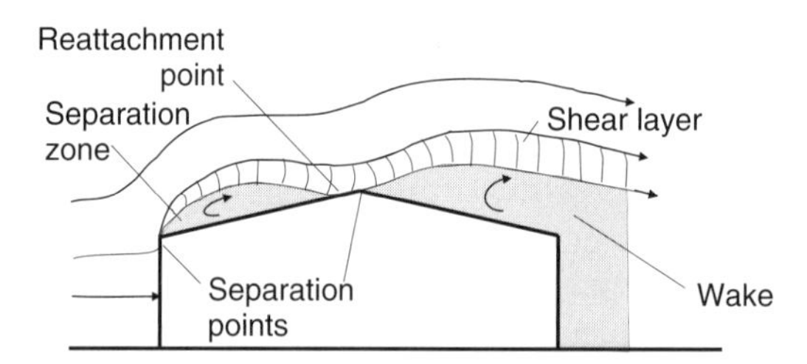

Figure 13.2.3. Flow about a building with sharp edges (after *Aérodynamique*, Centre Scientifique et Technique du Bâtiment, Nantes, France, 1980).

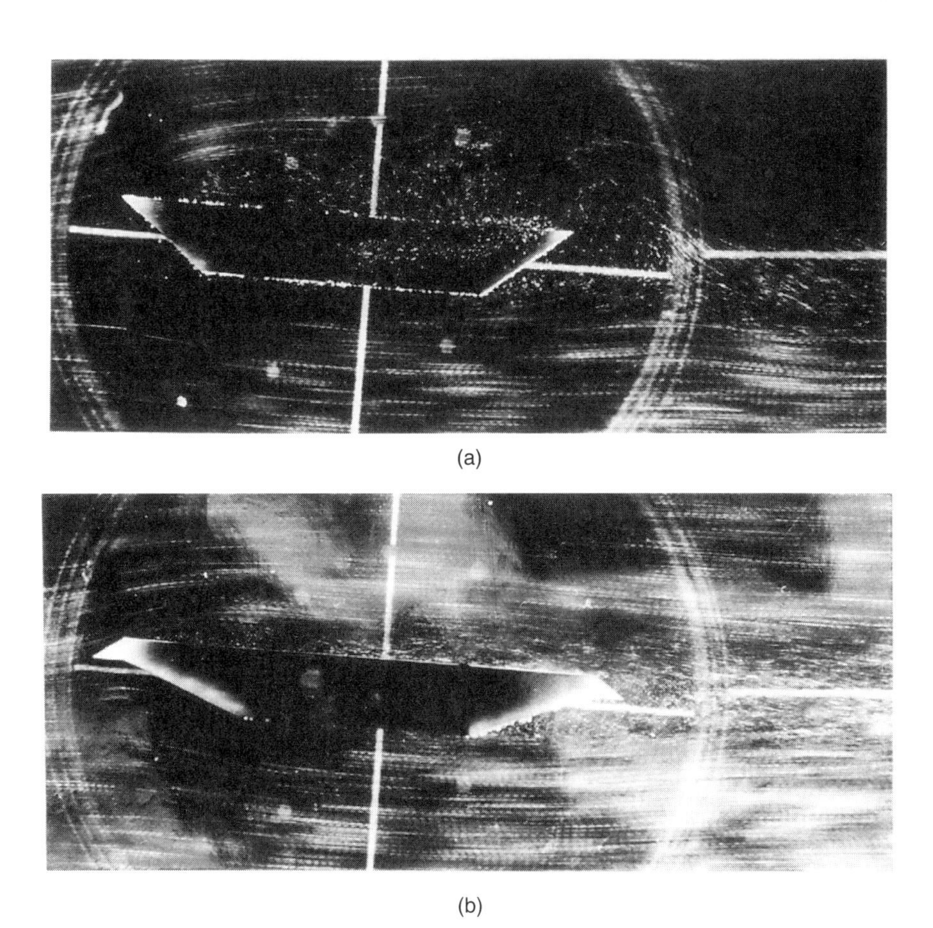

(a)

(b)

Figure 13.2.4. Visualization of water flow over (a) a model bridge deck section and (b) a partially streamlined model bridge deck section (courtesy of the National Aeronautical Establishment, National Research Council of Canada).

a clockwise spinning baseball moving from left to right. Figure 13.2.5b is a schematic of the forces acting on baseball with velocity V and angular velocity ω) [13-16].

The relative velocity of the flow with respect to the ball is directed from right to left. Entrainment of fluid due to friction at the surface of the spinning body increases the relative flow velocities with respect to the body near its top and decreases them near its bottom. By virtue of Bernoulli's equation, the static pressures are therefore lower near the top and higher near the bottom. The flow asymmetry induced by spinning therefore results in a net force denoted by F_M in Fig. 13.2.5b and called the Magnus force. In different aerodynamic contexts, flow asymmetries due to body motions can under certain conditions be the cause of galloping and other aeroelastic motions (Chapter 15).

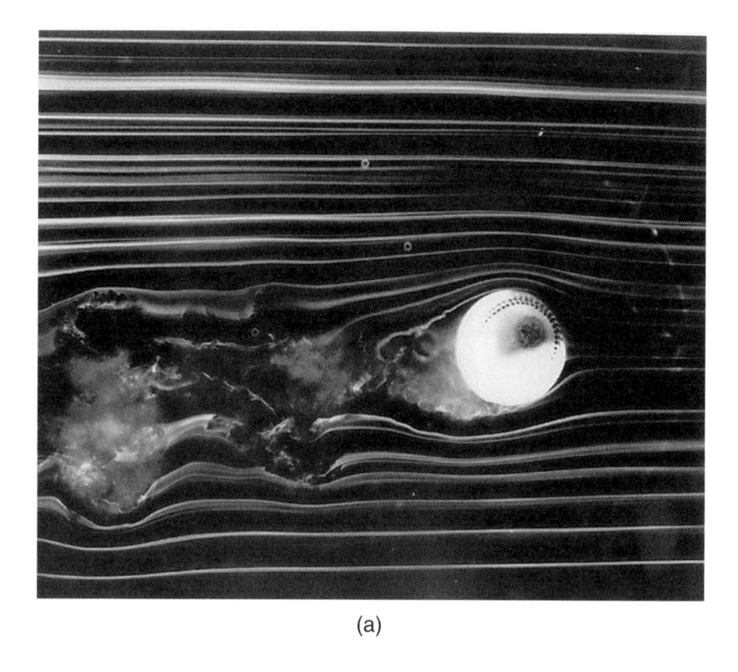

(a)

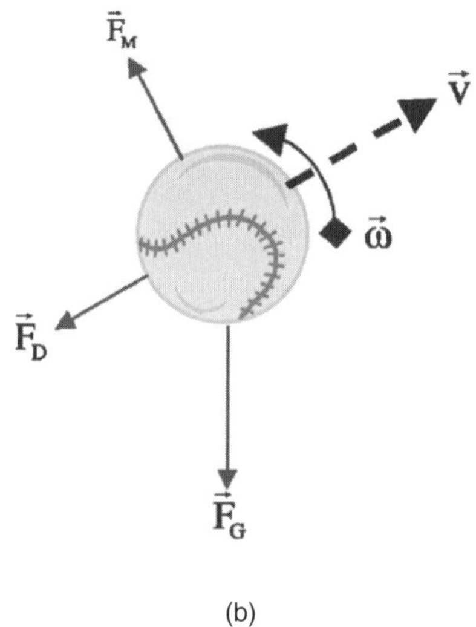

(b)

Figure 13.2.5. (a) Flow around a spinning baseball (courtesy of the National Institute of Standards and Technology); (b) schematic showing forces acting on baseball with velocity V and angular velocity ω (from A. M. Nathan, "The effect of spin on the flight of a baseball," *Am. J. Phys.* 76, 119–124, 2008, by permission of the author and the *American Journal of Physics*).

The increase in the downwind area through which the flow occurs is a function of the curvature of the boundary surface. For given Reynolds number *Re*, the flow is more prone to separation if the radius of curvature is small. At sharp edges, where the radius of curvature is very small or zero, separation will occur except in flows with very low *Re* (e.g., honey-like flows). Depending upon the flow and the boundary shape, a separated flow may experience *reattachment* (Fig. 13.2.3).

In the separation zone and in the wake, the flow velocity is vanishingly small (the flow in the wake is referred to in German as *Totwasser*, dead water). Just outside of the shear layers, the velocities are larger than the velocity V_0 far upwind (since streamlines must crowd around the body, much as they do around a hill, so that the same amount of fluid flows per unit time around the obstructing body as in the corresponding unobstructed flow far upwind). By virtue of Bernoulli's equation (Eq. 13.2.1), the pressures just outside the separation layers are therefore smaller than the pressures far upwind. Those pressures are impressed on the body (they are for practical purposes the same on the body as just outside the shear layer), meaning that in the separated flow region the body is subjected to negative pressures. In addition, local vortices can be responsible for the creation of large suctions.

The resultant of the forces due to positive and negative pressures acting on a body in the direction of the mean flow is called the drag force, or the *drag*.

13.2.3 Vortex Shedding, the Strouhal Number, Vortex-Induced Fluctuating Lift

Cylindrical or prismatic bodies immersed in uniform, smooth flow shed in their wake alternating vortices at a dominant frequency n_s given by the relation

$$n_s = S\frac{V}{D}, \tag{13.2.6}$$

where D is a characteristic dimension of the cross section, V is the constant flow velocity, and the *Strouhal number* S depends upon the cross section of the body and the Reynolds number. For circular cylinders $S \approx 0.2$ for $Re \approx 2 \times 10^5$, with values higher by up to 5–10% for $Re \approx 10^7$ [13-1]. For square cylinders $S \approx 0.12$ for $Re \approx 10^5$. Vortex shedding also occurs under three-dimensional conditions, that is, in nonuniform and turbulent flows, and for tapered and/or relatively short bodies. Examples of vortex shedding are shown in Figs. 13.2.6 and 13.2.7. In both cases, the vorticity is regular. At any given instant, the fluctuating flow in a prism's or cylinder's wake is *asymmetrical* about a line parallel to the oncoming flow (Fig. 13.2.7). Therefore, the pressures the flow induces on the cylinder are also asymmetrical. The effect of the asymmetrical pressures is to induce on the body a transverse fluctuating load *perpendicular* to the oncoming flow, called *lift*.

Figure 13.2.6. Flow around a rectangular cylinder ($Re = 200$). (From Y. Nakamura, "Bluff-body aerodynamics and turbulence," *J. Wind. Eng. Ind. Aerodynam.* 49, 65–68, 1993 (copyright 1993, with permission from Elsevier).

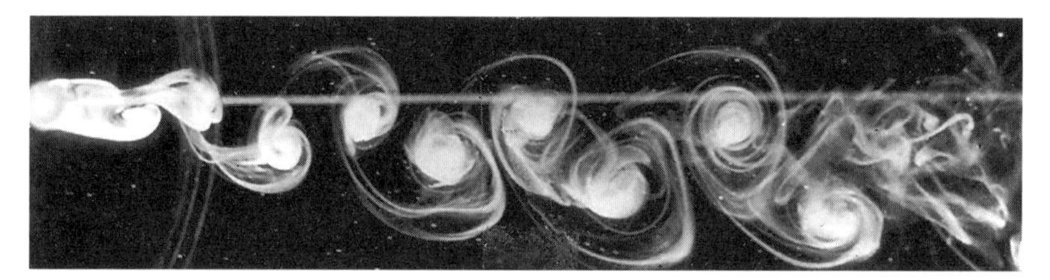

Figure 13.2.7. Vortices shed in the wake of a circular cylinder in a water tunnel (National Aeronautical Establishment, National Research Council of Canada).

13.2.4 Pressure and Force Coefficients, Internal Pressures

13.2.4.1 Nondimensional Pressure and Force Coefficients. Since the quantity $\frac{1}{2}\rho V^2$ has the dimensions of pressure, the ratio

$$C_p = \frac{p - p_0}{\frac{1}{2}\rho V^2}, \tag{13.2.7}$$

where V is a reference wind speed (e.g., the wind speed at the elevation of a building's eave or top), is a nondimensional quantity called the *pressure coefficient*. Similarly, drag or lift force coefficients may be defined as ratios of drag forces or lift forces to the reference force $\frac{1}{2}\rho V^2 A$, where A is a reference area. Moment coefficients are defined as ratios between aerodynamic moments and reference moments $\frac{1}{2}\rho V^2 AB$, where B is a reference linear dimension.

Pressures generally vary randomly in time and in space. Pressures induced on the windward face of a building by oncoming wind normal to that face are proportional to the product $\frac{1}{2}C_p(\overline{V} + V')^2$, where V' is the velocity fluctuation about the mean $\overline{V}$. Neglecting second-order terms, it follows that the fluctuating pressures on the windward face are approximately proportional to $C_p\overline{V}V'$.

The random variation of pressures in space (Fig. 13.2.8) means that pressures at different points do not act in tune (i.e., have imperfect spatial coherence), so peak total forces per unit tend to decrease as the area increases. This fact is reflected in the definition and use in the ASCE Standard of gust effect factors and of effective pressures that decrease as the tributary area increases.

13.2.4.2 Internal Pressures. Loads on exterior wall or roof components are determined by both external and internal pressures. If the building has an opening on the windward (leeward) side and is otherwise sealed, the wind flow will create a positive (negative) internal pressure, as shown in Fig. 13.2.9a (Fig. 13.2.9b). In most cases, the opening or the porosity distribution over the building envelope is not well known, and internal pressures could be positive

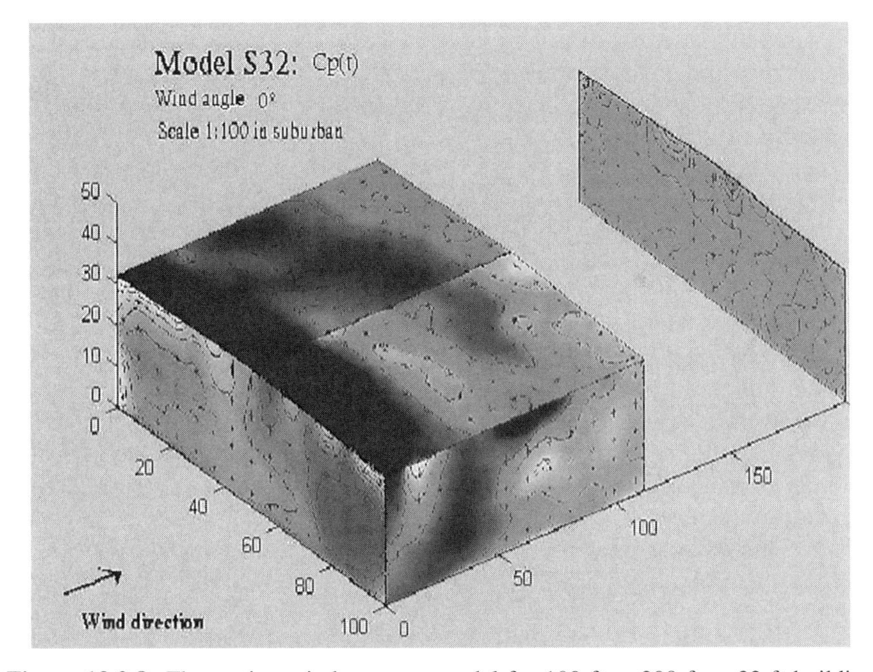

Figure 13.2.8. Fluctuating wind pressure model for 100 ft × 200 ft × 32 ft building in suburban terrain; gable roof with 1/24 slope. Mean wind speed is normal to end walls. Note asymmetry of pressures with respect to vertical plane containing ridge line. (Based on 1:100 model scale boundary-layer wind tunnel simulation, University of Western Ontario; created by Dr. A. Grazini, National Institute of Standards and Technology.)

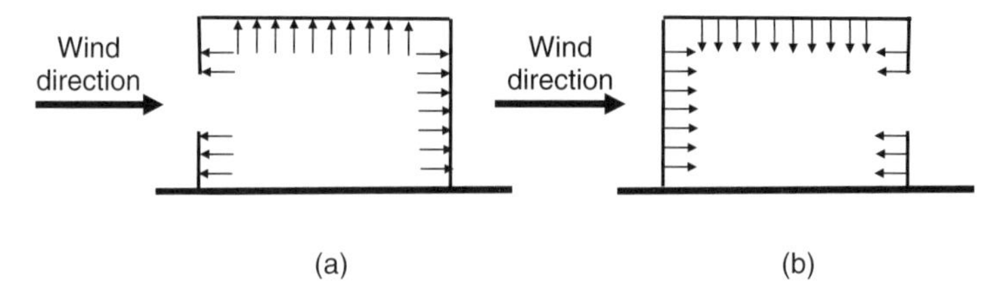

Figure 13.2.9. Mean internal pressures in buildings with openings.

or negative. The ASCE Standard provisions specify internal pressures that depend upon the size and distribution of openings in the building envelope (Sects. 3.3 and 4.2.1). For recent research on internal pressures, see [13-2].

13.2.5 Reynolds Number and Drag; Drag Reduction for Bodies with Rounded Shapes

Consider the case of the circular cylinder in uniform laminar flow. Figure 13.2.7 shows vorticity shed periodically in a cylinder's wake for $30 \leq Re < 5{,}000$ or so. Note the coherence of the vortex formations. For higher Re, the flow undergoes various changes, until for $Re > 10^5$ or so the wake becomes turbulent, no longer exhibiting strong coherence; the vortex shedding is irregular, although it has a dominant frequency [13-1] (Sect. 13.2.3). For $10^5 < Re < 2 \times 10^5$, the drag coefficient $C_D \approx 1.2$. At $Re \approx 2 \times 10^5$, C_D drops dramatically, from $C_D \approx 1.2$ to $C_D \approx 0.25$, then increases gradually with the Reynolds number until it reaches a value $C_D \approx 0.6$ for $Re = 10^7$.

The drop in the drag coefficient can be explained as follows. At relatively low velocities V, the flow in the boundary layer that forms near the cylinder surface is laminar (smooth). Higher velocities V cause the flow in the boundary layer upwind of the separation points to become unstable, that is, to change from laminar to turbulent. The turbulence causes an exchange of fluid particles between the boundary layer and the zone just outside it. The outside fluid that, owing to this exchange, penetrates into the boundary layer, has on average higher velocities than the velocities in the boundary layer, within which the flow is retarded. The effect of these higher velocities is that flow reversal and separation occur further downwind than would have been the case in the absence of the turbulent transport, and the wake becomes narrower. This reduces the cylinder's area subjected to negative pressures, and therefore reduces the drag. The cylinder with a narrower wake behaves as if it were aerodynamically more streamlined.

A similar reduction of the drag is effected by the presence of turbulence in the oncoming flow. Finally, turbulence promoting exchange of fluid between the boundary layer and the higher-velocities zone outside can be artificially created by roughening the surface of the body in various ways. For example, a shift of the flow separation downstream, with consequent narrowing of the wake and decrease in drag, is achieved in golf balls through the provision of dimples that cause turbulence.

13.3 AERODYNAMIC TESTING

To date, testing remains the only accepted means of obtaining aerodynamic data usable for the design of engineering structures. Flachsbart, working under Prandtl's supervision, discovered as early as 1932 that aerodynamic pressures on buildings can differ markedly in shear flows (i.e., flows in which the mean wind speeds increase with distance from the ground or from the wind tunnel floor) from pressures measured in uniform, smooth flows.[1] Nevertheless, until the 1960s, wind tunnel tests aimed at producing results used in building codes were conducted only in uniform, smooth flows. Such results are reproduced in Table 4.4.1, Table 4.5.1, and Appendix A4 of [7-1].

In the 1960s, Jensen rediscovered Flachsbart's finding. Attempts have subsequently been made to build a wind tunnel in which the shear flow developed naturally by friction at the wind tunnel floor over a sufficiently long fetch.

However, it turned out that the fetch required to achieve this goal would typically be too large. For this reason, spire-like passive devices are typically placed upwind of the roughness fetch. The spires and, depending upon application, additional passive devices, help to produce a flow with features resembling to some degree those of atmospheric boundary layer (ABL) flows (Fig. 13.3.1).

A rigorous simulation of atmospheric flows would require that the nondimensional form of the equations of fluid motion and their attendant boundary conditions be the same in the prototype and at model scale. As is shown in Sects. 13.3.1–13.3.4, this is not possible in practice, owing to the violation of the Reynolds number similarity requirement and, in the case of tall structures, the violation of the Rossby number as well. Wind tunnel testing is therefore an art that requires consideration of the errors inherent in imperfect simulations. Attempts to quantify those errors are made by performing full-scale aerodynamic measurements, a difficult endeavor owing to large uncertainties in the wind flow that often are encountered in practice.

Limitations of wind tunnel testing discussed in Sects. 13.3.1–13.3.4 have led to the development of alternative, larger testing facilities capable of testing models with dimensions and flow velocities closer or even equal to those of the prototype. Such facilities can provide aerodynamic data for use in design or for destructive testing in conventional structural testing laboratories. Facilities capable of producing sufficiently large flow speeds can also be used directly for destructive testing purposes, although issues remain with respect to the use of such facilities for destructive testing that accounts for safety margins inherent in design practice.

For *aeroelastic testing*, see Chapter 19.

[1]Having refused to divorce his Jewish wife, Flachsbart was prohibited by the Nazi authorities from publishing his results in the open literature [13-3]. His 1932 work was largely unknown until 1986, when it was rediscovered in the library of the National Bureau of Standards [7-1, p. 173].

Figure 13.3.1. Meteorological wind tunnel, Wind Engineering Laboratory, Colorado State University. Model and turntable are in the foreground, and spires are in the background (courtesy of Professor Bogusz Bienkiewicz; photo by Gregory E. Stace).

13.3.1 Basic Similarity Requirements[2]

Basic similarity requirements can be determined from dimensional analysis. For buildings, it may be assumed that the aerodynamic force F on a body is a function of flow density ρ, flow velocity V, a characteristic dimension D, a characteristic frequency n, and the flow viscosity μ; the following relation governing dimensional consistency holds:

$$F \stackrel{d}{=} \rho^{\alpha} V^{\beta} D^{\gamma} n^{\delta} \mu^{\varepsilon} \tag{13.3.1}$$

where α, β, γ, δ, and ε are exponents to be determined. Each of the quantities ρ, V, D, n, and μ can be expressed dimensionally in terms of the three fundamental quantities: mass M, length L, and time T, so Eq. 13.2.1 can be written as

$$\frac{ML}{T^{-2}} \stackrel{d}{=} \left(\frac{M}{L^3}\right)^{\alpha} \left(\frac{L}{T}\right)^{\beta} (L)^{\gamma} \left(\frac{1}{T}\right)^{\delta} \left(\frac{M}{LT}\right)^{\varepsilon} \tag{13.3.2}$$

[2]Sections 13.3.1 and 13.3.2 are taken from [7-1] and were prepared by Professor R.H. Scanlan.

(the dimensions of the viscosity follow from Eq. 13.2.4). Dimensional consistency requires that

$$\begin{aligned} 1 &= \alpha + \varepsilon \\ 1 &= -3\alpha - \beta + \gamma - \varepsilon \\ -2 &= -\beta - \delta - \varepsilon \end{aligned} \tag{13.3.3}$$

from which there follows, for example, that

$$\begin{aligned} \alpha &= 1 - \varepsilon \\ \beta &= 2 - \varepsilon - \delta \\ \gamma &= 2 - \varepsilon + \delta \end{aligned} \tag{13.3.4}$$

Substitution of these relations in Eq. 13.3.1 yields

$$F \stackrel{d}{=} \rho^{1-\varepsilon} V^{2-\varepsilon-\delta} D^{2-\varepsilon+\delta} n^{\delta} \mu^{\varepsilon} \tag{13.3.5}$$

or

$$F \stackrel{d}{=} \rho V^2 D^2 \left(\frac{Dn}{V}\right)^{\delta} \left(\frac{\mu}{\rho VD}\right)^{\varepsilon} \tag{13.3.6}$$

meaning that the dimensionless force coefficient $F/\rho V^2 D^2$ is a function of the dimensionless ratios Dn/V and $\mu/\rho VD$ (or of their reciprocals). In some wind engineering problems (e.g., the vibrations of suspended bridges), the aerodynamic forces are also functions of the acceleration of gravity g. By introducing g^{ζ} into Eq. 13.3.1, it can easily be shown that the force is also a function of the nondimensional ratio V^2/Dg, called the *Froude number*. The nondimensional ratio $\rho VD/\mu = VD/\nu$ is the well-known *Reynolds number*, and $\nu = \mu/\rho$ is the kinematic viscosity of the fluid (Sect. 13.2.1.2). The parameter nD/V is called the reduced frequency, while its reciprocal is the reduced velocity. If the frequency n being considered is the vortex-shedding frequency, the reduced frequency is called the *Strouhal number* (Sect. 13.2.3). If n is replaced by the Coriolis parameter (Sect. 10.1), the reduced velocity is called the *Rossby number*.

13.3.2 Basic Scaling Considerations

Similarity requires that the reduced frequencies and the Reynolds numbers be the same in the laboratory and in the prototype. This is true regardless of the nature of the frequencies involved (e.g., vortex-shedding frequencies, natural frequencies of vibration, frequencies of the turbulent components of the flow), or of the densities being considered (e.g., fluid density, density of the structure). For example, if the reduced frequency is the same in the prototype and in the laboratory (i.e., at model scale), applying this requirement to the vortex-shedding frequency n_v and to the fundamental frequency of vibration of the structure n_s, we have

$$\left(\frac{n_v D}{V}\right)_p = \left(\frac{n_v D}{V}\right)_m \tag{13.3.7}$$

and

$$\left(\frac{n_s D}{V}\right)_p = \left(\frac{n_s D}{V}\right)_m \tag{13.3.8}$$

It follows from Eqs. 13.3.7 and 13.3.8 that

$$\left(\frac{n_s}{n_v}\right)_p = \left(\frac{n_s}{n_v}\right)_m \tag{13.3.9}$$

This is also true of the ratios of all other relevant quantities (lengths, densities, velocities). Thus, for the density of the structure and the density of the fluid, it must be the case that

$$\left(\frac{\rho_s}{\rho_{air}}\right)_p = \left(\frac{\rho_s}{\rho_f}\right)_m \tag{13.3.10}$$

where ρ_f is the density of the fluid in the laboratory. For the same reason,

$$\left(\frac{V(z_1)}{V(z_2)}\right)_p = \left(\frac{V(z_1)}{V(z_2)}\right)_m \tag{13.3.11}$$

In particular, if in the prototype the velocities conform to a power law with exponent $\overline{\alpha}$, it follows from Eq. 13.3.11 that in the laboratory the velocities must conform to the power law with the same exponent $\overline{\alpha}$. To show this, Eq. 13.3.11 is rewritten as follows:

$$\left(\frac{z_1}{z_2}\right)_p^{\overline{\alpha}} = \left(\frac{V(z_1)}{V(z_2)}\right)_m \tag{13.3.12}$$

Since $(z_1/z_2)_{prot} = (z_1/z_2)_{model}$ by virtue of geometric similarity, it follows from the preceding equation that similarity is satisfied if

$$\left(\frac{z_1}{z_2}\right)_m^{\overline{\alpha}} = \left(\frac{V(z_1)}{V(z_2)}\right)_m \tag{13.3.13}$$

Since there are three fundamental requirements concerning mass, length, and time, three fixed choices of scale can be made. This choice determines all other scales. For example, let the *length scale*, the *velocity scale*, and the *density scale* be denoted by $\lambda_L = D_m/D_p$, $\lambda_V = V_m/V_p$, and $\lambda_\rho = \rho_m/\rho_p$. The reduced frequency requirement

$$\left(\frac{nD}{V}\right)_p = \left(\frac{nD}{V}\right)_m \tag{13.3.14}$$

controls the *frequency scale* λ_n for all pertinent test frequencies. From Eq. 13.3.14, it follows immediately that $\lambda_n = \lambda_V/\lambda_L$. The time scale λ_T is the reciprocal of λ_n.

13.3.3 Violation of Rossby Number Similarity

The violation of the Rossby number has no effects on the testing of low- or mid-rise buildings. However, even if a boundary layer of sufficient depth were developed in the wind tunnel by friction at the tunnel floor over a long distance upwind of the model, it would cause building models to be subjected to flow fluctuations that are dissimilar from the atmospheric flow fluctuations. To see this, recall that one of the effects of the Rossby number on the atmospheric flow is that the logarithmic law holds in a region, called the atmospheric boundary layer's surface layer, whose depth is proportional to the mean wind speed (Eq. 11.2.6; Appendix A2, Eq. A2.2.16). It was shown in Chapter 11 that the turbulence spectrum defined by Eq. 11.3.6 is valid in the region in which the logarithmic law holds. In strong winds, this region is a few hundred meters high. However, in a wind tunnel in which a 2 m deep boundary layer would develop naturally, the logarithmic law would hold only over about one-tenth of the boundary layer depth, that is, over at most about 0.2 m from the wind tunnel floor. For a 300 m tall building tested at a 1:300 length scale, this would amount to about one-fifth of the model height. Thus, according to similarity theory, Eq. 11.3.6 would be applicable over approximately the entire height of the prototype building, but over only one-fifth of the height of the building model.

It was mentioned earlier that, in conventional wind tunnels for civil engineering testing, the depth of the boundary layer developed naturally by friction at the floor turns out to be much less than 2 m or so. For this reason, turbulent shear flows are in practice created in wind tunnels by placing spires immediately upwind of the roughness fetch, with results that may vary across wind tunnels.

13.3.4 Violation of Reynolds Number Similarity

In principle, for similarity between prototype (i.e., full-scale) and wind tunnel flows to be achieved, the respective Reynolds numbers must be the same. This requirement is referred to as *Reynolds number similarity*. In conventional wind tunnels for testing models of structures, the fluid being used is air at atmospheric pressure, and Reynolds number similarity is unavoidably violated. For an illustration, see Numerical Example 13.2.1.

As follows from Section 13.2.5, the reproduction in the wind tunnel of the aerodynamic behavior of rounded bodies poses Reynolds number–related problems. Separation occurs farther downwind in the prototype (i.e., at full scale), where *Re* is larger than in the wind tunnel. Rendering the model surface rougher, or adding wires or other devices to the body surface to "trip" the flow, reduces this type of discrepancy between prototype and model-scale flows.

Unlike bodies with rounded shapes, bodies with sharp edges have fixed separation points (Fig. 13.2.3), whose position—at the edges—is assumed to be in practice independent of Reynolds number. It has therefore been hypothesized that flows around such bodies are similar at full scale and in the wind tunnel, even if Reynolds number similarity is violated. This hypothesis is not borne out by measurements. In the wind tunnel, friction forces are larger in relation to inertial forces than at full scale, so the counterparts of full-scale high-frequency flow fluctuations (i.e., smaller eddies) are damped out. This affects the local vorticity at edges and corners in the wind tunnel, resulting in local pressures typically weaker than at full scale [13-5]. An example is shown in Fig. 13.3.2 [13-6]. Also, the damping out in the wind tunnel

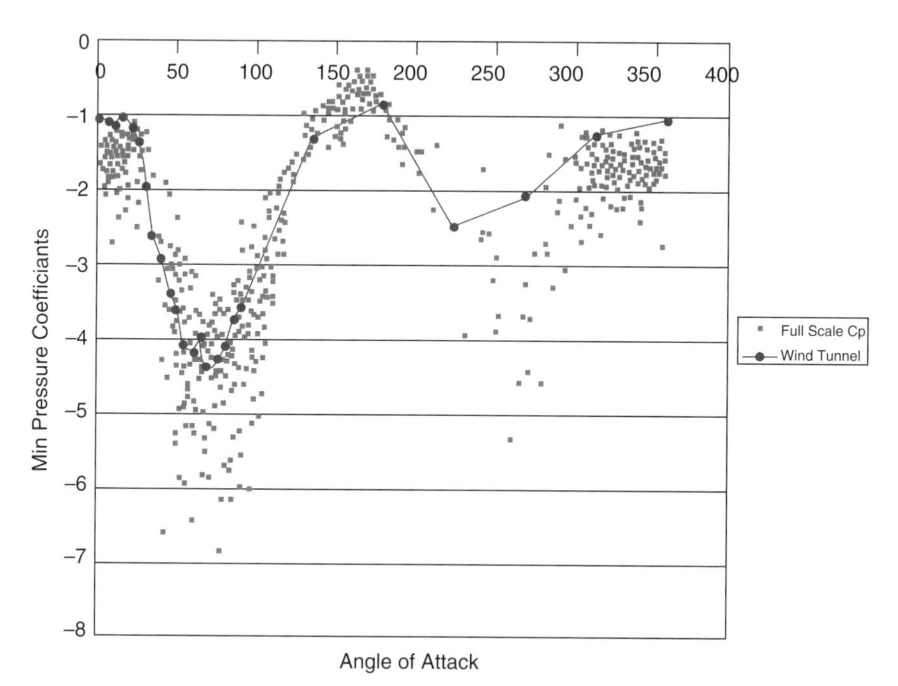

Figure 13.3.2. Minimum pressure coefficients at building corner, eave level, Texas Tech University experimental building, full-scale and wind tunnel measurements (from F. Long, *Uncertainties in pressure coefficients derived from full and model scale data*, report to the National Institute of Standards and Technology, Wind Science and Engineering Research Center, Texas Tech University).

of higher-frequency turbulence components, which are effective in bringing about the exchange of fluid particles between the boundary layer and the outer flow, results in weaker promotion of reattachment in the model than in the prototype.

Tests in a specialized wind tunnel capable of achieving smooth flows with high Reynolds numbers have shown that the flow around a body with sharp edges can be Reynolds number–dependent [13-4]. In the prototype, such smooth flows would correspond to relatively turbulence-free flows under stable stratification conditions, as can occur, for example, over bodies of water in winter.

For some tall buildings, the loss of high-frequency velocity fluctuations content in the laboratory also reduces the strength of the resonant fluctuations induced by the oncoming flow.

13.3.5 Pressure Measurements, Comparisons between Peak Pressures

Measuring simultaneously time histories of fluctuating pressures at up to hundreds of pressure taps is a novel capability that has transformed approaches to the definition and use of the aerodynamic loads for structural engineering purposes (see Chapters 18 and 19). Simultaneous pressure measurements automatically capture the spatial coherence among pressures at various taps.

An example of simultaneous pressure-measuring systems is the Electronic Pressure Scanning System developed by Scanivalve Corporation (www.scanivalve.com). This pressure-measuring system includes an Electronic Pressure Scanning Module (e.g., ZOC33, www.scanivalve.com/pdf/prod_zoc33_0512.pdf, with sensors arranged in blocks of eight, each of which has its calibration valve); a Digital Service Module (e.g., DSM3400, www.scanivalve.com/pdf/manual_dsm3400_hardware.pdf, which can service up to 16 Electronic Pressure Scanning Modules, i.e., up to 512 sensors, and contains an embedded computer, RAM memory, and a hard disk drive); a pressure calibration system; auxiliary instrumentation to regulate supply of clean, dry air; and data acquisition software.

The connection between the Electronic Pressure Scanning Module and the pressure taps is typically made through plastic tubes. A test model with tubes connecting the pressure taps to the scanning module is shown in Fig. 13.3.3. The distortions in amplitude and phase of the measured pressures are corrected using appropriate tubing transfer functions [13-18].

One problem that arises in comparing peak pressures obtained in different sets of measurements is that the peak pressure is a random variable with a distribution and a mean value. It is thus possible that the peak of one record is lower, while the peak of the other record is greater, than that mean value. To perform consistent comparisons, it is therefore necessary to estimate the probability distribution of the peak value and refer, in the comparisons, to a specified percentage point of the distribution. A procedure for doing so, based

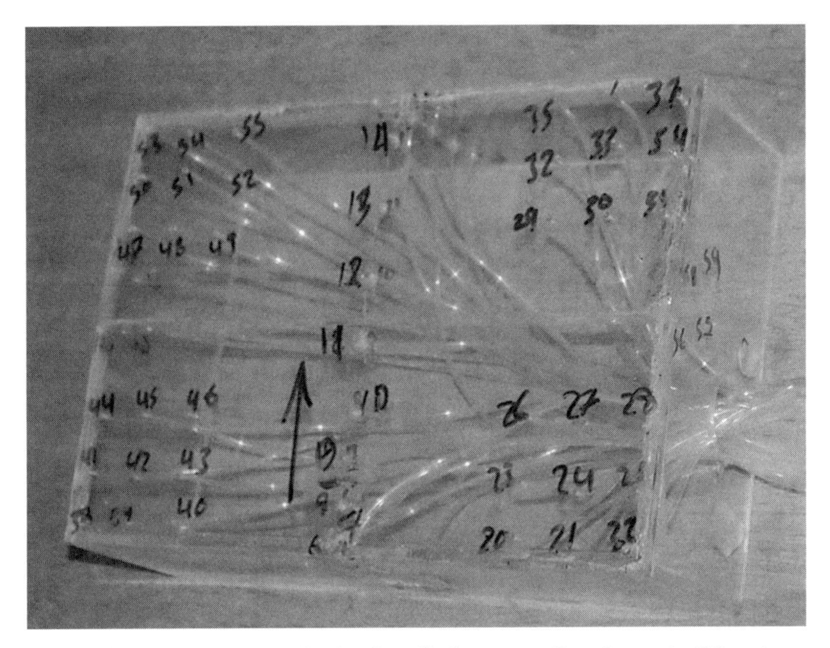

Figure 13.3.3. Pressure taps and tubes installed on a small-scale test building (courtesy of Dr. Aly Mousaad Aly Sayed Ahmed, Alexandria University, Egypt).

on random vibration theory, has been developed in [13-7]. The details of the procedure and the attendant software are available in www.nist.gov/wind (scroll down to III B).

13.3.6 Wind Tunnel Blockage

A body placed in the wind tunnel will partially obstruct the passage of air, causing the flow to accelerate and distorting the aerodynamic behavior of the model. This effect is called blockage (see, e.g., [7-1]). It may be assumed approximately that for 2% blockage ratios (i.e., ratios between the area of the obstructing body normal to the wind velocity and the wind tunnel cross-sectional area) the blockage corrections are about 5%, and that the magnitude of the blockage corrections is proportional to the blockage ratio.

13.3.7 Variation of Test Results from Laboratory to Laboratory and its Effect on the Reliability of ASCE 7 Standard Wind Pressures Provisions

Independent tests conducted by six prominent wind tunnel laboratories on models of two industrial buildings demonstrated that test results can vary significantly from laboratory to laboratory [13-8]. One of the metrics used in determining the outcome of the tests was the wind-induced bending moment at the knee of portal frames with 20 ft and 32 ft eave height. The ratios of largest to smallest reported estimates of the bending moments were in some cases higher than 2. The discrepancies were greater for the lower buildings in suburban terrain, and were attributed to differences in (1) mean wind profiles, for which the power law exponents varied among laboratories between 0.14 and 0.19 (target value 1/7) for open exposure, and between 0.16 and 0.23 (target value 0.22) for suburban exposure, and (2) turbulence intensities, which varied at a 20 ft elevation between 18 and 21% for open exposure and between 24 to 33% for suburban exposure [13-9].

Partly as a consequence of large errors inherent in the wind tunnel testing of low-rise buildings, differences exceeding in some instances 50% were found to exist between estimates of wind loads based on recent wind tunnel tests conducted at the University of Western Ontario and loads specified in the ASCE 7 Standard [13-10 to 13-13].

For taller buildings, the variability of results obtained in different wind tunnels appears to be lower, particularly in open terrain. However, discrepancies exceeding 40% have been reported between estimates of wind effects on tall buildings estimated by different laboratories (see Appendix A5). Such discrepancies are due to a variety of causes, of which the difference between characteristics of wind tunnel flows achieved in the respective wind tunnels is not necessarily the most important.

13.3.8 Large-Scale Testing Facilities

Similarity requirements applicable to wind tunnel facilities also apply to large-scale aerodynamic testing facilities, although different compromises with respect to flow simulation are required in the two cases. Testing is in some cases performed in both types of facilities to take advantage of their respective capabilities. A hybrid facility is described in [13-19].

13.3.8.1 IBHS Research Center. Figures 13.3.4 and 13.3.5 show an outside and inside view of the Institute for Business & Home Safety (IBHS) Research Center in South Carolina, a multi-peril facility capable of testing structures subjected to realistic Category 1, 2, and 3 hurricanes, extratropical windstorms, thunderstorm frontal winds, wildfires, and hailstorms.[3]

13.3.8.2 Wall of Wind Facilities. Smaller facilities of the type known as wall of wind have been developed in Florida. They are capable of producing up to Category 2 or 3 winds, including, as necessary, wind-driven rain and wind-borne debris. The transverse dimensions of the flows they generate are of the order of 5 m to 10 m (width) × 5 m (height). For a detailed description of the Florida International University (FIU) six-fan wall of wind (Fig. 13.3.6) and its capabilities, see [13-14]. As of this writing, a larger, twelve-fan facility is under construction at FIU.

[3]This description was kindly provided by IBHS.

Figure 13.3.4. Partial exterior view of IBHS Research Center (courtesy of the Institute for Business & Home Safety).

Figure 13.3.5. Interior view of IBHS Research Center with full-scale specimen, placed on the 55-foot diameter turntable with a surface area of 2,375 square feet. The 105-fan array with 300 hp motors is located to the left of the picture. (Courtesy of the Institute for Business & Home Safety.)

Figure 13.3.6. Six-fan wall of wind, Florida International University (courtesy of Drs. A. Gan Chowdhury and G. Bitsuamlak).

13.4 LOW-FREQUENCY TURBULENCE AND AERODYNAMIC PRESSURES ON RESIDENTIAL HOMES

A main cause for the differences among test results obtained in various laboratories is the fact that techniques for atmospheric flow simulation are not standardized. This is due in part to the variety of wind tunnel sizes and types used in wind engineering. In particular, low-frequency flow fluctuations, which contribute overwhelmingly to the turbulence intensity and the integral turbulence scale, may vary significantly from wind tunnel to wind tunnel.

For large buildings, the imperfect spatial coherence of the atmospheric flow fluctuations results in smaller overall wind effects than would be the case if the flows were perfectly coherent. However, for buildings with sufficiently small dimensions (e.g., residential homes), the effect of the imperfect spatial coherence is not significant. It may therefore be hypothesized that peak aerodynamic effects experienced by a small building subjected to flow whose velocities have significant low-frequency fluctuations are not substantially different from the peak aerodynamic effects induced by flows for which: (1) No low-frequency components are simulated (low-frequency components are defined as having reduced frequencies $nz/V(z) < 0.1$, say, where 0.1 is the commonly accepted practical lower limit of the inertial subrange; n = frequency, z = height above the surface, V = mean wind speed of the turbulent flow averaged over, say, 10 min or 1 hour); (2) the mean speed of the laboratory flow is augmented from $V(z)$ to $cV(z)$, where $c > 1$ is a factor such that $(c-1)V(z)$ is equal to the peak fluctuating velocity in the flow containing low-frequency fluctuations; (3) the vertical profiles of the simulated flow speeds $V(z)$ and $cV(z)$ are similar. This approach amounts in effect to replacing the low-frequency fluctuations of the flow with mean speed $V(z)$ by an incremental speed $(c-1)V(z)$ constant in time. This incremental speed may be viewed as a conceptual flow fluctuation with vanishing frequency, the spatial coherence of which is unity [13-15, 13-17].

In addition to eliminating a cause of discrepancies among measurements conducted in different laboratories, the proposed approach allows the use of considerably larger model scales than are possible in conventional testing. In boundary layer wind tunnels, similarity considerations impose the condition

$$\left(\frac{L_x}{D}\right)_{prot} = \left(\frac{L_x}{D}\right)_{model} \tag{13.4.1}$$

where L_x denotes the longitudinal integral turbulence scale, and D is a characteristic dimension of the structure. The difficulty of producing flows with large integral turbulence scales limits the geometric scale of the model. This limitation no longer exists in the absence of simulated low-frequency fluctuations [13-15], although blockage considerations must be taken into account.

A significant barrier to performing Computational Fluid Dynamics (CFD) calculations of aerodynamic pressures induced by atmospheric boundary layer flows is the difficulty of simulating numerically the imperfect spatial coherence of the turbulence in the oncoming flow. The approach to simplifying the oncoming flow described in this section is applicable not only to wind tunnel simulations, but to CFD calculations as well.

CHAPTER 14

STRUCTURAL DYNAMICS

14.1 INTRODUCTION

Under wind loads, flexible structures experience dynamic effects, that is, effects involving structural motions, including resonant amplification effects. A well-known example of resonant amplification is the effect on a bridge of a military formation marching in lockstep at a frequency equal or close to the bridge's fundamental frequency of vibration. The effects of successive steps are additive: a first step causes a deflection whose maximum is reached when the second step strikes. The second step causes an additional deflection, and subsequent steps keep adding to the response.

The mathematical model for the dynamic response is Newton's second law, a second-order differential equation of motion. The dynamic response can be obtained by solving the equation of motion in the frequency domain or in the time domain. Frequency domain approaches are useful, for example, when the fluctuating wind loads acting on the structure can be related linearly to the wind speed fluctuations in the atmosphere. Since the wind speed fluctuations are typically defined in terms of spectra and cross-spectra (Sect. 11.3.3), fluctuating loads can be similarly defined, and the response can then be conveniently obtained in the frequency domain. The requisite theory is presented in Sects. 14.2.3, 14.3.3–14.3.6, and 14.3.7.2.

Recent developments allow the synchronous measurement of pressure time histories at large numbers of points on building models or prototypes (Sect. 13.3.5, Chapter 18, and Chapter 19). Inherent in those measurements is phase information on pressure fluctuations at various points—and, therefore, information on the extent to which the respective pressures are spatially coherent. The requisite data on the wind loading acting on the structure (i.e., on the forcing function in the dynamic equations of motion) are thus available in the time domain from model tests.

An additional technological development that supports time domain solutions of wind-related structural dynamics problems is the availability of powerful computational capabilities. The requisite theory, applicable in the general three-dimensional case where mass centers do not coincide with elastic centers, is presented in Sect. 14.4.

14.2 THE SINGLE-DEGREE-OF FREEDOM LINEAR SYSTEM

The system of Fig. 14.2.1 consists of a particle of mass M concentrated at point B and of a member AB with linear elastic behavior and negligible mass. The particle is subjected to the force $F(t)$. Its displacement $x(t)$ is opposed by (1) a restoring force $-kx$ supplied by the member AB and (2) a damping force $-cdx/dt \equiv -c\dot{x}$, [1] where the stiffness k and the damping coefficient c are assumed to be constant. Newton's second law states that the product of the particle's mass by its acceleration, $M\ddot{x}$, equals the total force applied to the particle. The equation of motion of the system is

$$M\ddot{x} = -c\dot{x} - kx + F(t). \qquad (14.2.1)$$

Figure 14.2.1. Single-degree-of-freedom system.

[1] Here and elsewhere in the book the dot denotes differentiation with respect to time, that is, $\dot{x} \equiv \frac{dx}{dt}$.

With the notations $n_1 = \sqrt{k/M}/(2\pi)$ and $\zeta_1 = c/(2\sqrt{kM})$, where n_1 denotes the frequency of vibration of the oscillator[2], and ζ_1 is the damping ratio (i.e., the ratio of the damping c to the critical damping $\zeta_{cr} = 2\sqrt{kM}$ beyond which the system's motion would no longer be oscillatory), Eq. 14.2.1 becomes

$$\ddot{x} + 2\zeta_1(2\pi n_1)\dot{x} + (2\pi n_1)^2 x = \frac{F(t)}{M}. \tag{14.2.2}$$

For structures, ζ_1 is typically small (of the order of 1%).

14.2.1 Response to a Harmonic Load

In the particular case of a harmonic load $F(t) = F_0 \cos 2\pi nt$, it can be verified by substitution that the steady state solution of Eq. 14.2.2 is

$$x(t) = H(n)F_0 \cos(2\pi nt - \theta) \tag{14.2.3}$$

$$H(n) = \frac{1}{4\pi^2 n_1^2 M \left\{\left[1 - (n/n_1)^2\right]^2 + 4\zeta_1^2 (n/n_1)^2\right\}^{1/2}} \tag{14.2.4}$$

$$\theta = \tan^{-1} \frac{2\zeta_1(n/n_1)}{1 - (n/n_1)^2} \tag{14.2.5}$$

where θ is the *phase angle*, and $H(n)$ is the system's *mechanical admittance function* (or *mechanical amplification factor*). For $n = n_1$, that is, if the frequency of the harmonic forcing function coincides with the frequency of vibration of the oscillator, the amplitude of the response is largest, and is inversely proportional to the damping ratio ζ_1. In this case, the motion exhibits *resonance* (or *resonant amplification*). In the particular case $F(t) = F_0 \cos 2\pi nt$, the steady state response can be written as

$$x(t) = H(n)F_0 \sin(2\pi nt - \theta) \tag{14.2.6}$$

14.2.2 Response to an Arbitrary Load

Let the system described by Eq. 14.2.2 be subjected to the action of a load equal to the unit impulse function $\delta(t)$ acting at time $t = 0$, that is, to a load defined as follows (Fig. 14.2.2):

$$\delta(t) = 0 \text{ for } t \neq 0 \tag{14.2.7}$$

$$\lim_{\Delta t \to 0} \int_0^{\Delta t} \delta(t)\, dt = 1 \text{ for } t = 0. \tag{14.2.8}$$

The response of the system to the load $\delta(t)$ depends on time and is denoted by $G(t)$.

[2]The quantity $2\pi n$ is called circular frequency and is commonly denoted by ω.

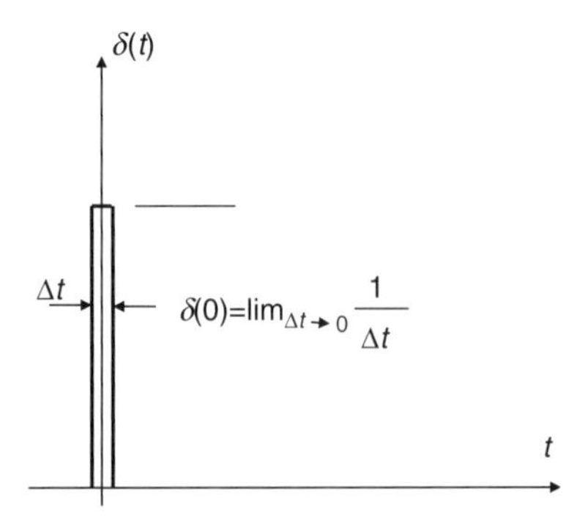

Figure 14.2.2. Unit impulse function.

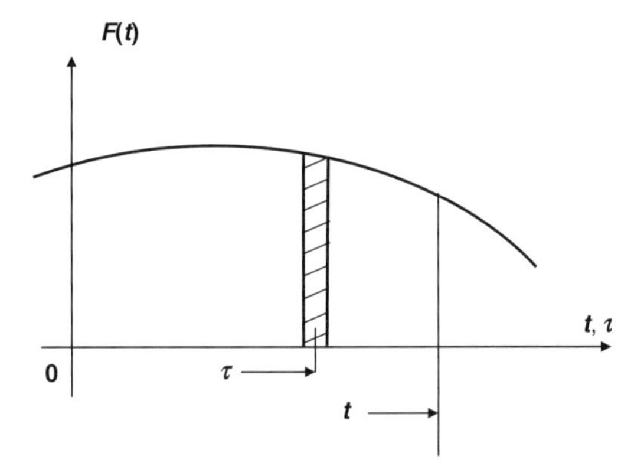

Figure 14.2.3. Load $F(t)$.

An arbitrary load $F(t)$ (Fig. 14.2.3) may be described as a sum of elemental impulses of magnitude $F(\tau')\,d\tau'$ acting each at time τ'. Since the system is linear, the response at time t to each such impulse is $G(t-\tau')F(\tau')\,d\tau'$. The total response is

$$x(t) = \int_{-\infty}^{t} G(t-\tau')F(\tau')\,d\tau' \tag{14.2.9}$$

The limits of the integral indicate that all the elemental impulses that have acted before time t have been taken into account. Denoting $\tau = t - \tau'$, Eq. 14.2.9 becomes

$$x(t) = \int_{0}^{\infty} G(\tau)F(t-\tau)\,d\tau \tag{14.2.10}$$

Let $F(t) = F_0 \cos 2\pi nt$. It follows from Eqs. 14.2.3 and 14.2.10 that

$$H(n)\cos\theta = \int_{0}^{\infty} G(\tau)\cos 2\pi n\tau\,d\tau \tag{14.2.11a}$$

$$H(n)\sin\theta = \int_{0}^{\infty} G(\tau)\sin 2\pi n\tau\,d\tau \tag{14.2.11b}$$

Equations 14.2.11a and b yield Eqs. 14.2.12a and b, whose summation yields Eq. 14.2.13:

$$H^2(n)\cos^2\theta = \int_0^\infty \int_0^\infty G(\tau_1)\cos 2\pi n\tau_1 G(\tau_2)\cos 2\pi n\tau_2\, d\tau_1\, d\tau_2 \quad (14.2.12a)$$

$$H^2(n)\sin^2\theta = \int_0^\infty \int_0^\infty G(\tau_1)\sin 2\pi n\tau_1 G(\tau_2)\sin 2\pi n\tau_2\, d\tau_1\, d\tau_2 \quad (14.2.12b)$$

$$H^2(n) = \int_0^\infty \int_0^\infty G(\tau_1)G(\tau_2)\cos 2\pi n(\tau_1 - \tau_2)\, d\tau_1\, d\tau_2 \quad (14.2.13)$$

14.2.3 Response to a Stationary Random Load

Now let the load $F(t)$ be a stationary process with spectral density $S_F(n)$. Using Eqs. A1.4.4, A1.4.5, and 14.2.10, we obtain the spectral density of the system response as follows:

$$\begin{aligned}
S_x(n) &= 2\int_{-\infty}^{\infty} R_x(\tau)\cos 2\pi n\tau\, d\tau \\
&= 2\int_{-\infty}^{\infty}\left[\lim_{T\to\infty}\frac{1}{T}\int_{-T/2}^{T/2} x(t)x(t+\tau)\, dt\right]\cos 2\pi n\tau\, d\tau \\
&= 2\int_{-\infty}^{\infty}\left\{\lim_{T\to\infty}\frac{1}{T}\int_{-T/2}^{T/2} dt\left[\int_0^\infty G(\tau_1)F(t-\tau_1)\, d\tau_1\right.\right. \\
&\quad \left.\left.\times\int_0^\infty G(\tau_2)F(t+\tau-\tau_2)\right]\right\}\cos 2\pi n\tau\, d\tau \\
&= 2\int_0^\infty G(\tau_1)\left\{\int_0^\infty G(\tau_2)\left[\int_{-\infty}^{\infty} R_F(\tau+\tau_1-\tau_2)\cos 2\pi n\tau\, d\tau\right]d\tau_2\right\}d\tau_1 \\
&= 2\int_0^\infty \int_0^\infty G(\tau_1)G(\tau_2)\cos 2\pi n(\tau_1-\tau_2)\, d\tau_1\, d\tau_2 \\
&\quad \times\int_{-\infty}^{\infty} R_F(\tau+\tau_1-\tau_2)\cos 2\pi n(\tau+\tau_1-\tau_2)d(\tau+\tau_1-\tau_2) \\
&\quad + 2\int_0^\infty \int_0^\infty G(\tau_1)G(\tau_2)\sin 2\pi n(\tau_1-\tau_2)\, d\tau_1\, d\tau_2 \\
&\quad \times\int_{-\infty}^{\infty} R_F(\tau+\tau_1-\tau_2)\sin 2\pi n(\tau+\tau_1-\tau_2)d(\tau+\tau_1-\tau_2)
\end{aligned} \quad (14.2.14)$$

where, in the last step, the following identity is used:

$$\cos 2\pi n\tau \equiv \cos 2\pi n[(\tau+\tau_1-\tau_2)-(\tau_1-\tau_2)] \quad (14.2.15)$$

From Eqs. A1.4.4, A1.4.7, and 14.12.13, there follows

$$S_x(n) = H^2(n)S_F(n) \tag{14.2.16}$$

This relation between frequency domain forcing and response is useful in applications.

14.3 CONTINUOUSLY DISTRIBUTED LINEAR SYSTEMS

14.3.1 Normal Modes and Frequencies; Generalized Coordinates, Mass and Force; Modal Equations of Motion

A linearly elastic structure with continuously distributed mass per unit length $m(z)$ and low damping can be shown to vibrate in resonance with the exciting force if the latter has certain sharply defined frequencies called the structure's *natural frequencies of vibration.* Associated with each natural frequency is a *mode*, or *modal shape*, of the vibrating structure. The first four normal modes $x_i(z)$ $(i = 1, 2, 3, 4)$ of a vertical cantilever beam with running coordinate z are shown in Fig. 14.3.1. The natural modes and frequencies are structural properties independent of the loads.

A deflection $x(z,t)$ along a principal axis of a continuous system, due to time-dependent forcing, can in general be written in the form

$$x(z,t) = \sum_i x_i(z)\xi_i(t) \tag{14.3.1}$$

where the functions $\xi_i(t)$ are called the *generalized coordinates* of the system, and $x_i(z)$ denotes the modal shape in the ith mode of vibration. For a building, similar expressions hold for deflections $y(z,t)$ in the direction of its second principal axis, and for horizontal torsional angles $\varphi(z,t)$. For structures whose centers of mass and elastic centers do not coincide, the x, y, and φ motions are coupled, as is shown in Sect. 14.4, which presents the development of the equations of motion for this general case.

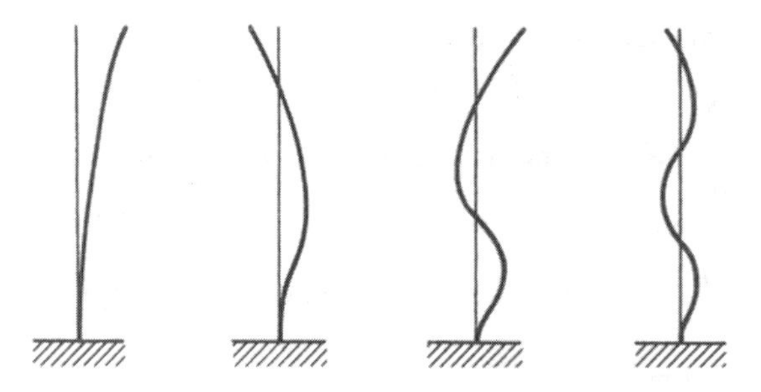

Figure 14.3.1. First four modal shapes of a cantilever beam.

In this section, we limit ourselves to presenting the modal equations of motion corresponding to the particular case of translational motion along a principal axis x:

$$\ddot{\xi}_i(t) + 2\zeta_i(2\pi n_i)\dot{\xi}(t) + (2\pi n_i)^2 \xi_i(t) = \frac{Q_i(t)}{M_i} \quad (i = 1, 2, 3, \ldots) \qquad (14.3.2)$$

where ζ_i, n_i, M_i, and Q_i are the ith mode damping ratio, natural frequency, generalized mass, and generalized force, respectively,

$$M_i = \int_0^H [x(z)]^2 m(z)\, dz \qquad (14.3.3)$$

$$Q_i = \int_0^H p(z,t) x_i(z)\, dz \qquad (14.3.4)$$

$m(z)$ is the mass of the structure per unit length, $p(z, t)$ is the load acting on the structure per unit length, and H is the structure's height. For a concentrated load acting at $z = z_1$,

$$p(z,t) = F(t)\delta(z - z_1) \qquad (14.3.5)$$

where $\delta(z - z_1)$ is defined, with a change of variable, as in Eq. 14.2.8,

$$\begin{aligned} Q_i(t) &= \lim_{\Delta z \to 0} \int_{z_1}^{z_1+\Delta z} p(z,t) x_i(z)\, dz \\ &= x_i(z_1) F(t) \end{aligned} \qquad (14.3.6)$$

14.3.2 Response to a Concentrated Harmonic Load

If a concentrated load

$$F(t) = F_0 \cos 2\pi n t \qquad (14.3.7)$$

is acting on the structure at a point of coordinate z_1, by virtue of Eq. 14.3.6 the generalized force in the ith mode is

$$Q_i(t) = F_0 x_i(z_1) \cos 2\pi n t \qquad (14.3.8)$$

and the steady state solutions of Eq. 14.3.2 are similar to the solution 14.2.3 of Eq. 14.2.2:

$$\xi_i(t) = F_0 x_i(z_1) H_i(n) \cos(2\pi n t - \theta_i) \qquad (14.3.9)$$

$$H_i(n) = \frac{1}{4\pi^2 n_i^2 M_i \left\{\left[1 - (n/n_1)^2\right]^2 + 4\zeta_i^2 (n/n_i)^2\right\}^{1/2}} \qquad (14.3.10)$$

$$\theta_i = \tan^{-1} \frac{2\zeta_i (n/n_i)}{1 - (n/n_i)^2} \qquad (14.3.11)$$

The response of the structure at a point of coordinate z is then

$$x(z,t) = F_0 \sum_i x_i(z)x_i(z_1)H_i(n)\cos(2\pi nt - \theta_i) \tag{14.3.12}$$

It is convenient to write Eq. 14.3.12 in the form

$$x(z,t) = F_0 H(z,z_1,n)\cos[2\pi nt - \theta(z,z_1,n)] \tag{14.3.13}$$

where, as follows immediately from Eqs. A1.4.8a and A1.4.8b,

$$H(z,z_1,n) = \Big\{\Big[\sum_i x_i(z)x_i(z_1)H_i(n)\cos\theta_i\Big]^2 + \Big[\sum_i x_i(z)x_i(z_1)H_i(n)\sin\theta_i\Big]^2\Big\}^{1/2} \tag{14.3.14}$$

$$\theta(z,z_1,n) = \tan^{-1}\frac{\sum_i x_i(z)x_i(z_1)H_i(n)\sin\theta_i}{\sum_i x_i(z)x_i(z_1)H_i(n)\cos\theta_i} \tag{14.3.15}$$

Similarly, the steady state response at a point of coordinate z to a concentrated load

$$F(t) = F_0 \sin 2\pi nt \tag{14.3.16}$$

acting at a point of coordinate z_1 can be written as

$$x(z,t) = F_0 H(z,z_1,n)\sin[2\pi nt - \theta(z,z_1,n)] \tag{14.3.17}$$

14.3.3 Response to a Concentrated Stationary Random Load

Let the response at a point of coordinate z to a concentrated unit impulsive load $\delta(t)$ acting at time $t = 0$ at a point of coordinate z_1 be denoted $G(z, z_1, t)$. Following the same reasoning that led to Eq. 14.2.10, the response $x(z,t)$ to an arbitrary load $F(t)$ acting at a point of coordinate z_1 is

$$x(z,t) = \int_0^\infty G(z,z_1,\tau)F(t-\tau)\,d\tau \tag{14.3.18}$$

Note the complete similarity of Eqs. 14.3.12, 14.3.17, and 14.3.18 to Eqs. 14.2.3, 14.2.6, and 14.2.10, respectively. Therefore, the same steps that led to Eq. 14.2.16 yield the relation between the spectra of the random forcing and the response:

$$S_x(z,z_1,n) = H^2(z,z_1,n)S_F(n) \tag{14.3.19}$$

14.3.4 Response to Two Concentrated Stationary Random Loads

Let $x(z,t)$ denote the response at a point of coordinate z to two stationary loads $F_1(t)$ and $F_2(t)$ acting at points with coordinates z_1 and z_2, respectively. The autocovariance of $x(z,t)$ is (see Eq. A1.5.1):

$$
\begin{aligned}
R_x(z,\tau) &= \lim_{T\to\infty} \frac{1}{T}\int_{-T/2}^{T/2} x(z,t+\tau)\,dt \\
&= \lim_{T\to\infty} \frac{1}{T}\int_{-T/2}^{T/2} \left[\int_0^\infty G(z,z_1,\tau_1)F_1(t-\tau_1)\,d\tau_1 \right.\\
&\quad \left. + \int_0^\infty G(z,z_2,\tau_1)F_2(t-\tau_1)\,d\tau_1\right] \\
&\quad \times \left[\int_0^\infty G(z,z_1,\tau_2)F_1(t+\tau-\tau_2)\,d\tau_2 \right.\\
&\quad \left. + \int_0^\infty G(z,z_2,\tau_2)F_2(t+\tau-\tau_2)\,d\tau_2\right] dt \qquad (14.3.20)\\
&= \int_0^\infty G(z,z_1,\tau_1)\left[\int_0^\infty G(z,z_1,\tau_2)R_{F1}(\tau+\tau_1-\tau_2)\,d\tau_2\right] d\tau_1 \\
&\quad + \int_0^\infty G(z,z_2,\tau_1)\left[\int_0^\infty G(z,z_2,\tau_2 R_{F_2}(\tau+\tau_1-\tau_2)\,d\tau_2\right] d\tau_1 \\
&\quad + \int_0^\infty G(z,z_1,\tau_1)\left[\int_0^\infty G(z,z_2,\tau_2)R_{F_1F_2}(\tau+\tau_1-\tau_2)\,d\tau_2\right] d\tau_1 \\
&\quad + \int_0^\infty G(z,z_2,\tau_1)\left[\int_0^\infty G(z,z_1,\tau_2)R_{F_1F_2}(\tau+\tau_1-\tau_2)\,d\tau_2\right] d\tau_1
\end{aligned}
$$

The spectral density of the displacement $x(z,t)$ is

$$
\begin{aligned}
S_x(z,n) &= 2\int_{-\infty}^{\infty} R_x(z,\tau)\cos 2\pi n\tau\,d\tau \\
&= 2\int_{-\infty}^{\infty} R_x(z,\tau)\cos 2\pi n[(\tau+\tau_1-\tau_2)-(\tau_1-\tau_2)]d(\tau+\tau_1-\tau_2)
\end{aligned}
\qquad (14.3.21)
$$

Substitute the right-hand side of Eq. 14.3.20 for $R_x(z,\tau)$ in Eq. 14.3.21. Using the relations

$$H(z,z_1,n)\cos\theta(z,z_1,n) = \int_0^\infty G(z,z_1,\tau)\cos 2\pi n\tau\,d\tau \qquad (14.3.22a)$$

$$H(z,z_1,n)\sin\theta(z,z_1,n) = \int_0^\infty G(z,z_1,\tau)\sin 2\pi n\tau\,d\tau \qquad (14.3.22b)$$

(which are similar to Eqs. 14.2.11a, b), and

$$H(z, z_1, n)H(z, z_2, n)\cos[\theta(z, z_1, n) - \theta(z, z_2, n)]$$
$$= \int_0^\infty \int_0^\infty G(z, z_1, \tau_1)_1 G(z, z_2, \tau_2)\cos 2\pi n(\tau_1 - \tau_2)\, d\tau_1\, d\tau_2 \quad (14.3.23a)$$
$$H(z, z_1, n)H(z, z_2, n)\sin[\theta(z, z_1, n) - \theta(z, z_2, n)]$$
$$= \int_0^\infty \int_0^\infty G(z, z_1, \tau_1)_1 G(z, z_2, \tau_2)\sin 2\pi n(\tau_1 - \tau_2)\, d\tau_1\, d\tau_2 \quad (14.3.23b)$$

which are derived from Eqs. 14.3.22a, b, and following the steps that led to Eq. 14.2.16, there results

$$\begin{aligned} S_x(z, n) = {} & H^2(z, z_1, n)S_{F_1}(n) + H^2(z, z_2, n)S_{F_2}(n) \\ & + 2H(z, z_1, n)H(z, z_2, n)\{S^C_{F_1F_2}(n)\cos[\theta(z, z_1, n) - \theta(z, z_2, n)] \\ & + S^Q_{F1F2}(n)\sin[\theta(z, z_1, n) - \theta(z, z_2, n)]\} \end{aligned} \quad (14.3.24)$$

where $S^C_{F1F2}(n)$ and $S^Q_{F1F2}(n)$ are the co-spectrum and quadrature spectrum of the forces $F_1(t)$ and $F_2(t)$, defined by Eqs. A1.5.5 and A1.5.6, respectively.

14.3.5 Effect of the Correlation of the Loads upon the Magnitude of the Response

Let two stationary random loads $F_1(t) \equiv F_2(t)$ act at points of coordinates z_1 and z_2, respectively. The loads $F_1(t)$ and $F_2(t)$ are *perfectly correlated.* By definition, in this case $S^C_{F_1F_2}(n) = S_{F_1}(n)$, and $S^Q_{F_1F_2}(n) = 0$ (Eqs. A1.4.5 and A1.5.1, A1.4.4 and A1.5.5; A1.4.7 and A1.5.6). From Eq. 14.3.24,

$$\begin{aligned} S_x(z, n) = {} & \\ & \{H^2(z, z_1, n) + H^2(z, z_2, n) + 2H(z, z_1, n)H(z, z_2, n)\cos[\theta(z, z_1, n) \\ & -\theta(z, z_2, n)]\}\, S_{F_1}(n) \end{aligned} \quad (14.3.25)$$

If $z_1 = z_2$,

$$S_x(z, n) = 4H^2(z, z_1, n)S_{F_1}(n) \quad (14.3.26)$$

Consider now two loads $F_1(t)$ and $F_2(t)$ for which the cross-covariance $R_{F_1F_2}(\tau) = 0$. Then, by Eqs. A1.5.5 and A1.5.6,

$$S^C_{F_1F_2}(n) = S^Q_{F_1F_2}(n) = 0 \quad (14.3.27)$$

and, if $S_{F_1}(n) \equiv S_{F_2}(n)$,

$$S_x(z,n) = [H^2(z,z_1,n) + H^2(z,z_2,n)]S_{F_1}(n) \tag{14.3.28}$$

If $z_1 = z_2$,

$$S_x(z,n) = 2H^2(z,z_1,n)S_{F_1}(n) \tag{14.3.29}$$

The spectrum of the response to the action of the two uncorrelated loads is in this case only half as large as in the case of the perfectly correlated loads.

14.3.6 Distributed Stationary Random Loads

The spectral density of the response to a distributed stationary random load can be obtained by generalizing Eq. 14.3.24 to the case where an infinite number of elemental loads, rather than two concentrated loads, are acting on the structure. Thus, if the load is distributed over an area A, and if it is noted that in the absence of torsion the mechanical admittance functions are independent of the across-wind coordinate y, the spectral density of the along-wind fluctuations may be written as

$$S_x(z,n) = \int_A \int_A H(z,z_1,n)H(z,z_2,n)\Big\{S^C_{p'_1p'_2}(n)\cos[\theta(z,z_1,n) - \theta(z,z_2,n)] + S^Q_{p'_1p'_2}(n)\sin[\theta(z,z_1,n) - \theta(z,z_2,n)]\Big\}\,dA_1\,dA_2 \tag{14.3.30}$$

where p'_1 and p'_2 denote pressures acting at points of coordinates y_1, z_1 and y_2, z_2, respectively. It can be verified that from Eq. 14.3.30 there follows[3]

$$\begin{aligned} S_x(z,n) = {} & \frac{1}{16\pi^4}\sum_i\sum_j \frac{x_i(z)x_j(z)}{n_in_jM_iM_j}\,\frac{1}{\{[1-(n/n_i)^2]^2 + 4\zeta_i^2(n/n_i)^2\}\{[1-(n/n_j)^2]^2 + 4\zeta_j^2(n/n_j)^2\}} \\ & \times \Bigg[\Bigg\{\left[1-\left(\frac{n}{n_i}\right)^2\right]\left[1-\left(\frac{n}{n_j}\right)^2\right] + 4\zeta_i\zeta_j\frac{n}{n_i}\frac{n}{n_j}\Bigg\}\int_A\int_A x_i(z_1)x_j(z_2)S^C_{p'_1p'_2}(n)\,dA_1\,dA_2 \\ & + \Bigg\{2\zeta_j\frac{n}{n_j}\left[1-\left(\frac{n}{n_i}\right)^2\right] - 2\zeta_i\frac{n}{n_i}\left[1-\left(\frac{n}{n_j}\right)^2\right]\Bigg\}\int_A\int_A x_i(z_1)x_j(z_2)S^Q_{p'_1p'_2}(n)\,dA_1\,dA_2\Bigg] \end{aligned} \tag{14.3.31}$$

[3]By using Eqs. 14.3.14 and 14.3.15, 14.3.10 and 14.3.11, and A1.24a, b. For a derivation of Eq. 14.3.31 in terms of complex variables, see [14-1].

If the damping is small and the resonant peaks are well separated, the cross-terms in Eq. 14.3.31 become negligible, and

$$S_x(z,n) = \sum_i \frac{x_i^2 \int_A \int_A x_i(z_1)x_i(z_2)S^C_{p'_1p'_2}(n)\,dA_1\,dA_2}{16\pi^4 n_i^4 M_i^2\{[1-(n/n_i)^2]^2 + 4\zeta_i^2(n/n_i)^2\}} \tag{14.3.32}$$

14.3.7 Example: Along-Wind Response

To illustrate the application of the material presented in this chapter, we consider the along-wind response of tall buildings subjected to pressures per unit area $p(y,z,t) = \overline{p}(z) + p'(y,z,t)$ (Fig. 14.3.2).

14.3.7.1 Mean Response. The along-wind deflection induced by the mean pressures $\overline{p}(z)$ is

$$\overline{x}(z) = B\sum_i \frac{\int_0^H \overline{p}(z)x_i(z)\,dz}{4\pi^2 n_i^2 M_i} x_i(z) \tag{14.3.33}$$

Consider the case of loading induced by wind with longitudinal speed $U(z,t) = \overline{V}(z) + u'(z,t)$ normal to a building face. The sum of the mean

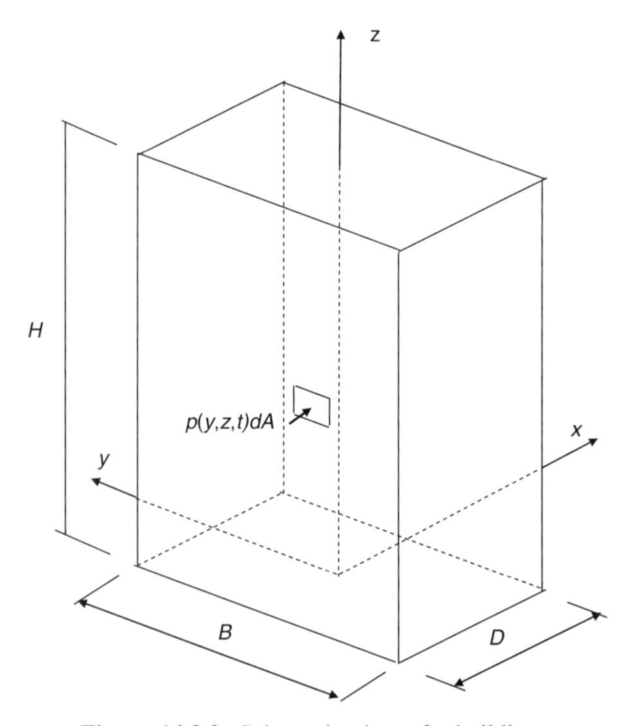

Figure 14.3.2. Schematic view of a building.

pressures $\overline{p}(z)$ acting on the windward and leeward faces of the building is then

$$\overline{p}(z) = \tfrac{1}{2}\rho(C_w + C_l)\overline{V}^2(z) \tag{14.3.34}$$

where ρ is the air density; C_w and C_l are the values, averaged over the building width B, of the mean positive pressure coefficient on the windward face and of the negative pressure coefficient on the leeward face, respectively; and $\overline{V}(z)$ is the mean wind speed at elevation z in the undisturbed oncoming flow. Equation 14.3.33 then becomes

$$\overline{x}(z) = \frac{1}{2}\rho(C_w + C_l)B \sum_i \frac{\int_0^H \overline{V}^2(z)x_i(z)\,dz}{4\pi^2 n_i^2 M_i} x_i(z) \tag{14.3.35}$$

14.3.7.2 Fluctuating Response: Deflections and Accelerations. The co-spectrum of the pressures at points M_1, M_2 of coordinates (y_1, z_1), (y_2, z_2), respectively, can be written as

$$S^C_{p'_1 p'_2} = S^{1/2}_{p'}(z_1, n)S^{1/2}_{p'}(z_2, n)\ \mathrm{Coh}(y_1, y_2, z_1, z_2, n)N(n) \tag{14.3.36}$$

where $S^{1/2}_{p'}(z_i, n)$ is the spectral density of the fluctuating pressures at point $P_i (i = 1, 2)$, Coh (y_1, y_2, z_1, z_2, n) is the coherence of pressures both acting on one of the building faces, and $N(n)$ is the coherence of pressures of which one is acting on the windward face and the other is acting on the leeward face of the building. By definition, if both P_1 and P_2 are on the same building face, $N(n) \equiv 1$. Since

$$p(z, t) \approx \tfrac{1}{2}\rho C[\overline{V}(z) + u(z)]^2 \tag{14.3.37}$$

where C (which is equal to C_w or C_l, depending upon whether the pressure acts on the windward or leeward face) is the average pressure coefficient,

$$S_{p'}(z_i, n) \approx \rho^2 C^2 \overline{V}^2(z) S_u(z_i, n) \tag{14.3.38}$$

where we used the fact that u^2 is small in relation to $2\overline{V}(z)u(z)$.

Equation 14.3.32 then becomes

$$\begin{aligned} S_x(z_i, n) \approx{}& \frac{\rho^2}{16\pi^4} \sum_i \frac{x_i^2(z)\left[C_w^2 + 2C_w C_l N(n) + C_l^2\right]}{n_i^4 M_i^2\{\left[1 - (n/n_i)^2\right]^2 + 4\zeta_i^2(n/n_i)^2\}} \\ &\times \int_0^B \int_0^B \int_0^H \int_0^H x_i(z_1)x_i(z_2)U(z_1)U(z_2) \\ &\times S_u^{1/2}(z_1)S_u^{1/2}(z_2)\ \mathrm{Coh}(y_1, y_2, z_1, z_2, n)\,dy_1\,dy_2\,dz_1\,dz_2 \end{aligned} \tag{14.3.39}$$

The coherence $\mathrm{Coh}(y_1, y_2, z_1, z_2, n)$ may be expressed as in Eq. 11.3.14 (Sect. 11.3.3.2). A simple, tentative expression for the function $N(n)$, a measure of the coherence between pressures on the windward and leeward faces, is:

$$N(n) = 1 \text{ for } n\overline{V}(z)/D < 0.2 \tag{14.3.40a}$$

$$N(n) = 0 \text{ for } n\overline{V}(z)/D \geq 0.2 \tag{14.3.40b}$$

The mean square value of the fluctuating along-wind deflection is (Eq. A1.3.5)

$$\sigma_x^2(z) = \int_0^\infty S_x(z, n)\, dn \tag{14.3.41}$$

From Eq. A1.3.6b it follows that the mean square value of the along-wind acceleration is

$$\sigma_{\ddot{x}}^2(z) = 16\pi^2 \int_0^\infty n^4 S_x(z, n)\, dn \tag{14.3.42}$$

The expected value of the largest peak deflection occurring in the time interval T is

$$x_{\max} = K_x(z)\sigma_x(z) \tag{14.3.43}$$

where (see Eqs. A1.7.12 and A1.7.6)

$$K_x(z) = [2 \ln \nu_x(z)T]^{1/2} + \frac{0.577}{[2 \ln \nu_x(z)T]^{1/2}} \tag{14.3.44}$$

$$\nu_x(z) = \left[\frac{\displaystyle\int_0^\infty n^2 S_x(z, n)\, dn}{\displaystyle\int_0^\infty S_x(z, n)\, dn} \right]^{1/2} \tag{14.3.45}$$

Similarly, the largest peak of the along-wind acceleration is, approximately,

$$\ddot{x}_{\max} = K_{\ddot{z}}(z)\sigma_{\ddot{x}}(z) \tag{14.3.46}$$

$$K_{\ddot{x}}(z) = [2 \ln \nu_{\ddot{x}}(z)T]^{1/2} + \frac{0.577}{[2 \ln \nu_{\ddot{x}}(z)T]^{1/2}} \tag{14.3.47}$$

$$\nu_{\ddot{x}}(z) = \left[\frac{\displaystyle\int_0^\infty n^6 S_x(z, n)\, dn}{\displaystyle\int_0^\infty n^4 S_x(z, n)\, dn} \right]^{1/2} \tag{14.3.48}$$

It can be shown that the mean square value of the deflection may be written, approximately, as a sum of two terms: the "background term" that entails no resonant amplification, and is due to the quasi-static effect of the fluctuating pressures, and the "resonant term," which is associated with resonant amplification due to force components with frequencies equal or close to the fundamental natural frequency of the structure, and is inversely proportional to the damping ratio [7-1, p. 212].

14.4 TIME DOMAIN SOLUTIONS FOR THREE-DIMENSIONAL DYNAMIC RESPONSE[4]

In general, the dynamic response of flexible buildings subjected to wind loads entails translational motions along their principal axes, and torsional motions. The torsional motions are due to the eccentricity of the aerodynamic loads with respect to the elastic center. If the mass center is eccentric with respect to the elastic center, additional torsional effects occur owing to inertial force effects. An example of torsional deformations induced by wind is shown in Fig. 19.3.2.

14.4.1 Dynamic Modeling

14.4.1.1 Natural Frequencies and Modes of Vibration. We now describe a procedure for obtaining modal frequencies and shapes. The total kinetic energy of the system is

$$T = \frac{1}{2}\sum_{n=1}^{n_f}(m_n\dot{x}_n^2 + m_n\dot{y}_n^2 + I_n\dot{\varphi}_n^2) \qquad (14.4.1)$$

where x_n, y_n are the displacements of the center of mass of the nth floor in the x and y direction, respectively; $\varphi(z_n)$ is the torsional rotation of the nth floor; and n_f is the total number of floors.

The total strain energy of the system is

$$V = \frac{1}{2}\{q\}^T[k]\{q\}, \qquad (14.4.2)$$

where

$$\{q\}^T = \{x_1, x_2, \ldots, x_{nf}, y_1, y_2, \ldots, y_{nf}, \phi_1, \phi_2, \ldots \phi_{nf}\}, \qquad (14.4.3)$$

$$[k] = [a]^{-1}, \qquad (14.4.4)$$

$$[a] = \begin{bmatrix} [x_{i1,x}x_{i2,x}, \ldots, x_{in_f,x}] & [x_{i1,y}x_{i2,y}, \ldots, x_{in_f,y}] & [x_{i1,\varphi}, x_{i2,\varphi}, \ldots, x_{in_f,\varphi}] \\ [y_{i1,x}y_{i2,x}, \ldots, y_{in_f,x}] & [y_{i1,y}y_{i2,y}, \ldots, y_{in_f,y}] & [y_{i1,\varphi}, y_{i2,\varphi}, \ldots, y_{in_f,\varphi}] \\ [\varphi_{i1,x}\varphi_{i2,x}, \ldots, \varphi_{in_f,x}] & [\varphi_{i1,y}\varphi_{i2,y}, \ldots, \varphi_{in_f,y}] & [\varphi_{i1,\alpha}, \varphi_{i2,\alpha}, \ldots, \varphi_{in_f,\alpha}] \end{bmatrix}, \qquad (14.4.5)$$

$[k]$ is the system's stiffness matrix, and $[a]$ is its inverse, that is, the flexibility matrix. The nine component matrices of matrix $[a]$ are represented in the right-hand side of Eq. 14.4.5 by their respective i^{th} rows. The size of matrix $[a]$ is $3n_f \times 3n_f$. The terms of matrix $[a]$ are displacements in the x or y direction or torsional rotations at the center of mass of floor $i (i = 1, 2, \ldots, n_f)$ due to

[4]Professor M. Grigoriu's contribution to this section is acknowledged with thanks.

a unit horizontal force in the x or y direction or a unit torsional moment at the center of mass of floor $j\,(j = 1, 2, \ldots, n_f)$, and can be obtained by using standard structural analysis programs. (For example, the term $y_{i2,x}$ is the y displacement of the center of mass of floor i due to a unit horizontal force acting at the center of mass of floor 2 in direction x.) In the stiffness matrix $[k]$, the restoring force $k_{ix,2y}$ represents the horizontal force in the x-direction at the center of mass of floor i induced by a unit horizontal displacement in the y direction at the center of mass of floor 2. The positions of the centers of mass and elastic centers can vary from floor to floor, and are implicitly accounted for by the terms of the matrix $[k]$.

The displacements of the mass center and the torsional rotation at the elevation z_i of the ith floor for the freely vibrating structure form a vector $\{w(t)\}$ of dimension $3n_f$. Its terms are:

$$w_1(t) = x_1(t), w_2(t) = x_2(t), \ldots, w_{n_f}(t) = x_{n_f}(t);\; w_{n_f+1}(t) = y_1(t), \ldots, w_{2n_f}(t) = y_{n_f}(t);\; w_{2n_f+1}(t) = \varphi_1(t), \ldots, w_{3n_f}(t) = \varphi_{n_f}(t) \qquad (14.4.6)$$

The equations of motion of the undamped, freely vibrating system are obtained from Lagrange's equations, and can easily be shown to be

$$[M]\{\ddot{w}(t)\} + [k]\{w\} = \{0\} \qquad (14.4.7)$$

where $[M]$ is a diagonal matrix of the floor masses and mass moments of inertia. These equations are coupled owing to the cross-terms of the matrix $[k]$. The natural frequencies of vibration $\omega_i\,(i = 1, 2, \ldots, 3n_f)$ are obtained by solving the secular equation numerically and are ordered so that $\omega_1 < \omega_2 < \ldots < \omega_{3nf}$. (The secular equation is obtained by substituting in Eq. 14.4.7 the vector $\{w \cos \omega t\}$ for the vector $\{w(t)\}$.) The corresponding modes of vibration are the $3n_f$ eigenvectors corresponding to the eigenvalues obtained by solving the secular equation, and constitute a matrix $[\mu]$ of dimension $3n_f \times 3n_f$. The eigenvector $\{\mu_j\}$ corresponds to the eigenvalue ω_j.

14.4.1.2 *Equations of Motion under Excitation by Wind from Direction θ.* Wind with mean velocity $\overline{V}_\theta(H)$ blowing from direction θ induces aerodynamic forces in the x and y directions at the elevation z_l of pressure tap l, equal to the projection on those directions of the pressures $p_{l\theta}(t)$ times the pressure tap tributary area A_l. The aerodynamic torsional moment about the mass center due to the tap l is equal to the pressures at that tap times the respective tributary area A_l, times the respective horizontal distance from the tap to the mass center at the tap level.

The equations of motion of the forced system are

$$[M]\{\ddot{w}_\theta(t)\} + [k]\{w_\theta\} = \{F_\theta(t)\} \qquad (14.4.8)$$

where $\{F_\theta(t)\}$ is the vector of the wind forces (torsional moments) $F_{x_1\theta}(t)$, $F_{x_2\theta}(t), \ldots, F_{x_{n_f}\theta}(t), F_{y_1\theta}(t), \ldots, F_{y_{n_f}\theta}(t), M_{\varphi_1\theta}(t), \ldots, M_{\varphi_{n_f}\theta}(t)$. acting at the centers of mass of floors $1, 2, \ldots, n_f$. The subscript θ indicates that the forcing and response vectors are associated with the direction θ from which the wind blows. Substituting in Eq. 14.4.8 the transformed variables

$$\{w_\theta(t)\} = [\mu]\{\xi_\theta(t)\} \tag{14.4.9}$$

where $[\mu]$ is the matrix consisting of the $3n_f$ eigenvectors $\{\mu_j\}(j = 1, 2, \ldots, 3n_f)$, yields

$$[M][\mu]\{\ddot{\xi}_\theta(t)\} + [k][\mu]\{\xi_\theta(t)\} = \{F_\theta(t)\}. \tag{14.4.10}$$

Pre-multiplying Eq. 14.4.10 by $[\mu]^T$, where the superscript T denotes transpose,

$$[\mu]^T[M][\mu]\{\ddot{\xi}_\theta(t)\} + [\mu]^T[k][\mu]\{\xi_\theta(t)\} = [\mu]^T\{F_\theta(t)\} \tag{14.4.11}$$

Owing to the orthogonality of the eigenvectors, Eq. 14.4.11, to which modal viscous damping terms proportional to the modal damping ratios ζ_m are added, can be written:

$$M_m\ddot{\xi}_{m\theta}(t) + 2M_m\omega_m\zeta_m\dot{\xi}_{m\theta}(t) + M_m\omega_m^2\xi_{m\theta}(t)$$
$$= \{[\mu]^T\{F_\theta(t)\}]\}_m(m = 1, 2 \ldots, 3n_f). \tag{14.4.12}$$

In structural dynamics the quantities M_m and the quantities in the right-hand side of Eqs. 14.4.12 are called generalized masses and generalized forces, respectively. It follows from the unforced equation of motion of the system that $M_m\omega_m^2 = [[\mu]^T[k][\mu]]_m$. Once Eqs. 14.4.12 are solved numerically, the physical coordinates $\{w_\theta(t)\}$ (i.e., the coordinates $x_1(t), x_2(t), \ldots x_{nf}(t), y_1(t), \ldots, y_{nf}(t), \varphi_1(t), \ldots, \varphi_{nf}(t)$ due to winds from direction θ, see Eq. 14.4.6) are given by Eq. 14.4.9. Accelerations $\{\ddot{w}_\theta(t)\}$ are obtained by differentiating Eq. 14.4.9 twice, the second derivatives of the generalized coordinates being known once Eq. 14.4.12 is solved.

From the requirement that the reduced frequency $nD/\overline{V}$ be the same for model and prototype, it follows that the time interval Δt between successive discrete values of the forcing $\{F_\theta(t)\}$ is

$$\Delta t = \frac{D}{D_{\text{mod}}} \frac{\overline{V}_{\text{mod}}(H_{\text{mod}})}{\overline{V}(H)} \Delta t_{\text{mod}}, \tag{14.4.13}$$

where prototype quantities are unsubscripted, and the subscript "mod" pertains to the model.

14.4.1.3 *Wind Effects on Structural Members.* One advantage of the time domain approach is its clarity and simplicity. In particular, spatial coherence is accounted for automatically by the simultaneously measured pressure data. This allows the convenient and accurate calculation of wind effects needed for the design of individual structural members. To show this, we assume that the structural system consists of beams, columns, and trusses, but the procedure we describe can be easily adapted to other systems.

The structure is subjected to time-dependent applied aerodynamic and inertial forces and torsional moments. The resultant aerodynamic and inertial forces are applied at the center of mass of each floor (or, to save computation time, of each group of several floors) and act in the building's principal directions. To their action is added the action of the torsional moments. The wind effects at a cross section of any member, or any other wind effect, can then be calculated at each time t by summing up algebraically the products of the aerodynamic and inertial forces and moments by the respective influence coefficients that convert the forces into wind effects, including bending moments, shears, and axial forces. The influence coefficients are obtained by using standard structural analysis programs.

For the design of steel structures, it is not individual internal forces that need to be considered, but rather weighted sums of internal forces governed by interaction equations (equations wherein sums of demand-to-capacity ratios are required not to exceed unity) and by wind and gravity load combinations. A similar approach is applicable to reinforced concrete structures. Within the framework of estimates based on the frequency domain approach, designers are required to consider as many as 20 separate combinations of wind effects due, for example, to motion in the principal directions and in torsion. These combinations are obtained by using a variety of generic guessed-at combination factors.

In reality, rather than being generic, the combined wind effects depend on the known aerodynamic and inertial loading, and on the influence coefficients specifically applicable to the wind effects of concern. In the past, this dependence could not be accounted for in a rigorous manner owing to limited computational capabilities, and because the aerodynamic information obtained in the wind tunnel was incomplete and was not recorded for later use by the structural engineer or other parties. Additional details on time domain approaches are presented in Chapter 19.

CHAPTER 15

AEROELASTICITY

15.1 INTRODUCTION

Flow-induced structural motions can modify the flow in a manner that, in turn, affects the structural motions. Structural motions that modify the aerodynamic action of the flow on the structure are called *self-excited*, and the behavior associated with them is termed *aeroelastic*. The flutter of the Brighton Chain Pier Bridge (termed in the 1800s "undulation") (Fig. 15.1.1) and, more than one century later, the flutter of the original Tacoma Narrows Bridge (Fig. 15.1.2) are notorious examples of aeroelastic behavior. Tall chimneys and buildings can also respond aeroelastically.

To describe the interaction between aerodynamic forces and the structural motions they induce, it is in principle necessary to solve the full equations of motion describing the flow, with time-dependent boundary conditions imposed by the moving structure. For bluff bodies in turbulent flow, this problem defies analytical capabilities and, in spite of continual progress, it remains difficult to solve dependably by computational fluid dynamics methods, especially for structures with complex shapes. Therefore, the aeroelastic characterization of civil engineering structures relies largely on empirical modeling and laboratory testing. The applicability to the prototype of laboratory test results and associated empirical models needs to be assessed carefully, owing to the violation of the Reynolds number similarity criterion. However, for carefully modeled structures, aeroelastic test results are generally assumed to yield reasonably realistic results.

The flow-body interaction is destabilizing or stabilizing if an energy transfer occurs from the flow into the body or from the body into the flow, respectively.

SKETCH Shewing the manner in which the 3rd Span of the CHAIN PIER at BRIGHTON undulated
just before it gave way in a storm on the 29th of November 1836.

255 feet

SKETCH Shewing the appearance of the 3rd Span after it gave way.

Figure 15.1.1. Brighton chain pier failure, 1836. (From J. Russel, "On the Vibration of Suspension Bridges and Other Structures, and the Means of Preventing Injury from This Cause," *Transactions of the Royal Scottish Society of Arts*, 1841).

Figure 15.1.2. Flutter of Tacoma Narrows suspension bridge, 1940 (from F. B. Farquarson, ed., *Aerodynamic Stability of Suspension Bridges*, Part 1, Bulletin 116, University of Washington Engineering Experimental Station, Seattle, WA, 1949–1954).

By analogy with viscous damping forces, which tend to reduce a body's oscillatory energy, aeroelastic forces are called *negative aerodynamic damping* forces if they tend to *increase* the oscillatory energy, that is, if they are destabilizing. Stabilizing aerodynamic forces are called positive aerodynamic damping forces.

The purpose of this section is to provide an introduction to basic aeroelastic phenomena of interest in the design of flexible structures. Section 15.2 discusses aeroelasticity phenomena associated with vortex lock-in. Section 15.3 is devoted to galloping. Galloping is a relatively simple aeroelastic behavior and therefore provides a useful starting point for understanding the basic flow-structure interactions that produce flutter. Fundamentally, galloping occurs in bodies with certain cross-sectional shapes because, as the body moves, the angle of attack of the relative flow velocity with respect to the body changes, causing flow asymmetries that excite body motions. In this sense, the aeroelastic phenomena that govern flutter motion are similar. They are more complicated, however, for two reasons. First, they involve effects due to vorticity, which entail dependencies on reduced frequencies, and are absent in the galloping case. Second, they are associated with motions in several degrees of freedom, instead of just one degree of freedom. Section 15.4 considers flutter, as well as buffeting in the presence of flutter effects. For material on aeroelastic testing, see Chapter 19. For additional material on the aeroelasticity of civil engineering structures, see [7-1, 15-1, 15-2].

15.2 VORTEX-INDUCED OSCILLATIONS

15.2.1 Vortex-Induced Lock-in

The shedding of vortices in the wake of a body gives rise to fluctuating *lift forces*. If the body is flexible, or if it has elastic supports, it will experience motions due to aerodynamic forces and, in particular, to the fluctuating lift force. As long as the motions are sufficiently small, they do not affect the vortex shedding, and Eq. 13.2.6 remains valid. If the vortex-shedding frequency n_s, and therefore the frequency of the associated lift force, is equal to a natural frequency of vibration of the body n_1, then resonant amplification can occur. Experiments show that this is the case not only at the flow speed $n_s D/S$ (Eq. 13.2.6), but also at any speed V within an interval $n_s D/S - \Delta V < V < n_s D/S + \Delta V$, where $\Delta V/V$ is of the order of a few percent and depends upon cross-sectional shape and the mechanical damping. Within that interval, the vortex-shedding frequency no longer conforms to Eq. 13.2.6, but aligns itself to the body's frequency n_s.

This is an aeroelastic effect: While the flow affects the body motion, the body motion in turn affects the flow insofar as it produces *lock-in*, that is, a synchronization of the vortex-shedding frequency with the frequency of vibration of the body.

15.2.2 Vortex-Induced Oscillations and Lift Force Along-Span Correlations

A second aeroelastic effect due to a cylinder's vibration is of interest in practice. If the cylinder is infinitely rigid, vortex-induced lift forces per unit span at different stations along the cylinder are imperfectly correlated—that is, they are not perfectly in phase with each other. However, if the cylinder oscillates under excitation by vortex-induced lift forces, the oscillations bring about an increase in the correlation among those forces. This in turn results in increased overall lift forces and oscillation amplitudes.

15.2.3 Modeling of Vortex-Induced Oscillations as a Single-Degree-Of-Freedom Motion

A variety of vortex-induced oscillation models are available in which the aeroelastic forces depend upon adjustable parameters fitted to match experimental results. By construction, those models provide a reasonable description of the observed aeroelastic motions. However, the empirical models may not be valid as a motion predictor for conditions other than those of the experiments.

We note Scanlan's simple but useful single-degree-of-freedom model applicable to long, elastically-supported cylinders in uniform smooth flow [7-1]:

$$m\left[\ddot{y} + 2\pi\zeta n_1\dot{y} + (2\pi n_1)^2 y\right] = \frac{1}{2}\rho V^2 D\left[Y_1(K)\left(1 - \varepsilon\frac{y^2}{D^2}\right)\right]\frac{\dot{y}}{V}, \quad (15.2.1)$$

where m is the body mass per unit length, ζ is the damping ratio, n_1 is the frequency of vibration of the body, D is the cylinder's diameter, V is the flow velocity, ρ is the density of the fluid, $K = 2\pi Dn/V$, and the vortex-shedding frequency n satisfies the Strouhal relation $n = SV/D$; Y_1 and ε are adjustable parameters that must be fitted to experimental results. This model allows for negative and positive aerodynamic damping at low and high body displacements, respectively, that is, for the aeroelastic transfer of energy from the flow to the body and from the body to the flow.

15.3 GALLOPING

Galloping is a large-amplitude aeroelastic oscillation (one to ten or more cross-sectional dimensions of the body) that can be experienced by cylinders or prisms with certain types of cross section (e.g., rectangular section, D-section, ice-laden power cables). The oscillations occur in a plane normal to the oncoming flow velocity, at frequencies that are much lower than the vortex-shedding frequencies for the same sections. The flow speeds that cause galloping motions are typically considerably larger than those that cause vortex lock-in. Flow reattachment, which is present at vortex lock-in and flutter, does not occur in the galloping case; fully separated flows are thus a feature of galloping motions, and result in the absence of vortex-induced pressures on the body.

Fundamental to the galloping phenomenon is the fact that the angle of attack of the relative flow velocity with respect to the body changes as the body experiences an incipient motion away from its position of equilibrium. This changed relative velocity creates in bodies with certain cross-sectional shapes asymmetrical pressure distributions that enhance that incipient motion, rather than suppress it, as would be the case if the body were aeroelastically stable.

15.3.1 Glauert-Den Hartog Necessary Condition for Galloping Motion

We consider a square cylinder immersed in a flow with velocity V in the x-direction (Fig. 15.3.1). The positive y-coordinate in Fig. 15.3.1 is *downwards*. It is assumed that the body is elastically restrained and mechanically damped in the direction normal to the x-axis. Let a small perturbation cause the cylinder to move *downward* from its position of equilibrium. Associated with the motion is a downward velocity $\dot{y}$ of the body with respect to the flow. The relative vertical velocity of the flow with respect to the body is then $v = -\dot{y}$ (*upward*), and the total relative velocity, V_r, of the flow with respect to the body is the vector sum of the velocities V and $-\dot{y}$ (Fig. 15.3.1). The flow has an effective angle of attack with respect to the body $\alpha \approx -\dot{y}/V$, that is, *the relative velocity V_r makes an angle α with the axis x*, and the flow around the body is therefore *asymmetrical* with respect to the x-axis. As indicated earlier, if the aerodynamic properties of the body are such that, for small α, the force induced on the body by the relative velocity V_r tends to push the downward-moving body further downward, rather than bringing it back to its original undeformed position (i.e., if V_r causes the average pressure on the upper side of the body to be larger than on the lower side), then the body is unstable—it will experience galloping motion. We now obtain the necessary condition for this to be the case.

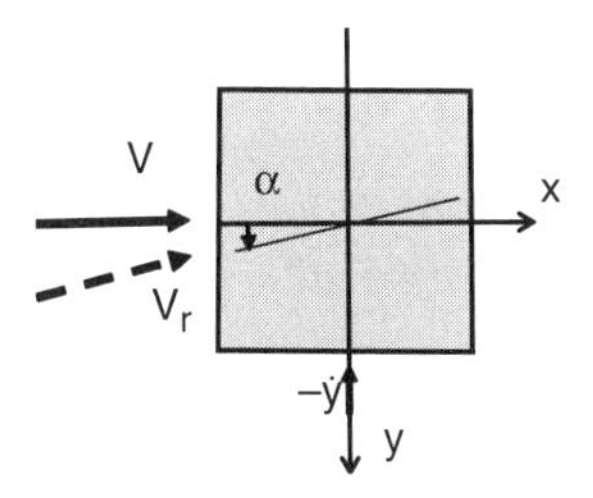

Figure 15.3.1. Flow velocity V and relative flow velocity V_r with respect to a square cylinder moving downwards with velocity $\dot{y}$.

The drag along, and the lift force normal to, the direction α may be written as

$$D(\alpha) = \frac{1}{2}\rho V_r^2 B C_D(\alpha), \tag{15.3.1}$$

$$L(\alpha) = \frac{1}{2}\rho V_r^2 B C_L(\alpha) \tag{15.3.2}$$

The projection of these components on the direction y is

$$F_y(\alpha) = -D(\alpha)\sin\alpha - L(\alpha)\cos\alpha. \tag{15.3.3}$$

We now write $F_y(\alpha)$ in the alternative form

$$F_y(\alpha) = \frac{1}{2}\rho V^2 B C_{F_y}(\alpha) \tag{15.3.4}$$

Since $V = V_r \cos\alpha$, it follows from Eqs. 15.3.1 to 15.3.4 that

$$C_{F_y}(\alpha) = \frac{-[C_L(\alpha) + C_D(\alpha)\tan\alpha]}{\cos\alpha}. \tag{15.3.5}$$

The differentiation of Eq. 15.3.5 with respect to α and the evaluation of the result at $\alpha = 0$ yield

$$\left.\frac{dC_{F_y}}{d\alpha}\right|_{\alpha=0} = -\left(\frac{dC_L}{d\alpha} + C_D\right)_{\alpha=0}. \tag{15.3.6}$$

For $\alpha = 0$, $d\alpha = \dot{y}/V$, and

$$F_y \approx \left.\frac{dF_y}{d\alpha}\right|_{\alpha=0} d\alpha, \tag{15.3.7}$$

$$F_y \approx \left.\frac{dF_y}{d\alpha}\right|_{\alpha=0} \frac{\dot{y}}{V} \tag{15.3.8}$$

$$= -\frac{1}{2}\rho V^2 B\left(\frac{dC_L}{d\alpha} + C_D\right)_{\alpha=0} \frac{\dot{y}}{V} \tag{15.3.9}$$

The equation of motion of the body in the y-direction is

$$m\left[\ddot{y} + 2\pi\zeta n_1 \dot{y} + (2\pi n_1)^2 y\right] = F_y$$
$$= -\frac{1}{2}\rho V^2 B\left(\frac{dC_L}{d\alpha} + C_D\right)\frac{\dot{y}}{V} \tag{15.3.10}$$

The sum of the mechanical and aerodynamic damping affecting the system is

$$2m\omega_1\zeta + \frac{1}{2}\rho VB\left(\frac{dC_L}{d\alpha} + C_D\right)_{\alpha=0} = d \qquad (15.3.11)$$

The second term in the left-hand side of Eq. 15.3.11 can be interpreted as an aerodynamic damping term. For galloping to be possible, the aerodynamic damping must be negative—that is, the quantity between parentheses in Eq. 15.3.11 (in which C_D is positive) must be negative. This condition is known as the *Glauert-Den Hartog necessary condition for galloping*. If the mechanical damping is vanishingly small, galloping can in theory occur for any flow velocity, however small. It follows from the necessary condition for galloping that circular cylinders cannot gallop, since for such bodies $\frac{dC_L}{d\alpha} = 0$. A plot for which the second term in the left-hand side of Eq. 15.3.11 is negative is shown in Fig. 15.3.2.

Tests have shown that the derivatives $\frac{dC_{F_y}}{d\alpha}$ do not depend on the frequency of the body motion and can be obtained from aerodynamic force measurements on the *fixed* body. $\frac{dC_{F_y}}{d\alpha}$ are called *steady state aerodynamic lift coefficient derivatives* or, for short, steady state aerodynamic derivatives. We note in the next section that, in the case of flutter, the aeroelastic behavior is characterized by quantities of a similar nature, called flutter aerodynamic derivatives, which, unlike the steady state derivatives that characterize galloping motion, depend upon the oscillation frequency.

15.3.2 Modeling of Galloping Motion

To describe the galloping motion requires the development of the lift coefficient C_{F_y} in powers of $\frac{\dot{y}}{V}$. The following expression was proposed in [15-3]:

$$C_{F_y}(\alpha) = A_1\left(\frac{\dot{y}}{V}\right) - A_2\left(\frac{\dot{y}}{V}\right)^2\frac{\dot{y}}{|\dot{y}|} - A_3\left(\frac{\dot{y}}{V}\right)^3 + A_5\left(\frac{\dot{y}}{V}\right)^5 - A_7\left(\frac{\dot{y}}{V}\right)^7 \qquad (15.3.12)$$

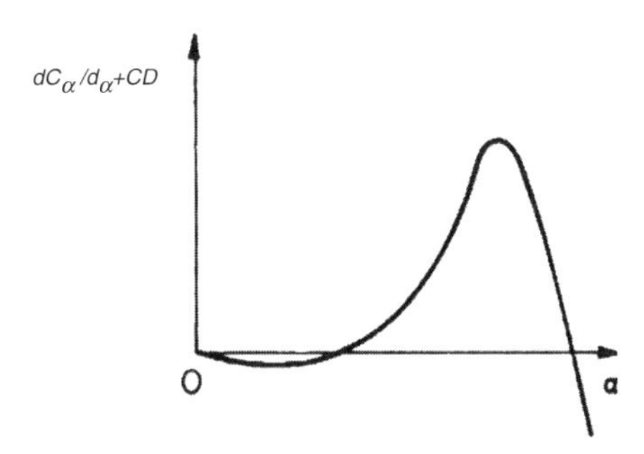

Figure 15.3.2. Dependence of $dC_\alpha/d\alpha + C_D$ on α.

If the dependence of C_D and C_L upon α is known, C_{F_y} (Eq. 15.3.5) is plotted against $\tan\alpha = \frac{\dot{y}}{V}$. The coefficients in Eq. 15.3.12 can then be estimated on the basis of this plot, for example by using a least squares technique. The solution of the equation of motion can be of three types, depending upon whether the coefficient A_1 is less than, equal to, or larger than zero [7-1].

Efforts are being made to model analytically aeroelastic across-wind motions of buildings at velocities higher than those at which vortex-induced resonance occurs. It was suggested in [15-4] that such motions may in some instances be of the galloping type, but it is not clear that this is indeed the case.

15.4 FLUTTER

Flutter is an aeroelastic phenomenon that occurs in flexible bodies, such as bridge decks, with relatively flat shapes in plan. It involves oscillations with amplitudes that grow in time and can result in catastrophic structural failure. Like other aeroelastic phenomena, flutter entails the solution of equations of motion involving inertial, mechanical damping, elastic restraint, and aerodynamic forces (including self-excited forces) that depend upon the ambient flow and the shape and motion of the body.

Assume that the mechanical damping is negligible. The motion of the body is aeroelastically *stable* if, following a small perturbation away from its position of equilibrium, the body will revert to that position owing to stabilizing self-excited forces associated with the perturbation. As the flow velocity increases, the aerodynamic forces acting on the body change, and for a certain critical value of the flow velocity the self-excited forces cause the body to be neutrally stable. For velocities larger than the critical velocity, the oscillations initiated by a small perturbation from the position of equilibrium will grow in time. The self-excited forces that cause these growing oscillations can be viewed as producing a *negative aeroelastic damping* effect.

The main difficulty in solving the flutter problem for bridges is the development of expressions for the self-excited forces. For thin airfoil flutter in incompressible flow, self-excited forces due to small oscillations have been derived from basic aerodynamic theory. However, for bridge sections, the airfoil solutions are in general not applicable.

To date, perhaps the most influential contribution to solving the flutter problem for bridges is a simple conceptual framework wherein the self-excited forces due to relatively small bridge deck oscillations can be characterized by fundamental functions called flutter aerodynamic derivatives. As pointed out earlier, in the galloping case the self-excited forces depend on the steady state derivatives $\frac{dC_{F_y}}{d\alpha}$, which may be obtained from measurements on the *fixed* body; in contrast, flutter derivatives depend upon oscillation frequency, and must be obtained from measurements on the *oscillating* body.

Although it is accompanied at all times by vortex shedding with frequency equal to the flutter frequency, *flutter is a phenomenon distinct from vortex-induced oscillation*. The latter entails aeroelastic flow-structure interactions only for flow velocities at which the frequency of the vortex shedding is equal or close to the structure's natural frequency. For velocities higher than those at which lock-in occurs, the oscillations are much weaker than at lock-in. In contrast, for velocities higher than those at which flutter sets in, the strength of the oscillations increases monotonically with velocity.

Some aerodynamicists have expressed the belief that the across-wind oscillations of the John Hancock tower in Boston may have been induced by flutter, rather than by vortex shedding. No documentation supporting this belief appears to exist, however, and flutter typically does not occur in buildings. Nevertheless, the flutter phenomenon as it affects bluff bodies is of potential interest to all structural engineers, and therefore warrants an introductory discussion in this chapter. In Sect. 15.4.1, we consider two-dimensional bridge deck behavior in smooth flow. Section 15.4.2 is concerned, in a two-dimensional context, with bridge deck flutter in turbulent flow and the consequent buffeting of the bridge deck.

15.4.1 Formulation of the Two-Dimensional Bridge Flutter Problem in Smooth Flow

The dependence of flutter derivatives upon the oscillation frequency n of the fluttering body can be expressed in terms of the nondimensional *reduced frequency*

$$K = 2\pi Bn/V,$$

where B is the width of the deck, and V is the mean wind flow velocity. If the horizontal displacement p of the deck is also taken into account, the equations of motion of a two-dimensional section of a symmetrical bridge deck with linear viscous damping and elastic restoring forces in smooth flow can be written as

$$m\ddot{h} + c_h\dot{h} + C_h h = L_h \tag{15.4.1a}$$

$$I\ddot{\alpha} + c_\alpha\dot{\alpha} + C_\alpha\alpha = M_\alpha \tag{15.4.1b}$$

$$m\ddot{p} + c_p\dot{p} + C_p p = D_p \tag{15.4.1c}$$

where h, α, and p are the vertical displacement, torsional angle, and horizontal displacement, respectively (Fig. 15.4.1). A unit span is acted upon by the aerodynamic lift L_h, moment M_α, and drag D_p; has mass m; mass moment of inertia I; vertical, torsional, and horizontal restoring forces with stiffness C_h, C_α, and C_p, respectively; and viscous damping coefficients c_h, c_α, and c_p.

It was seen that a galloping body experiences a single-degree-of-freedom motion, and that, for small displacements, the aeroelastic force acting on the body is linear with respect to the time rate of change of the across-wind

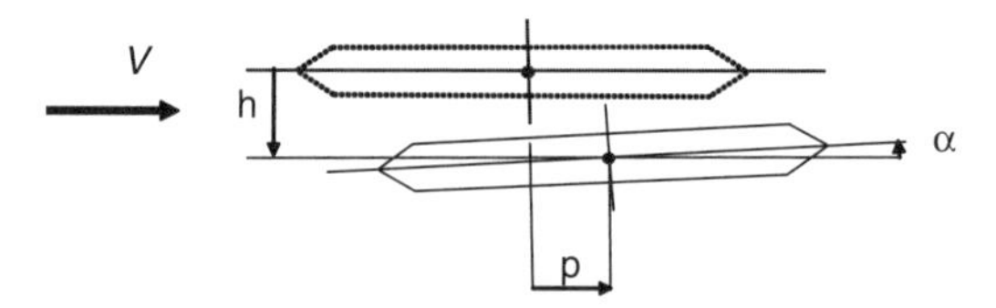

Figure 15.4.1. Notations.

displacement y, the proportionality factor being a function of aerodynamic origin (Eq. 15.3.9). Flutter entails motions with three degrees of freedom (h, α, and p), and the expressions for the aeroelastic forces acting on the body are therefore lengthier than in the galloping case, although conceptually they are related. Indeed, for small displacements, the aeroelastic forces can be written as sums of terms that, like their galloping counterparts, are linear with respect to the rates of change of h, α, and p, the factors of proportionality being also functions of aerodynamic origin. However, unlike in the case of galloping, terms proportional to h, α, and p come into play as well. (For example, it is intuitively clear that the aerodynamic effects on the body of Fig. 15.4.1 depend upon the angle of attack α.) In mathematical terms, the expressions for the aeroelastic forces can therefore be written as sums of terms as follows:

$$L_h = \frac{1}{2}\rho V^2 B\left[KH_1^*(K)\frac{\dot{h}}{V} + KH_2^*(K)\frac{B\dot{\alpha}}{V} + K^2H_3^*(K)\alpha + K^2H_4^*(K)\frac{h}{B} + KH_5^*(K)\frac{\dot{p}}{V} + K^2H_6^*(K)\frac{p}{B}\right] \tag{15.4.2a}$$

$$M_\alpha = \frac{1}{2}\rho V^2 B\left[KA_1^*(K)\frac{\dot{h}}{V} + KA_2^*(K)\frac{B\dot{\alpha}}{V} + K^2A_3^*(K)\alpha + K^2A_4^*(K)\frac{h}{B} + KA_5^*(K)\frac{\dot{p}}{V} + K^2A_6^*(K)\frac{p}{B}\right] \tag{15.4.2b}$$

$$D_p = \frac{1}{2}\rho V^2 B\left[KP_1^*(K)\frac{\dot{p}}{V} + KP_2^*(K)\frac{B\dot{\alpha}}{V} + K^2P_3^*(K)\alpha + K^2P_4^*(K)\frac{p}{B} + KP_5^*(K)\frac{\dot{h}}{V} + K^2P_6^*(K)\frac{h}{B}\right] \tag{15.4.2c}$$

Terms proportional to $\ddot{h}$, $\ddot{\alpha}$ and $\ddot{p}$ (i.e., so-called added mass terms, reflecting the forces due to the body motion that result in fluid accelerations around the body) do not appear in Eq. 15.4.2, as these terms are negligible in wind engineering applications. The role of the terms in h and p is to account for changes in the frequency of vibration of the body due to aeroelastic effects, while the terms in α reflect the role of the angle of attack noted earlier. The quantities $\dot{h}/U$ and $B\dot{\alpha}/V$ are effective angles of attack (e.g., the ratio $\dot{h}/V$ has the same significance as in the case of galloping, i.e., it represents the angle of attack of the relative velocity of the flow with respect to the moving body). Those quantities are nondimensional. The coefficients H_i^*, A_i^*, and P_i^* are known as *Scanlan flutter derivatives*,[1] and are also nondimensional. The original form of Eqs. 15.4.2a–c was inspired by Theodorsen's classical mathematical derivations of aeroelastic forces and moments on an airfoil [15-6], of which it is an empirical counterpart.

To summarize, each term in Eqs. 15.4.2 can be viewed as similar in form to terms of the type

$$L = \frac{1}{2}\rho V^2 B C_L = \frac{1}{2}\rho V^2 B \frac{dC_L}{d\alpha}\alpha \tag{15.4.3}$$

for small angles of attack α. Quantities such as KH_i^* and $K^2A_i^*$ are therefore analogous to aerodynamic lift coefficient derivatives $\frac{dC_{F_y}}{d\alpha}$ that arise in the galloping case. Unlike in the galloping case, where, owing to the absence of vortex-induced pressures on the body, the derivatives can be obtained experimentally from static tests (that is, tests in which the body is at rest), for the flutter case the coefficients of the displacements and their time rate of change must be obtained experimentally from measurements on the oscillating body, which, owing to its elongated shape, is affected by vortex-induced pressures. For this reason, those coefficients are called *motional* aerodynamic derivatives, which go over into *steady state* aerodynamic derivatives for $K \to 0$.

The solution of the flutter equations can be obtained if plots of the flutter derivatives H_i^*, A_i^*, and P_i^* are available from measurements as functions of K. It is then assumed that the expressions for h, α, and p are proportional to $e^{i\omega t}$; these expressions are inserted into Eqs. 15.4.1a–c, and the determinant of the amplitudes of h, α, and p is set to zero.

For each value of K, a complex equation in n is obtained. The flutter velocity is the velocity

$$V_c = \frac{2\pi B n_c}{K_c}. \tag{15.4.4}$$

to which there corresponds a value of K, denoted by K_c, that yields a real solution $n = n_c$. In Eqs. 15.4.2a–c, the terms containing first derivatives of the displacements are measures of *aerodynamic damping*. If, among these terms, only those associated with the coefficients H_1^*, A_2^*, and P_1^* are significant, the total (structural plus aerodynamic) damping can be written as

$$c_h - {}^1\!/\!_2\rho U^2 BKH_1^*, \quad c_\alpha - {}^1\!/\!_2\rho U^2 BKA_2^*, \quad c_p - {}^1\!/\!_2\rho U^2 BKP_1^*, \qquad (15.4.5\text{a, b, c})$$

for the vertical, torsional, and horizontal degree of freedom, respectively.

[1]Equations 15.4.2a–c are formulated in terms of real variables, viewed by many practitioners to be best suited for structural engineering purposes. An alternative approach wherein the aeroelastic forces and the displacements they induce in the bridge are expressed in terms of complex variables is preferred by others, insofar as it may offer insights into phase relationships among various aeroelastic forces and displacements.

The original Tacoma Narrows bridge (Fig. 15.1.2) had negligible H_1^* values for all K, meaning that the total damping (Eq. 14.4.5a) for motion in the h-direction was positive, thus precluding flutter in the vertical degree of freedom. However, A_2^* was positive for $K > 0.16$ or so. As the effect of horizontal deck motions appears to have been negligible, for sufficiently high flow velocity the total damping given by Eq. 15.4.5b was negative, so flutter involving only the torsional degree of freedom occurred in wind with mean velocity of about 20 m/s. The bridge's susceptibility to flutter was due to the use of an "H" section (the horizontal line in the "H" representing the deck, and the vertical lines representing the girders supporting it). Owing to their inherent instability, "H" bridge sections are no longer used.

15.4.2 Bridge Section Response to Turbulent Wind in the Presence of Aeroelastic Effects

The aerodynamic forces induced on a bridge deck by turbulent wind are due to: (1) aeroelastic forces associated with flutter derivatives, and (2) buffeting forces induced by turbulent flow.

The expressions for the aeroelastic forces have the same form as for the smooth flow case (Eqs. 15.4.2). However, the aerodynamic coefficients H_i^*, A_i^*, P_i^* should be obtained from measurements in turbulent flow, since turbulence may affect the aerodynamics of the bridge deck by changing the configuration of the separation layers and the position of reattachment points. Through complex aerodynamic mechanisms, turbulence can affect the flutter derivatives and, therefore, the flutter velocity—in many instances favorably, but possibly also unfavorably. The buffeting forces per unit span may be written as follows:

$$\begin{aligned} L_b &= \frac{1}{2}\rho V^2 B\left[2C_L\frac{u(t)}{V} + \left(\frac{dC_L}{d\alpha} + C_D\right)\frac{w(t)}{V}\right] \\ M_b &= \frac{1}{2}\rho V^2 B\left[2C_M\frac{u(t)}{V} + \left(\frac{dC_M}{d\alpha}\right)\frac{w(t)}{V}\right] \\ D_b &= \frac{1}{2}\rho V^2 B\left[2C_D\frac{u(t)}{V}\right]. \end{aligned} \qquad (15.4.6a, b, c)$$

For example, Eq. 15.4.6c is derived from the expression for the total (mean plus fluctuating) drag force D, where

$$D = \overline{D} + D_b = \frac{1}{2}\rho C_D B\,[V + u(t)]^2, \qquad (15.4.7)$$

V is the mean flow velocity, $u(t)$ is the along-wind (longitudinal) component of the turbulent velocity fluctuation at time t, the mean drag force is defined as

$$\overline{D} = \frac{1}{2}\rho C_D B V^2, \qquad (15.4.8)$$

and the drag coefficient C_D is measured in turbulent flow. For the two-dimensional case, the solution of the buffeting problem in the presence of aeroelastic effects is obtained from Eqs. 15.4.1a–c, in which the right-hand sides consist of the sums $L_h + L_b$, $M_\alpha + M_b$, $D_p + D$, respectively, rather than just L_h, M_α, and D_p. In solving the equations, the angle of attack at each time step must be taken into account.

Even though the two-dimensional case can provide useful insights into the behavior of a bridge, to be useful in applications to actual bridges the solution must be obtained for the three-dimensional case, in which the bridge displacement and the aerodynamic forces are functions of position along the span [7-1, 15-2, 15-5].

For an interesting application to an aeroelastic study of the Golden Gate Bridge, see Sect. 13.1.4 of [7-1].

CHAPTER 16

STRUCTURAL RELIABILITY UNDER WIND LOADING

16.1 INTRODUCTION

Structures are designed so that specified *limit states* have sufficiently small *probabilities of being exceeded*—or, equivalently, sufficiently long *mean recurrence intervals* (MRIs). Examples of such limit states are:

- A demand-to-capacity index (DCI)[1] equal to unity.
- A specified inter-story drift, dependent on type of cladding and/or partitions, and on insurance considerations.
- A specified peak or r.m.s. of the top-floor accelerations.

Other limit states may be specified, depending upon the building, its contents, and its functions. Associated with each limit state is a minimum allowable MRI. The more severe the consequences of exceeding the limit state, the larger are the minimum allowable MRIs.

Building codes specify limit states related to life safety (i.e., to building integrity, the loss of which might jeopardize human lives), and associated

[1]For members experiencing only one type of internal force (e.g., tension), the DCI is equal to the demand-to-capacity ratio, that is, to the ratio between the internal force and a measure of the nominal capacity of the member to resist it. For steel member cross sections experiencing two or more types of internal force (e.g., column cross sections experiencing a compressive axial force, a bending moment about one of the cross section's principal axes, and a bending moment about the second principal axis), the DCI is the *sum* of the ratios between the internal forces and the respective measures of nominal capacities, that is, the DCI is identical to the left-hand side of the pertinent *design interaction equation*. A similar definition of the DCI applies to reinforced concrete members.

MRIs of the demand that causes them. For example, the ASCE Standard 7-10 specifies a nominal 700-year MRI for the strength design of typical structures (as opposed to such critical structures as, e.g., hospitals, fire stations, or structures whose failure entails significant danger of loss of human life, for which the Standard specifies a nominal 1,700-year MRI). This increase is not based on explicit structural reliability or other calculations, but rather on professional consensus based on experience, intuition, or belief.

Limit states and associated MRIs not related to life safety may be established by agreement among the owner, the designer, and the insurer, although some nonstructural limit states may require compliance with regulatory requirements. The concept of *performance-based design* has been developed with a view to allowing the development and implementation of performance requirements agreed upon by various stakeholders or imposed by regulation. For an example, consider the hypothetical case of a hospital that collapsed during an earthquake and was rebuilt to conform to upgraded code requirements. During a subsequent earthquake, the building performed well from a structural viewpoint. However, the seismic motions compromised the piping system, rendering the hospital unusable. Performance-based design is largely meant to define limit states and the associated MRIs aimed to prevent nonstructural failures, which can have serious effects on functionality, community resilience, and, via insurance or the lack of it, the financial health of individuals, institutions, or communities.

The primary goal of structural reliability is to help develop design criteria ensuring that risks of structural failure are acceptably small (risk being defined as the probability of occurrence of an undesirable event times the consequence of the occurrence of that event). In the early phases of its development, it was believed that structural reliability could accomplish this goal for any structural system by performing the following steps: (1) clear and unambiguous definition of failure limit states, (2) calculation of probabilities of exceedance of those limit states, and (3) specification of the maximum acceptable values of those probabilities. The clear definition of limit states can be a difficult task, however, as are calculations of probabilities of failure. For structures consisting of systems, such calculations are rarely possible in practice. In addition, probability distribution tails, which determine failure probabilities, are largely unknown. Finally, the specification of the acceptable failure probability for any limit state can be a complex economic or political issue that exceeds the bounds of structural engineering.

Improved forecasting capabilities, which allow sufficient time for evacuation, have resulted in massively reduced loss of life due to hurricanes, particularly in developed countries. This is not the case for earthquakes, which strike suddenly, and for which the motivation to perform research into failure limit states has been far stronger than for winds. The ASCE 7 Standard specifies seismic design criteria based on nonlinear analyses, consistent with the requirement that the structure not collapse under a Maximum Considered Earthquake with a 2,500-year MRI. No similar design criteria have been developed, and little research into nonlinear structural behavior has been performed for structures subjected to wind loads. What constitutes ultimate post-elastic structural behavior under wind loads? How different are the probabilities of ultimate limit states defined as incipient collapse of structural systems from those of limit state defined by first yield in structural members—that is, what is the effective strength reserve of a structure designed for the latter limit state? Except in a few particular cases, no quantitative answers to these questions are available for structures under wind loads.

In view of the insuperable difficulties inherent in the original goals of structural reliability, the discipline has settled for more modest goals. Under the demand inherent in the wind and gravity loads affected by their respective load factors, each member cross section must experience demand-to-capacity indexes (DCIs) not greater than unity. Recall that DCIs are sums (or other functions) of demand-to-capacity ratios, the capacity being affected by resistance factors smaller than unity. This approach is called Load and Resistance Factor Design (LRFD). In the ASCE 7-05 Standard, the MRI $\overline{N}$ of the wind load was typically assumed to be 50 years, and the wind load factor was assumed to be 1.6. In the ASCE 7-10 Standard, for formal rather than substantive reasons, the wind load factor is specified to be 1.0, but $\overline{N}$ is assumed to be 700 years, 1,700 years, or 300 years, depending upon the risk category of the structure, so that the resulting wind effects are similar to those of the ASCE 7-05 Standard.

Past experience with wind effects on buildings suggests that the member-by-member approach just described is safe,[2] even though it does not provide any explicit indication of the structural system's incipient collapse limit state and the probability of exceedance of that state. However, complacency is not in order, as is shown by the example of criteria for design against the 100-year storm surge, proven by Hurricane Katrina effects to be inadequate even though they were considered to be based on past experience. As is shown in Sect. 16.2, the LRFD approach is not rigorous probabilistically; therefore, its applicability to novel types of structure needs to be considered with caution.

The following sections are concerned with topics associated with the LRFD approach, including the limitations of this approach (Sect. 16.2), the dependence of design MRIs on wind directionality (Sect. 16.3), structural strength reserve (Sect. 16.4), design MRIs for multi-hazard regions (Sect. 16.5), individual uncertainties and overall uncertainty in the estimation of wind effects (Sect. 16.6), and the calibration of design MRIs for structures experiencing significant dynamic effects or for which errors in the estimation of extreme wind effects are significantly larger than the typical errors accounted for in the ASCE 7-10 Standard (Sect. 16.7).

[2]The degree to which is the case depends upon the structure's strength reserve, see Sect. 16.4.

16.2 FIRST-ORDER SECOND-MOMENT APPROACH, LOAD AND RESISTANCE FACTORS

The so-called first-order second-moment (FOSM) approach considered in this section was developed, primarily in the 1970s, as a substitute for a structural reliability theory based on explicit estimation of failure probabilities.

16.2.1 Failure Region, Safe Region, and Failure Boundary

Consider a member subjected to a load Q, and let the load that induces a given limit state (e.g., first yield) be denoted by R. Both Q and R are random variables that define the *load space*. Failure occurs for any pair of values for which

$$R - Q < 0 \tag{16.2.1}$$

The *safe region* is defined by the inequality

$$R - Q > 0 \tag{16.2.2}$$

The *failure boundary* separates the failure and the safe regions, and is defined by the relation

$$R - Q = 0 \tag{16.2.3}$$

Relations similar to Eqs. 16.2.1–16.2.3 hold in the *load effect space*, defined by the variables Q_e and R_e, where Q_e is an effect induced in the structure by the loading Q (e.g., the stress induced by the load Q), and R_e is the corresponding limit state. The failure boundary is then

$$R_e - Q_e = 0 \tag{16.2.4}$$

Henceforth, we use for simplicity the notations Q, R for both the load and the load effect space. In general, Q and R are functions of independent random variables $X_1, X_2, \ldots, X_n$ (e.g., terrain roughness, aerodynamic coefficients, wind speeds, natural frequencies, damping ratios, strength) called *basic variables*, say

$$Q = Q(X_1, X_2, \ldots, X_m) \tag{16.2.5}$$

$$R = R(X_{m+1}, X_{m+2}, \ldots, X_n) \tag{16.2.6}$$

Substitution of Eqs. 16.2.5 and 16.2.6 into Eq. 16.2.3 yields the failure boundary in the space of the basic variables, which is defined by the equation

$$g(X_1, X_2, \ldots, X_n) = 0 \tag{16.2.7}$$

It can be useful in applications to map the failure region, the safe region, and the failure boundary onto the space of variables $Y_1, Y_2, \ldots, Y_r$, defined by transformations

$$Y_i = Y_1(X_1, X_2, \ldots, X_n) \quad (i = 1, 2, \ldots, n) \tag{16.2.8}$$

One example is the frequently used set of transformations

$$Y_1 = \ln R \tag{16.2.9a}$$

$$Y_2 = \ln Q \tag{16.2.9b}$$

On the failure boundary, $R = Q$, so in the coordinates Y_1, Y_2 the failure boundary is $Y_1 = Y_2$.

16.2.2 Safety Indexes

Denote by S the failure boundary in the space of the reduced variables $x_{i\ \mathrm{red}} = (X_i - \overline{X}_i)/\sigma_{xi}$ where the variables X_i are mutually independent, and $\overline{X}_i$ and σ_{xi} are, respectively, the mean and standard deviation of X_i. (The subscript "red" stands for "reduced.") The reliability index, denoted by β, is defined as the shortest distance in this space between the origin (i.e., the image in the space of the reduced variables of the point with coordinates $\overline{X}_i$) and the boundary S. The point on the boundary S that is closest to the origin, and its image in the space of the original basic variables X_i, are called the *design point*. For any given structural problem, *the numerical value of the safety index depends upon the set of variables in which the problem is formulated*, as can be seen in subsequent examples.

16.2.2.1 Safety Indexes: Example 1. Assume that the resistance is deterministic, that is, $R \equiv \overline{R}$. The mapping of the failure boundary

$$Q - \overline{R} = 0$$

onto the space of the reduced variate $q_{\mathrm{red}} = (Q - \overline{Q})/\sigma_Q$ is a point q^*_{red} such that $Q = \overline{R}$, that is, $q^*_{\mathrm{red}} = (\overline{R} - \overline{Q})/\sigma_Q$. The origin in that space is the point for which $Q = \overline{Q}$, and is therefore $q_{\mathrm{red}} = 0$. The distance between the origin and the failure boundary is the safety index $\beta = (\overline{R} - \overline{Q})/\sigma_Q$ (Fig. 16.2.1). The larger the safety index β (i.e., the larger the difference $\overline{R} - \overline{Q}$, and the smaller the standard deviation σ_Q), the smaller is the probability of failure. It is seen that the reliability index provides some indication of a member's safety. However, this indication is largely qualitative, unless information is available on the probability distribution of the variate Q.

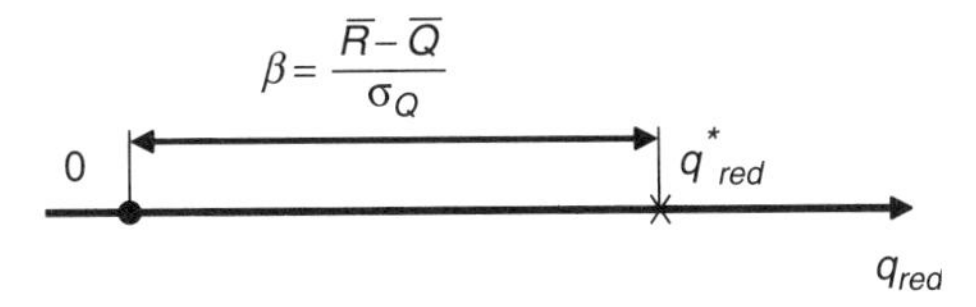

Figure 16.2.1. Index β for member with random load and deterministic resistance.

16.2.2.2 Safety Indexes: Example 2. Consider the failure boundary in the load space, and assume that both R and Q are random variables. The mapping of the failure boundary (Eq. 16.2.3) onto the space of the reduced variables $q_{\text{red}} = (Q - \overline{Q})/\sigma_Q$ and $r_{\text{red}} = (R - \overline{R})/\sigma_R$ yields

$$\sigma_Q q_{\text{red}} - \sigma_R r_{\text{red}} - (\overline{R} - \overline{Q}) = 0 \tag{16.2.10}$$

(Fig. 16.2.2). The distance between the origin and the line (16.2.10) is

$$\beta = \frac{\overline{R} - \overline{Q}}{\left(\sigma_R^2 + \sigma_Q^2\right)^{1/2}} \tag{16.2.11}$$

16.2.2.3 Safety Indexes: Example 3. Instead of operating in the load space R, Q, we consider the failure boundary in the space defined by the transformation (16.2.9a, b). Following exactly the same steps as in Sect. 16.2.2.2, but applying them to the variables Y_1 and Y_2, the safety index is in this case

$$\beta = \frac{\overline{Y}_1 - \overline{Y}_2}{\left(\sigma_{Y_1}^2 + \sigma_{Y_2}^2\right)^{1/2}} \tag{16.2.12}$$

Expansion in a Taylor series yields the expression

$$Y_1 = \ln \overline{R} + (R - \overline{R})\frac{1}{\overline{R}} - \frac{1}{2}(R - \overline{R})^2 \frac{1}{\overline{R}^2} + \cdots \tag{16.2.13}$$

and a similar expression for Y_2. Averaging these expressions, neglecting second- and higher-order terms, and using the notations $\sigma_R/\overline{R} = V_R, \sigma_Q/\overline{Q} = V_Q$,

$$\beta \approx \frac{\ln(\overline{R}/\overline{Q})}{\left(V_R^2 + V_Q^2\right)^{1/2}} \tag{16.2.14}$$

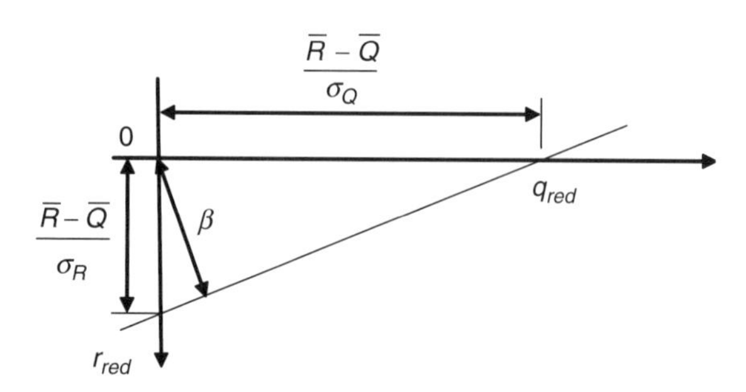

Figure 16.2.2. Index β for member with random load and random resistance.

16.2.3 Reliability Indexes and Failure Probabilities

If the variates R and Q are *normally distributed*, the probability of their difference is normal, and the probability of failure is

$$P_f = 1 - \Phi\left(\frac{\overline{R} - \overline{Q}}{\sqrt{\sigma_R^2 + \sigma_Q^2}}\right) \tag{16.2.15}$$

where the quantity between parentheses is equal to the safety index defined in Eq. 16.2.11, and Φ is the standardized normal cumulative distribution function. If the variates R and Q are *lognormally distributed*, it can be shown that the probability of failure is

$$P_f \approx 1 - \Phi\left(\frac{\ln(\overline{R}/\overline{Q})}{\sqrt{V_R^2 + V_Q^2}}\right) \tag{16.2.16}$$

where the quantity between parentheses is equal to the safety index defined in Eq. 16.2.14. The usefulness of Eq. 16.2.15 and Eq. 16.2.16 is limited by the fact that typically neither the load nor the resistance is normally or lognormally distributed.

16.2.4 Partial Safety Factors, Load and Resistance Factors

Consider a structure characterized by a set of variables with means and standard deviations $\overline{X}_i$ and σ_i, and design points X_i^* $(i = 1, 2, \ldots, n)$ (see Sect. 16.2.2) in the space of the original and the reduced variables, respectively. By definition

$$X_i^* = \overline{X}_i + \sigma_{\overline{X}_i} x_{i\,\mathrm{red}}^* \tag{16.2.17}$$

Equation 16.2.17 can be written in the form

$$X_i^* = \gamma_{\overline{X}_i} \overline{X}_i \tag{16.2.18}$$

where

$$\gamma_{\overline{X}_i} = 1 + V_{\overline{X}_i} x_{i\ \mathrm{red}}^* \tag{16.2.19}$$

The quantity $\gamma_{\overline{X}_i}$ is termed the *partial safety factor* applicable to the mean of the variable X_i.

In design applications, the means $\overline{X}_i$ are seldom used. Instead, nominal design values—such as the wind effect with an $\overline{N}$-yr MRI, or the factored nominal yield stress—are used instead. Let these nominal values be denoted by $\tilde{X}_i$. Equation 16.2.18 can then be written as

$$X_i^* = \gamma_{\tilde{X}_i} \tilde{X}_i \tag{16.2.20}$$

where

$$\gamma_{\tilde{X}_i} = \frac{\overline{X}_i}{\tilde{X}_i} \gamma_{\overline{X}_i} \tag{16.2.21}$$

The factor $\gamma_{\tilde{X}_i}$ is the *partial safety factor* applicable to the nominal design value of the variable X_i. In the particular case in which the variables of concern are the load Q and the resistance R, the partial safety factors are called *load and resistance factors*. For the resistance factor, the letter φ, rather than γ, is typically used.

From the definition of the partial safety factor $\gamma_{\overline{X}_i}$ (Eq. 16.2.19), and the definition of the checking point in the space of the reduced variables corresponding to $Y_1 = \ln R$ and $Y_2 = \ln Q$ (Eq. 16.2.14), it follows that if higher-order terms are neglected,

$$\varphi_{\overline{R}} \approx \exp(-\alpha_R \beta V_R) \tag{16.2.22}$$

$$\gamma_{\overline{Q}} \approx \exp(-\alpha_Q \beta V_Q) \tag{16.2.23}$$

$$\alpha_R = \cos\left[\tan^{-1}(V_Q/V_R)\right] \tag{16.2.24}$$

$$\alpha_Q = \sin\left[\tan^{-1}(V_Q/V_R)\right] \tag{16.2.25}$$

where β is the safety index given by Eq. 16.2.14.

The following linear approximation to Eq. 16.2.23 has been developed for use in standards:

$$\gamma_{\overline{Q}} = 1 + 0.55\beta V_Q \tag{16.2.26}$$

where β is given by Eq. 16.2.14 [16-1].

16.2.4.1 Effects of Approximate Load Factor Estimation. Consider, for example, two members, I and II, for which the following statistics apply:

Member	$\overline{Q}$ (ksi)	σ_Q (ksi)	$\overline{R}$ (ksi)	σ_R (ksi)	β (Eq. 16.2.14)	$\gamma_{\overline{Q}}$ (Eq. 16.2.23)	$\gamma_{\overline{Q}}$ (Eq. 16.2.26)
I	13.3	3.27	35.27	3.39	3.69	2.32	1.50
II	18.0	2.79	35.27	3.39	3.69	1.63	1.31

It is seen that differences between the load factors based on Eqs. 16.2.23 (the "exact" expression) and Eq. 16.2.26 (the "approximate" expression) can be significant. Even the validity of Eq. 16.2.23 as a measure of relative risk is problematic, as members with the same safety index β can have significantly different calculated failure probabilities, depending upon the extent to which the load and/or resistance distributions differ from the lognormal distributions inherent in Eq. 16.2.14. Nevertheless, the index β can serve as an approximate indicator of the relative degree of safety of different members, provided that those members' respective loading and resistance characteristics do not differ significantly from each other.

16.2.4.2 Calibration of Safety Index β, Limitations of the Load and Resistance Factor Approach. Values of the index β have been the object of calibration efforts. It appears that, based on engineering practice as reflected in standards and codes, they were determined to be typically approximately 3.0 for members subjected to gravity loads, although to our knowledge such determination is not clearly traceable. Similar efforts for members subjected predominantly to wind loads suggested that those values are lower than for gravity loads, "at least according to the methods used for structural checking in conventional design. These are methods which are simplified representations of real building behavior and they have presumably given satisfactory performance in the past" [16-11, p. 6]. Load factors for combinations involving wind loading used in the American National Standard A58 and in its successor, the ASCE 7 Standard, are based on the decision

> "to propose load factors for combinations involving wind ... loads that will give calculated β values which are comparable to those existing in current practice, and not to attempt to raise these values to those for gravity loads by increasing the nominal loads or the load factors for wind ... loading ... The profession may well feel challenged (1) to justify more explicitly ... why current simplified wind ... calculations may be yielding conservative estimates of ... safety; (2) to justify why current safety levels for gravity loads are higher than necessary if indeed this is true; (3) to explain why lower safety levels are appropriate for wind ... vis-à-vis gravity loads, or (4) to agree to raise wind ... loads or load factors to achieve a similar reliability as that inherent in gravity loads. While the writers feel that arguments can be cited in favor or against all four options, they decided that this report was not the appropriate forum for what should be a profession-wide debate." [16-11, p. 7].

Reference [16-11] further points out that the results presented therein are not applicable to loads that are considered to be outside the scope of the A 58 Standard (and, hence, of the ASCE 7 Standard).

For a confirmation that, as stated in [16-11], important issues pertaining to load factors for wind still need to be addressed, see Appendix A5. Section 16.7 was developed in response to this need.

16.3 DEPENDENCE OF WIND EFFECTS ON WIND DIRECTIONALITY

The ASCE 7-10 Standard assumes that, following multiplication by a blanket wind directionality factor (see Sect. 4.2.3), wind effects induced in structural members by $\overline{N}$-yr nondirectional wind speeds have $\overline{N}$-yr MRIs. The approximation inherent in this assumption can be unacceptably large. For this reason, users of the wind tunnel method typically estimate MRIs of wind effects by accounting for wind directionality explicitly, rather than via a wind directionality factor (for details, see Sects. 18.3 and Steps 6 and 7 of Sect. 19.4.2).

For the purpose of wind-induced loss estimation, building orientation, which is a factor in determining wind directionality effects, is another source of uncertainty, and therefore needs to be considered in the overall error calculations [16-2]. This uncertainty does not come into play for structures whose orientation is known. Orientation is also not an issue if nondirectional wind speeds and pressure coefficients are used to determine wind effects.

16.4 STRUCTURAL STRENGTH RESERVE

The purpose of designing structural members by using Allowable Stress Design (ASD) or Load and Resistance Factor Design (LRFD) is to ensure that, as they attain the respective limit states, their behavior is acceptable. For unacceptable behavior to occur, the wind speeds must be larger than the speeds that induce those states. How much larger? The answer to this question yields the MRI of the structure's limit state associated with the onset of unacceptable behavior. A structure with large strength reserve is one for which that MRI is considerably larger than the MRI inducing the limit states associated with ASD and LRFD. Unlike for structures subjected to seismic loads, the issue of the strength reserve available in structures subjected to wind loads has not been the object of systematic study. In Sect. 16.4.1, we consider the assessment of strength reserve as a ratio of ultimate to allowable wind load acting on portal frames by considering one wind direction at a time. In Sect. 16.4.2, we consider a similar problem, where the overall effect of wind directionality is taken into account.

16.4.1 Strength Reserve Assessment: Ratios of Ultimate to Allowable Wind Loads

For low-rise industrial steel buildings with gable roofs and portal frames, nonlinear push-over studies have been conducted in which the buildings were subjected to two sets of wind pressures [16-3]. One set consisted of wind pressures based on aerodynamic information specified for low-rise structures in the ASCE 7 Standard. The second set consisted of simultaneous wind pressures measured and recorded in the wind tunnel at a large number of taps on the building model's surface. The structural design of the frames was based on ASCE 7 Standard loads and the ASD approach. The purpose of the studies was twofold. First, to compare the strength reserve estimated by using the simplified wind loads inherent in the ASCE Standard on the one hand and recorded wind tunnel data on the other. Second, to examine the degree to which the strength reserve is changed by the adoption of alternative designs. The following alternative features of the lateral bracing and joint stiffening were considered:

1. Lateral bracing at bottom flanges of rafters with (a) 2.5 m spacing, or (b) 6 m spacing.
2. Knee joints: (a) Horizontal and vertical stiffeners, (b) horizontal, vertical, and diagonal stiffeners.
3. Vertical stiffener at ridge.

Ultimate strength analyses were performed for the pertinent load combinations involving wind. Factors λ were calculated, representing the ratio between ultimate and allowable wind load for each load combination being considered, the ultimate wind load corresponding to incipient failure through local or global instability as determined by using a finite element analysis program.

Reducing the distance between bracings of the rafter's lower flanges increased the strength reserve more effectively than providing diagonal stiffeners in the knee joint. Significant differences were found between the values of λ obtained under loading by pressures specified in the ASCE 7 Standard provisions and loading by the more realistic pressures measured in the wind tunnel. Failure modes for the frame knee are shown in Fig. 16.4.1. See also [16-4].

16.4.2 Strength Reserve Assessment: MRIs of Ultimate Wind Effects Estimated by Accounting for Wind Directionality

The following methodology applicable to rigid buildings was developed for the estimation of MRIs of ultimate wind effects by accounting for wind directionality [16-5]:

1. Using recorded wind tunnel pressure data, obtain the loads that induce peak internal forces (axial forces, bending moments, shear forces) in a number of cross sections deemed to be critical. Obtain loads corresponding to a unit wind speed at 10 m above ground over open terrain for, say, 16 or 36 wind directions spanning the 360° range. These loads,

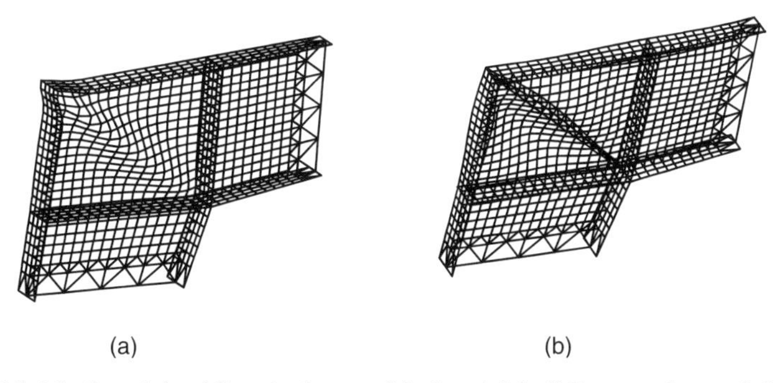

Figure 16.4.1. Local buckling in knee of industrial building steel portal frame: (a) knee without diagonal stiffener, (b) knee with diagonal stiffener.

multiplied by the square of wind speeds V considered in design, are used in Step 2.

2. Using nonlinear finite element analyses, determine the wind speed from each direction θ_i that causes the frame to experience incipient failure, defined as the onset of deformations that increase so fast under loads that implicit nonlinear finite element analyses fail to converge to a solution.
3. From available wind climatological data, create, by simulation, time series of directional wind speeds with length t_d that exceeds the anticipated MRIs of the failure events (see Sects. 12.5, 12.6).
4. Count the number n_f of cases in which directional wind speeds in the time series created in Step 3 exceed the directional wind speeds determined in Step 2 to produce incipient failure events. The MRI in years of the failure event is estimated as $\overline{T} = t_d/n_f$.

This methodology was applied to an industrial low-rise building portal frame located in a hurricane-prone region. The frame was strengthened by triangular stiffeners at the column supports and by haunches and horizontal, vertical, and diagonal stiffeners at the knee joints. Owing to such strengthening, the estimated failure MRI was in this case very high (100,000 years, corresponding to a 1/2000 probability that the frame will fail during a 50-year lifetime).

16.5 DESIGN CRITERIA FOR MULTI-HAZARD REGIONS

16.5.1 Strong Winds and Earthquakes

Structures in regions subjected to both strong earthquakes and strong winds are currently designed by considering separately loads induced by earthquakes and by winds, and basing the final design on the more demanding of those loads. The rationale for this approach has been that the probability of simultaneous occurrence of both earthquakes and high winds is negligibly small. This argument is incorrect. In fact, implicit in this approach, which is used in the ASCE 7 Standard, are risks of failure that can be greater by a factor of up to two than risks for structures exposed to wind only or to earthquakes only.

An intuitive illustration of this statement follows. Assume that a motorcycle racer applies for insurance against personal injuries. The insurer will calculate a rate commensurate with the risk that the racer will be injured in a motorcycle accident. Assume now that the motorcycle racer is also a high-wire artist. In this case, the insurance rate would be increased because the risk of injury within a specified period of time, either in a motorcycle or high-wire accident, will be larger than the risk due to only one of these types of accident. This is true even though the nature of the injuries in the two types of event may differ. This argument is expressed formally as

$$P(s_1 \cup s_2) = P(s_1) + P(s_2) \tag{16.5.1}$$

where $P(s_1)$ = annual probability of event s_1 (injury due to a motorcycle accident) and $P(s_2)$ = annual probability of event s_2 (injury due to a high-wire accident), and $P(s_1 \cup s_2)$ = probability of injury due to a motorcycle *or* a high-wire accident.

Equation 16.5.1 is applicable to structures as well, particularly to members experiencing large demands under lateral loads (e.g., columns in lower floors). For details and case studies, see [16-6, 16-7].

16.5.2 Winds and Storm Surge

Unlike earthquakes and windstorms, winds and storm surge are not independent events. Therefore, for some applications, it is necessary to consider their joint effects. This entails the following steps: (1) select a stochastic set of hurricane storm tracks in the region of interest, (2) use the selected storm tracks to generate time histories of wind speeds and corresponding time histories of storm surge heights at sites affected by those wind speeds, (3) use those time histories to calculate time series of wind and storm surge effects, (4) obtain from those time series estimates of joint effects of wind and storm surge with the mean recurrence intervals of interest [16-8, 16-9]. Note that in this approach the calculations are performed in the load effects space (Sect. 16.2.1).

One basic element in performing estimates of storm surge heights is the bathymetry at and near the site of interest. As was noted in more detail in Sect. 10.2.1, to be realistic, storm surge hazard scales must consider local bathymetry, which plays a crucial role in determining actual surge impacts.

16.6 INDIVIDUAL UNCERTAINTIES AND OVERALL UNCERTAINTY IN THE ESTIMATION OF WIND EFFECTS

Sections 16.6.1 to 16.6.3 recapitulate uncertainties due to micrometeorological, aerodynamic, and wind climatological factors. Section 16.6.4 considers overall uncertainties obtained by compounding uncertainties due to those factors. In addition to errors discussed in Sects. 16.6.1–16.6.3, for structures that experience dynamic effects, errors associated with natural frequencies, modes of vibration, and damping should be taken into account (see Sects. 16.6.4 and 16.7).

16.6.1 Errors Associated with Micrometeorological Modeling

It is necessary to achieve, to within reasonable bounds, flow simulations that are replicable across testing facilities and, *a fortiori*, within the same facility. At present this requirement is typically not satisfied, owing to the lack of general agreement on adequate and enforceable performance standards for generating laboratory flows, and differences among aerodynamic test results obtained on the same model in different wind tunnels were found to be as high as 100% (Sect. 13.3.7). Flows achieved in the laboratory should be considered adequate if they induce on models of structures pressures and internal forces that do not differ by more than, say, 15% from corresponding averages obtained across laboratories. Such a criterion would be far more effective in ensuring the quality of flow simulations on buildings than are the weak and incomplete criteria on the wind tunnel procedure incorporated in Chapter 31 of the ASCE Standard 7-10.

In addition to errors in the laboratory simulation of target simulation flows, it is necessary to account for uncertainties in the target flow models themselves, which can be of the order of, say, 10–20%.

16.6.2 Errors Associated with Aerodynamic Modeling

Aerodynamic factors are closely intertwined with micrometeorological factors. In particular, the Reynolds number affects the features of both the oncoming flow turbulence and the aerodynamic pressures due to flow-structure interaction. In both cases, the internal friction effects are stronger in the wind tunnel than in the prototype. This results in reduced high-frequency turbulent eddy transport of particles with large momentum from the free flow zone into the separation bubble and, therefore, in flow reattachment further downstream than in the prototype (Fig. 13.2.3). The suppression in the wind tunnel of high-frequency turbulence components causes the reduction of local pressure fluctuations in zones of high suction (Fig. 13.3.2). The distortion of aerodynamic effects due to violation of Reynolds number similarity is stronger for bodies with curved shapes, for which the deficit in the high-frequency flow fluctuations achieved in the laboratory causes earlier flow separation, a larger wake, and stronger drag than in the prototype. In practice full- or large-scale test results are used in some cases in attempts to reduce errors in wind tunnel test results.

16.6.3 Errors Associated with Wind Climatological Modeling

Nontropical storm wind speeds acting on buildings are typically estimated from observed wind speed data over open terrain at meteorological stations. The observations are typically affected by *observation errors*. The data are not necessarily measured at 10 m elevation, and for the sake of uniformity they are typically converted into wind speeds at that elevation. The data then need to be converted to wind speeds in terrain with the appropriate exposure at the elevation of the top of the building, that is, the elevation to which pressure, force, or moment coefficients are referenced in the wind tunnel. Both conversions (see Sect. 11.2) are affected by *conversion errors*. In addition, because the number of measurements on the basis of which wind speeds are estimated is limited, the estimates are affected by *sampling errors* (Sect. 12.7). Finally, the estimates are likely affected by *probabilistic modeling errors* (Sect. 12.8).

Tropical storm wind speeds (including hurricane wind speeds) are affected by similar errors. In particular, the climatological parameters on the basis of which the wind speeds are estimated are based on data samples of limited size (of the order of 50 to 100 years). Their probabilistic models are therefore affected by sampling errors, as well as by probabilistic modeling errors. In addition, significant errors may be present in the physical modeling of the tropical storms (Sect. 12.8).

16.6.4 Overall Uncertainties in the Estimation of Wind Effects

One of the useful features of a reliability-based framework for estimating uncertainties and safety margins is that the errors and uncertainties associated with the various parameters that determine wind effects are considered *collectively*, rather than individually. If the uncertainty with respect to an individual factor has a modest relative contribution to the overall uncertainty of the wind effect of interest, a design decision based solely on that uncertainty would be unwarranted.

For example, the issue of how much the size of aerodynamic time histories measured in the laboratory may be reduced without significant penalty from a structural safety viewpoint is best resolved within a structural reliability framework. Even though a shorter length of the time histories increases the uncertainty with respect to the magnitude of the peak aerodynamic coefficients, that increase can be negligible in relation to the *total* (overall) uncertainty associated with the wind effect of interest [16-10].

As a second example, the results of aerodynamic tests on low-rise building models discussed in Sect. 13.3.7 have variabilities across laboratories that are significantly larger in suburban than in open terrain. Taking these variabilities into account would result in larger wind load factors for buildings in suburban than in open terrain. This suggests that the advantage inherent in the lower wind loads specified for low-rise buildings in suburban terrain may be less significant than indicated by conventional calculations that do not take those variabilities into account.

To estimate, albeit approximately, overall uncertainties in the estimation of the wind effects, it may be assumed, for example, that: (1) pressures are affected by an uncertainty factor a that reflects experimental errors in wind tunnel measurements and depends upon the quality of the test facility; (2) wind speeds are affected by the product bcd, where (i) b reflects observation errors and errors in the conversion of 3-s (or 1 min) speeds to hourly (or 10-min) mean speeds at 10 m above ground in open terrain, (ii) c reflects uncertainties in the conversion of mean hourly (or 10-min) speeds at 10 m above ground in open terrain to mean hourly (or 10-min) wind speeds at the top of the building, and (iii) d reflects modeling and sampling errors in the estimation of the design wind speeds, which increase as the sample size of the observations decreases; (3) natural frequencies are affected by an uncertainty factor e; and (4) damping ratios are affected by an uncertainty factor f.

One possible assumption is that a, b, c, e have normal distributions with, say, 5–10% coefficients of variation. The factor d may be assumed to have a normal distribution with coefficient of variation dependent on the size of the data sample on which estimates of the design wind speeds are based and, possibly, on the degree of confidence in the distributional model of the extreme wind speeds. The factor f has been assumed to have a lognormal distribution [19-21]. It is seen that the estimates are affected by assumptions based at least in part on belief. Nevertheless, such estimates are useful insofar as they help improve estimates of the response in a manner commensurate with available information and engineering judgment on the magnitude of relevant errors and uncertainties.

16.7 CALIBRATION OF DESIGN MRIs IN THE PRESENCE OF DYNAMIC EFFECTS OR OF LARGE KNOWLEDGE UNCERTAINTIES

This section shows how larger than typical knowledge uncertainties in the micrometeorological, wind climatological, aerodynamic, and dynamic parameters of the wind-induced demand can be accounted for so that the MRI of the wind effects specified for design will be consistent, with respect to nominal risk, with the MRI specified in the ASCE Standard (see ASCE Sect. 31.4.1).

The estimated peak wind effect (e.g., the estimated $\overline{N}_1$-yr peak displacement occurring in a storm with, say, a one-hour duration) has an estimated value commonly denoted for convenience by $F_{pk}(\overline{N}_1)$. This value is called the $\overline{N}_1$*-yr point estimate* of the wind effect. It should be remembered, however, that the value of the peak effect is uncertain, owing to the various uncertainties discussed in Sect. 16.6, and that it has, therefore, a probability distribution $P[F_{pk}(\overline{N}_1)]$. For the purposes of this section, it is appropriate to change the notation of the $\overline{N}_1$-yr point estimate of the wind effect from the commonly used shorthand notation $F_{pk}(\overline{N}_1)$ to the more precise notation mean$[F_{pk}(\overline{N}_1)]$.

For Allowable Stress Design (ASD), $\overline{N}_1$ is typically required by the ASCE 7-05 Standard to be 50 years, a value deemed on the basis of past experience to be safe for a certain class of structures. However, for LRFD purposes, it is necessary to specify a point estimate of the $\overline{N}_2$-year wind effect $F_{pk}(\overline{N}_2)$, where $\overline{N}_2 > 50$ years. The value of $\overline{N}_2$ depends upon type of structure and the uncertainties in the estimation of the wind effect being considered.

The wind load factor is commonly defined as the ratio

$$LF_w = \frac{\text{mean}[F_{pk}(\overline{N}_2)]}{\text{mean}[F_{pk}(\overline{N}_1)]} \tag{16.7.1}$$

Calculations based on a set of typical knowledge uncertainties with respect to micrometeorological, climatological, aerodynamic, and dynamic response parameters led to a value of the wind load factor $LF_w = 1.6$,[3] which was incorporated in successive versions of the ASCE 7 Standard. An equivalent alternative to specifying a 50-year MRI for the basic wind speeds and multiplying those wind speeds by $\sqrt{LF_w}$ is to specify directly the MRI $\overline{N}_2$ estimated from Eq. 16.7.1, in which $LF_w = 1.6$.

However, for structures with knowledge uncertainties larger than those that led to the adoption of the load factor 1.6, it is necessary to make allowance for those larger uncertainties by using for design a load factor larger than 1.6 or, equivalently, a value for the MRI $\overline{N}_2$ commensurate with that larger load factor and, therefore, larger than the corresponding ASCE 7 Standard value. This can be achieved by a calibration process that entails the following steps:

1. Estimate the probability distributions $P[F_{pk,s}(\overline{N}_1)]$ and $P[F_{pk,ns}(\overline{N}_1)]$ and the point estimates inherent in those distributions, that is, mean$[F_{pk,s}(\overline{N}_1)]$ and mean$[F_{pk,ns}(\overline{N}_1)]$. The distributions are functions of the knowledge uncertainties that affect the respective response estimates, and are measures of the variability of the $\overline{N}_1$-yr peak effects due to the uncertainties in the parameters of the response. The subscripts $_s$ and $_{ns}$ indicate that the response corresponds to the uncertainties inherent in the Standard and to the nonstandard, larger uncertainties applicable to the building of interest, respectively. The probability distributions are estimated by calculating responses $[F_{pk,s}(\overline{N}_1)]_i$ and $[F_{pk,ns}(\overline{N}_1)]_i$ ($i = 1, 2, \ldots, m$, where m is sufficiently large, e.g., $m = 2000$), each corresponding to a set of the uncertain parameter values obtained by random sampling from the parameters' assumed distributions.[4] The distributions of the uncertain parameter values are generally not known, but may be assigned expressions believed to be reasonable (see Sect. 16.6.4).
2. Determine the percentile p of the distribution $P[F_{pk,s}(\overline{N}_1)]$ for which the peak effect corresponding to that percentile, $[F_{pk,s}(\overline{N}_1)]_p$, is equal to the point estimate of the peak wind effect with an $\overline{N}_{2s}$-year MRI considered for Load and Resistance Factor Design (LRFD). We denote that point estimate by mean$[F_{pk,s}(\overline{N}_{2s})]$. In accordance with ASCE Standard 7-10, $\overline{N}_{2s} = 700$, 1,700, or 300 years, depending upon the structure's risk category (see Sect. 3.1).

   Figure 16.7.1 shows the probability distribution $P[F_{pk,s}(\overline{N}_1)]$, the corresponding point estimate of the $\overline{N}_1$-yr peak effect, mean$[F_{pk,s}(\overline{N}_1)]$, and the value of $[F_{pk,s}(\overline{N}_1)]_p =$ mean$[F_{pk,s}(\overline{N}_{2s})]$. It is clear from the figure that p can be determined for given $\overline{N}_1$, $P[F_{pk,s}(\overline{N}_1)]$, and $\overline{N}_{2s}$. It follows that, for given $\overline{N}_1$ and p, $\overline{N}_{2s}$ is a function of the shape of the

[3]In earlier versions of ASCE 7 Standard the specified value of LF_w was 1.5.

[4]For tall buildings the responses $[F_{pk}(\overline{N}_1)]_i$ can be conveniently calculated by using specialized software; see www.nist.gov/wind, II A, and Chapter 19.

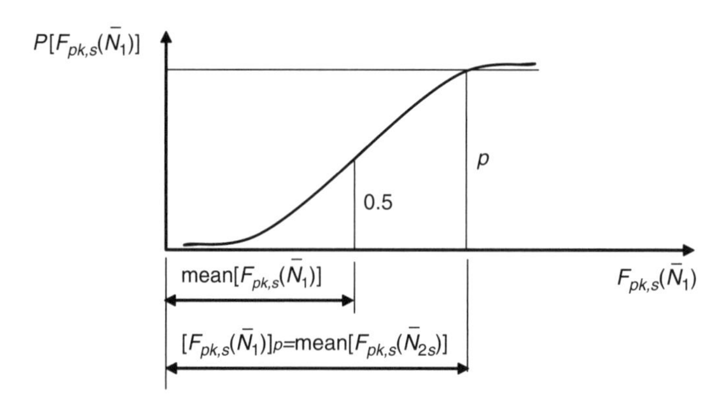

Figure 16.7.1. Probability distribution $P[F_{pk,s}(\overline{N}_1)]$, corresponding point estimate of the $\overline{N}_1$-yr peak effect mean$[F_{pk,s}(\overline{N}_1)]$, and value of $[F_{pk,s}(\overline{N}_1)]_p = \text{mean}[F_{pk,s}(\overline{N}_{2s})]$.

probability distribution $P[F_{pk,s}(\overline{N}_1)]$. In particular, *if the upper tail of the probability distribution is longer owing to greater uncertainties, so is the MRI of the peak wind effect's point estimate considered for LRFD.*

3. For an acceptable level of nominal risk to be ensured, the design peak wind effect for the nonstandard case must be equal to $[F_{pk,ns}(\overline{N}_1)]_p$, that is, to the value of $F_{pk,ns}(\overline{N}_1)$ corresponding in the distribution $P[F_{pk,ns}(\overline{N}_1)]$ to the percentile p (or, if the building owner or the building official so chooses, to a percentile $p_{ns} \geq p$). The mean recurrence interval $\overline{N}_{2ns}$ of the design wind effect is then obtained from the equality

$$\text{mean}[F_{pk,ns}(\overline{N}_{2ns})] = [F_{pk,ns}(\overline{N}_1)]_p \tag{16.7.2}$$

Inherent in the procedure just outlined is a calibration of the point estimate mean$[F_{pk,ns}(\overline{N}_{2ns})]$ with respect to current ASCE 7 Standard design practice. The calibration procedure takes into account the fact that the probability distribution $P[F_{pk,ns}(\overline{N}_1)]$ has a longer tail than $P[F_{pk,s}(\overline{N}_1)]$.

The ASCE 7 Standard wind tunnel procedure does *not* provide for such a calibration—that is, it does not ensure the requisite nominal risk consistency between the standard case and nonstandard cases, which typically arise when the wind tunnel procedure is deemed necessary.

In particular, the calibration can be performed for buildings experiencing dynamic effects, for which the nonstandard case is characterized by the presence, among other uncertainties, of uncertainties in the dynamic parameters (damping ratios, natural frequencies of vibration, modal shapes). Some wind engineering consultants have suggested that using for the design of such buildings the MRI $\overline{N}_{2s}$ used for rigid buildings (e.g., 1,700 years) will automatically ensure the requisite nominal risk consistency. It follows from the remark in italics at the end of Step 2 that, in general, this suggestion is not correct, meaning that the nominal risk inherent in current practice for the design of tall or super-tall buildings can be higher than the nominal risk inherent in the ASCE 7 Standard for ordinary buildings.

For either rigid or flexible structures, the procedure just outlined may also be used, for example, to determine the requisite value of $\overline{N}_{2ns}$ if the wind effects are based on short wind speed records (e.g., 10 years), rather than on records of typical length (e.g., 40 years).

CHAPTER 17

LOSS ESTIMATION

17.1 INTRODUCTION

The primary aims of wind-induced loss estimates are to help establish appropriate insurance premiums and reinsurance costs, and to assess benefits inherent in the use of techniques for reducing wind-induced damage in new and existing buildings. Loss estimates can also be used for emergency management and planning purposes.

Two basic approaches have been developed for producing such estimates. The traditional approach entails the application of statistical tools to the number of claims and recorded loss data. Owing primarily to the sparsity of such data, this approach was typically found to be inadequate. For this reason, a number of proprietary loss estimation models have been developed that make use of physical and probabilistic information. Where the available information is insufficient, the models are based largely on engineering judgment. Differences among loss estimates produced by the various loss models are due to differences among those models' elements, including:

- Meteorological, micrometeorological, and wind climatological models based on physical and/or probabilistic models of hurricane wind fields and landfall frequencies; hurricane paths, intensity decay over land, and terrain roughness; or on similar information for different types of storm.
- Exposure data[1] (i.e., data on numbers of buildings of various types at various geographical locations).

[1]In this context exposure is a term used by the insurance industry, distinct from the term exposure used in a micrometeorological sense, which pertains to the surface roughness upwind of the structure.

- Estimates of building performance under strong winds in the presence of wind-borne missiles and rain, as functions of type and quality of construction, building code requirements, and the presence or absence of shutters or other mitigation devices.
- Insurance practices (e.g., clauses concerning deductibles and limits).
- Assumptions on climate change.

A study reported in [17-1] examined data on claims provided by Citizens Property Insurance Corporation in conjunction with estimates of average annual losses (AAL) based on several proprietary loss models. The loss models were applied to single-family dwellings with insurance policies specifically focused on wind-induced damage. Loss estimates were found to differ among the models by factors as large as three, particularly in inland areas. The study traced the differences among individual elements of the loss models that caused the differences in the loss estimates; for example, estimates of annual loss reductions per building due to the use of shutters differed between two models from $1,040 to $78.

For the insurance and reinsurance processes to function in a fair, stable and economically rational manner, a reliable, transparent (i.e., *auditable* and *traceable*) loss estimation model is necessary. With this need in mind, the Florida Office of Insurance Regulation has supported the development, updating, and expansion of the Florida Public Hurricane Loss Model (FPHLM), designed to serve as a system open to public scrutiny. The FPHLM consists of three modules: (1) a meteorological model defining the wind field and its micrometeorological and wind climatological properties, (2) an engineering model yielding damage estimates, and (3) a loss estimation (actuarial) model for providing estimates of either (a) expected losses of a specific insurance portfolio caused by a specific hurricane, or (b) expected annual losses of specified insurance portfolios. The output from FPHLM is also intended to serve for the validation of other models, for rate making,[2] and for the analysis of the effectiveness of mitigation measures. The FPHLM has been initially developed for the most common types of single-family homes in Florida. For details on the engineering model, see [17-2].

17.2 ELEMENTS OF DAMAGE ESTIMATION PROCEDURES

Damage procedures contain, in principle, the following elements:

- Structural classification definitions and exposure databases.
- Definitions of basic damage and sub-damage states.
- Development of attendant damage matrices.
- Estimation of combined damage states and combined damage matrices.

[2]In insurers' terminology, rate making is the process by which insurance premiums are established.

17.2.1 Structural Classification and Exposure

Damage estimation methodologies include the development of structural classification and attendant data for specified geographical regions. Such data are usually not available from insurance portfolio files, which are typically focused on fire resistance. A simplified definition of structural types is required, which can be based on features such as: *method of construction* (e.g., manufactured, site-built), *roof* type (e.g., gable, hip), *roof cover* type (e.g., shingle, tile), *exterior wall* type (e.g., wood frame, masonry), *number of stories* (one or two), and footprint area of the building. Various combinations of those features then define basic building types. The simplifications in the classification of buildings, and the statistics on the various types of buildings, differ from loss model to loss model, and therefore contribute to differences among the respective loss estimates.

Exposure data provide information on numbers of structures of various types in the geographical regions of interest.

17.2.2 Basic Damage and Sub-Damage States

For loss estimation purposes, structures may be represented as assemblies of sets of components. For example, the FPHLM model includes five components: walls (W), openings (including garage doors) (O), roof-to-wall connections (C), roof cover (T), and roof sheathing (S). Damage to each component is called a *basic damage mode.* Each basic mode may be divided into *sub-damage modes* in accordance with the extent of damage. For example, basic damage to openings (lights) can be subdivided into four classes O_i $(i = 0, 1, 2, 3)$ corresponding, respectively, to no damage, light damage (loss of less than 25% of the openings), loss of 25 to 50% of the openings, and loss in excess of 50% of the openings. Sub-damage modes denoted by W_i, C_i, T_i, and S_i $(i = 0, 1, 2, 3)$ can be similarly defined for other basic damage modes.

For each component, a probabilistic estimate is required of both the wind-induced demand as a function of wind speed (including demand due to impacts by wind-borne missiles), and the capacity available to withstand the demand. In estimating the demand, it should be kept in mind that wind loading estimates based on building code specifications can differ widely from estimates based on ad hoc tests. In some instances, no adequate information on loading is available from any source, and engineering judgment is required to supply the missing information. This is also the case for estimates of capacity, pertaining as they do to limit states associated with failure, which typically involve post-elastic states.

17.2.3 Vulnerability Curves for Basic Damage or Sub-Damage States

Statistical estimates of capacity can be represented by *vulnerability* curves. Consider the following thought experiment. A set of n tests on one type of component is performed under the same loading L, or a storm with the same wind speed V affects in identical fashion a set of n buildings of the same type. The component damage state of interest will occur a number m_1 of times. If s sets of such n tests are performed under the loading L, or if s such sets of n buildings are subjected to the wind speed V, owing to the variability of the component strength, the component damage state in set i will occur m_i times ($i = 1, 2, \ldots, s$). The estimated mean value of m_i is denoted by $\overline{m}$. For wind loss estimation purposes, a vulnerability curve is sometimes defined in wind engineering as a representation of $\overline{m}$ as a function of wind speed (e.g., the 3-s peak speed at 10 m above ground over open terrain).

Assume, for example, that the probability that a damage state (e.g., total loss of roofing) will occur for wind speeds lower than or equal to 120 mph is 0.1 (meaning that, on average, one house in ten will experience total loss of roofing under winds of up to 120 mph), and the probability that the damage state will occur for wind speeds of up to 200 mph wind is 1 (every single building will experience total loss of roofing). It follows from this example that the vulnerability curve can be defined as a cumulative distribution function (CDF).

Alternatively, the vulnerability curve can be defined as the cost of repair for the damage state being considered, times the probability that the damage state will occur, divided by the total replacement cost of the building. If the total replacement cost of the building is $200,000 and the repair cost for the damage state being considered (say, total loss of roofing) is $10,000, it follows for our example that the ratio of the repair to the total replacement cost of the building is $10{,}000 \times 0.1/200{,}0000 = 0.005$ for 120 mph winds, and $10{,}000 \times 1/200{,}000 = 0.05$ for 200 mph winds. Similar definitions may be applied to combined damage states (Sect. 17.2.4).

In some applications, it is desirable to develop curves in which the ordinates, rather than corresponding to the mean of the damage across specimens, would correspond to some percentile p of the distribution of the damage. Such a representation is sometimes called a *fragility curve*. A vulnerability curve for which the probability density function of the damage induced by a specified wind speed is symmetrical corresponds to the particular case of a fragility curve with $p = 50\%$.

Note that in this section we considered two kinds of distribution. The first kind of distribution, $P(\text{damage state}|V)$, is conditional on a fixed wind speed V. Its mean was denoted by $\overline{m}$, and its other ordinates depended on the percentile p. The second kind of distribution represents the dependence of $\overline{m}$ on the variable V, in which case the distribution is defined as a vulnerability curve; alternatively, as pointed out earlier, a curve can be constructed representing the dependence on V of the pth quantile of the conditional distribution $P(\text{damage state}|V)$, in which case the distribution is called a fragility curve. Note, however, that there appears to be no standard definition of either type of curve.

17.2.4 Combined Damage States

Combined damage states are combinations of basic damage states (e.g., loss of roof cover *and* roof sheathing). The development of matrices of combined damage state probabilities (or, for short, damage matrices) can be based on a set-theoretical approach (Sect. 17.2.4.1), or on a strictly empirical engineering approach (Sect. 17.2.4.2).

The damage matrix is a property of the structure, independent of the extreme wind climate. While its calculation by an empirical engineering approach is appealing from a practical point of view, a set-theoretical framework offers a potential advantage if transparent—that is, auditable and traceable—comparative assessments of alternative procedures are performed by independent parties. If the damage matrix is developed within a rigorous, standardized set-theoretical framework, it is possible to describe and compare the engineering content of all procedures clearly and systematically. Since it is that content that determines the loss estimates, comparisons performed on this basis allow the best possible assessment of the procedures' respective merits or shortcomings.

17.2.4.1 Calculations Performed within a Set-Theoretical Framework [17-3]. A prerequisite for the calculations is the definition of the relevant combined damage states. It is then possible to develop the corresponding damage matrices.

17.2.4.1.1 Definition of Combined Damage States. Consider, for illustration purposes, Table 17.2.1. Its entries are ordinates of the conditional probabilities of two basic damage states, R (partial loss of roof) and F (total building collapse), as functions of wind speeds v. For example, for $v = 120$ mph, Prob($R|120$ mph) $= 0.8$. Curves such as those defined by Table 17.2.1 can be obtained fully or in part from laboratory tests, from analytical studies involving simulations and engineering judgment, and/or from inferences from post-disaster investigations.

Combined damage states must be chosen to ensure that no possible damage state is omitted, and no multiple counting of damage states occurs. For example, the following theoretically possible combinations of events are associated with the basic limit states R and F:

- Combinations of 4 states taken by 0: This set is empty.
- Combination of 4 states taken by 1: R, F, not R, not F.

TABLE 17.2.1. Conditional Probabilities of Basic Damage States R and F

v (mph)		80	90	100	110	120	130	140
R	$P(R\|v)$	0	0.03	0.2	0.5	0.8	1.0	1.0
F	$P(F\|v)$	0	0	0	0.02	0.05	0.2	1.0

- Combination of 4 taken by 2: R and F; R and not R (this set is empty); R and not F; F and not R; F and not F (this set is empty); not R and not F.
- Combination of 4 taken by 3: R and F and not R (this set is equivalent to F); R and F and not F (this set is equivalent to R); R and not R and not F (this set is equivalent to not F); F and not F and not R (this set is equivalent to not R).
- Combination of 4 taken by 4: R and F and not R and not F (this set is empty).

The total number of sets is $2^4 = 16$. Sixteen similar states involving "or" instead of "and" exist. Not all possible states are of interest in practice.

17.2.4.1.2 Probabilities of Combined Damage States. The probabilities of combined states can be calculated from probabilities of basic states and probabilistic information on their mutual dependence. We focus on the following damage states:

Case 1: R and not F (partial roof loss but no collapse). The probability of this state allows the estimation of the cost of repair of roofs for structures that did not collapse and need roof repair. The probability $P(R)$ can be written as:

$$P(R) = P(R \text{ and } F) + P(R \text{ and not } F) \qquad (17.2.1)$$

Therefore

$$P(R \text{ and not } F) = P(R) - P(R \text{ and } F) \qquad (17.2.2)$$

(see Fig. 17.2.1a).

Case 2: F and not R (structure collapsed but roof intact). This largely unrealistic state is included for illustration purposes, but would not be included in an engineering analysis of loss.

$$P(F) = P(F \text{ and } R) + P(F \text{ and not } R) \qquad (17.2.3)$$

$$P(F \text{ and not } R) = P(F) - P(F \text{ and } R) \qquad (17.2.4)$$

(see Fig. 17.2.1b).

Case 3: R and F (roof damaged and structure collapsed). The probability of this state (represented in a Venn diagram by the intersections of the regions R and F, Fig. 17.2.1c) allows the estimation of the cost of replacement of the entire structure, including the cost of repair of the roof. The joint probability $P(R \text{ and } F)$ can be written as

$$P(R \text{ and } F) = P(R|F)P(F) \qquad (17.2.4a)$$

$$= P(F|R)P(R) \qquad (17.2.4b)$$

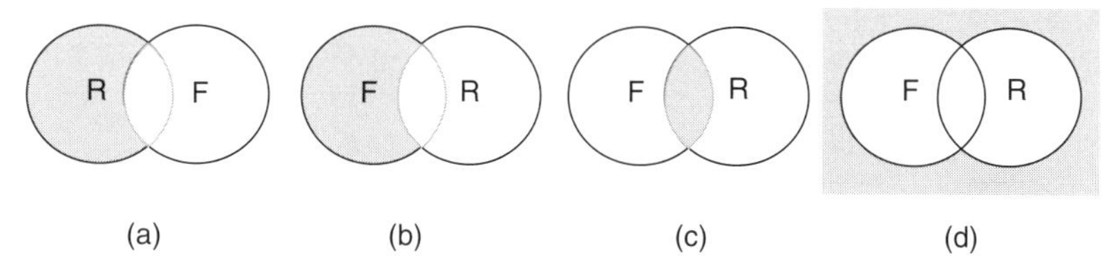

Figure 17.2.1. Venn diagrams: (a) R and not F, (b) F and not R, (c) R and F, (d) not R and not F.

where $P(R|F)$ is the probability of state R given that state F has occurred. A similar definition applies to $P(F|R)$.

The conditional probabilities are based on engineering considerations and are functions of the dependence between the states R and F. The specification of this dependence is an engineering input to the problem, and is a function of specific damage mechanisms. If R and F are independent, $P(R|F) = P(R)$ and $P(F|R) = P(F)$.

Case 4: not R and not F. This state designates "no damage" (Fig. 17.2.1d). We have

$$P(\text{not } R \text{ and not } F) = 1 - P(R \text{ or } F) \tag{17.2.5a}$$

$$= 1 - [P(R) + P(F) - P(R \text{ and } F)] \tag{17.2.5b}$$

We consider the following sub-cases.

(a) *100% dependence.* The probability of R given that F has occurred is unity (i.e., $P(R|F) = 1$). Physically, for our problem, this means that if the structure has collapsed, then the roof has necessarily been damaged. For this case

$$P(R \text{ and } F) = P(F) \tag{17.2.6}$$

(b) *Independence (0% dependence).* Physically, for our problem, independence means that collapse does not entail damage to the roof (collapse may be due to overturning while the roof remains intact). In this case $P(R|F) = P(R)$, that is,

$$P(R \text{ and } F) = P(R)P(F) \tag{17.2.7}$$

For example, for the two conditional probability curves defined by Table 17.2.1 and $v = 120$ mph, Table 17.2.2 shows the values of the joint distribution $P(R$ and $F)$ and the probability of no damage $P(\text{not } R$ and not $F)$ corresponding to various values of the conditional distribution $P(R|F)$.

Tables 17.2.3 and 17.2.4 show damage matrices corresponding to sub-cases A and B.

The cases R and not F, F and not R, R and F, not R and not F are represented in Venn diagrams in Figs. 17.2.1a, b, c, and d, respectively.

TABLE 17.2.2. Probabilities $P(R \text{ and } F)$ and $P(\text{not } R \text{ and not } F)$ Corresponding to $P(R) = 0.80$ and $P(F) = 0.05$ and Various Values of $P(R|F)$

$P(R\|F)$	$P(R$ and $F)$ (Eq. 17.2.4a)	$P(\text{not } R$ and not $F)$ (Eq. 17.2.5b)
0[a]	0	0.15
0.2	0.01	0.16
0.4	0.02	0.17
0.6	0.03	0.18
$0.8 = P(R)$[b]	0.04	0.19
1.0[c]	0.05	0.20

[a] R and F are mutually exclusive states.
[b] R and F are independent states.
[c] 100% dependence of R on F.

TABLE 17.2.3. Damage Matrix, Sub-Case A (100% Dependence)

v(mph)	80	90	100	110	120	130	140
R and not F	0	0.03	0.02	0.48	0.75	0.80	0
R and F	0	0	0	0.02	0.05	0.20	1.00
F and not R	0	0	0	0	0	0	0
No damage (not R and not F)	1	0.97	0.80	0.50	0.20	0	0

TABLE 17.2.4. Damage Matrix, Sub-Case B (Independence)

v(mph)	80	90	100	110	120	130	140
R and not F	0	0.03	0.02	0.49	0.76	0.80	0
R and F	0	0	0	0.01	0.04	0.20	1.00
F and not R	0	0	0	0.01	0.01	0	0
No damage (not R and not F)	1	0.97	0.80	0.49	0.19	0	0

A similar methodology has been applied for the case of five, rather than two, basic damage states in [17-4]. In this case, 457 mathematically possible combinations were found to exist, of which only 217 are both mutually exclusive and physically possible. For example, a structure cannot experience both the state $O_i T_j S_k W_0 C_0$ and $O_i T_j S_k W_0 C_1$. Because of its complexity, the methodology requires specialized mathematical techniques for its implementation.

17.2.4.2 Engineering Calculations Based on Monte Carlo Simulations. Some analysts view the set-theoretical methodology as impractical, and eschew its formalism and structure by using instead an engineering approach to loss estimation [17-2]. Simplifications in the description of the combined states reduce the number of combinations to fewer than 217 [17-2]. A simulation is performed by assuming a storm of given intensity, which may or may not cause breakage of openings and consequent changes in the aerodynamic behavior of the structure. The simulation is then repeated for successively increased wind speeds. For a given type of building, the output of the simulations is a damage matrix whose elements are, for each damage state, the probabilities of that state as functions of wind speed. The internal, utilities, and contents damage to the building are then extrapolated from the external damage by using empirical relations. Included are: damage due to water penetration (through broken windows and lost roof sheathing); damage to interior systems (electrical, plumbing, mechanical) and fixtures (fixed cabinets, carpeting, partitions, doors); and damage to contents. The damage model includes, in addition, estimates of damage to appurtenant structures (pool, deck, unattached garage), and additional living expenses (ALE). The resulting estimates of total damage result in the formulation of damage matrices and vulnerabilities for each statistically significant building type in the Florida building stock.

17.3 LOSS ESTIMATION

Denote by $\text{Damage}_{\text{type } m}(v_j - \Delta v_j/2 \leq v_j < v_j + \Delta v_j/2)$ the percentage of the total replacement cost of a building of type m in a specified region required to repair the damage to that building due to winds speeds $v_j - \Delta v_j/2 \leq v_j < v_j + \Delta v_j/2$. Type m may represent, for example, a typical manufactured home built in a specified region after the latest change in standard provisions on manufactured homes.

Further, denote by $C[(\text{damage state } i)_{\text{type } m}]$ the percentage of the total replacement cost of a building of type m required to repair the damage state i. We have

$$\begin{aligned} &\text{Damage}_{\text{type } m}(v_j - \Delta v_j/2 \leq v_j < v_j + \Delta v_j/2) \\ &\quad = \sum_i C(\text{damage state } i)_{\text{type } m} P[(\text{damage state } i)_{\text{type } m} \\ &\quad\quad |v_j - \Delta v_j/2 \leq v_i < v_j + \Delta v_j/2)] \end{aligned} \tag{17.3.1}$$

where $P[(\text{damage state } i)_{\text{type } m}|v_j - \Delta v_j \leq /2/v_j < v_j + \Delta v_j/2]$ is the probability of occurrence of $(\text{damage state } i)_{\text{type } m}$ given that $v_j - \Delta v_j/2 \leq v_j < v_j + \Delta v_j/2$. For example, (damage state i) may consist of 5–10% loss of roofing of the manufactured home, the cost of replacing such loss of roofing may be $C(\text{damage state } i)_{\text{type } m} = 4\%$ of the total cost of replacement of the building, and for $v_j = 100$ mph and $\Delta v_j = 10$ mph, the probability of 5–10% loss of roofing may be

$$P[(\text{damage state } i)_{\text{type } m}|v_j - \Delta v_j/2 \leq v_i < v_j + \Delta v_j/2] = 10\%$$

(i.e., under 95 to 105 mph winds, one manufactured home in 10 experiences 5–10% loss of roofing). The summation in Eq. 17.3.1 is performed over all mutually exclusive damage states that are relevant from an engineering viewpoint.

The average annual damage for a building of type m as a percentage of the total cost of replacement of the building can be written as

$$\begin{aligned}(\text{Average annual damage})_{\text{type } m} &= \sum_j \text{Damage}_{\text{type } m}(v_j - \Delta v_j/2 \leq v_j < v_j + \Delta v_j/2) \\ &\quad P(v_j - \Delta v_j/2 \leq v_j < v_j + \Delta v_j/2)\end{aligned} \tag{17.3.2}$$

Let the total cost of replacement of a building of type m be denoted by $(\text{Cost})_{\text{type } m}$. The average annual cost of damage (the average annual loss, or AAL) for one building of type m is then

$$(\text{AAL})_{\text{type } m} = (\text{Cost})_{\text{type } m} \times (\text{Average annual damage})_{\text{type } m} \tag{17.3.3}$$

For example, if $(\text{Average annual damage})_{\text{type } m} = 4\%$ and $(\text{Cost})_{\text{type } m} =$ \$200,000, $(\text{AAL})_{\text{type } m} = \$8{,}000$.

Denote the number of buildings of type m by n_m. For the entire stock of buildings of all types being considered in the area of interest, the total average annual cost of the damage is

$$(\text{AAL})_{\text{total}} = \sum_m n_m \times (\text{AAL})_{\text{type } m} \tag{17.3.4}$$

PART IV

WIND EFFECTS ON BUILDINGS

CHAPTER 18

RIGID BUILDINGS

18.1 INTRODUCTION

The total response of buildings to wind loads consists of three parts: (1) the mean response induced by the wind load (i.e., the *static* part of the response), (2) the fluctuating response due to wind force components whose frequencies differ significantly from the building's natural frequencies of vibration (i.e., the *quasi-static* or *background* part of the response), and (3) the fluctuating response due to wind force components with frequencies equal or close to the building's natural frequencies of vibration (i.e., the response associated with dynamic amplification effects, called the *resonant* part of the response).

Flexible buildings are discussed in Chapter 19. To estimate wind effects on flexible buildings, it is necessary to conduct dynamic analyses, that is, analyses that account for the resonant response. For *rigid buildings*, to which this chapter is devoted, the wind-induced resonant response is negligible. For practical purposes, buildings may be considered rigid if their fundamental natural frequency of vibration is at least 1 Hz (ASCE Sect. 26.9).[1] The codification of pressures on rigid buildings aims to achieve the best possible balance between the conflicting aims of developing simple, easy-to-use design tools and representing faithfully the actual aerodynamic pressures. The ASCE 7 Standard's analytical procedures are used mostly for routine building designs (see Part II of this book).

[1]The ASCE 7 Standard applies the specialized term *low-rise buildings* to a class of rigid buildings for which the mean roof height $h \leq 60$ ft and $h \leq$ least horizontal building dimension (ASCE Sect. 26.2). The term is tied to specific ASCE 7 Standard provisions, and is therefore used primarily within their context. In this section we use the more generic term "rigid buildings," which includes but is not limited to the low-rise buildings as defined in the ASCE 7 Standard.

Inherent in the ASCE 7 Standard provisions for wind loads on rigid buildings are large errors—in some cases, larger than 50% on the unconservative side [13-11 to 13-13]. The errors are due in part to the reduction of vast amounts of aerodynamic data to a few numbers contained in tables and plots. For buildings that warrant the use of more accurate procedures than those based on standard tables and plots, such errors can be significantly reduced by using database-assisted design (DAD) procedures. In such procedures, wind effects are determined by using large sets of electronically recorded aerodynamic pressure data measured simultaneously in the wind tunnel at large numbers of points on the building surfaces (Figs. 18.1.1–18.1.3; Fig. 13.3.3).

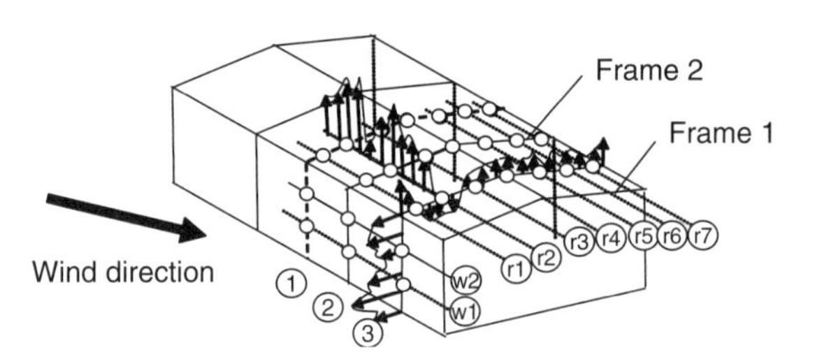

Figure 18.1.1. Building model with pressure taps tributary to Frame 2.

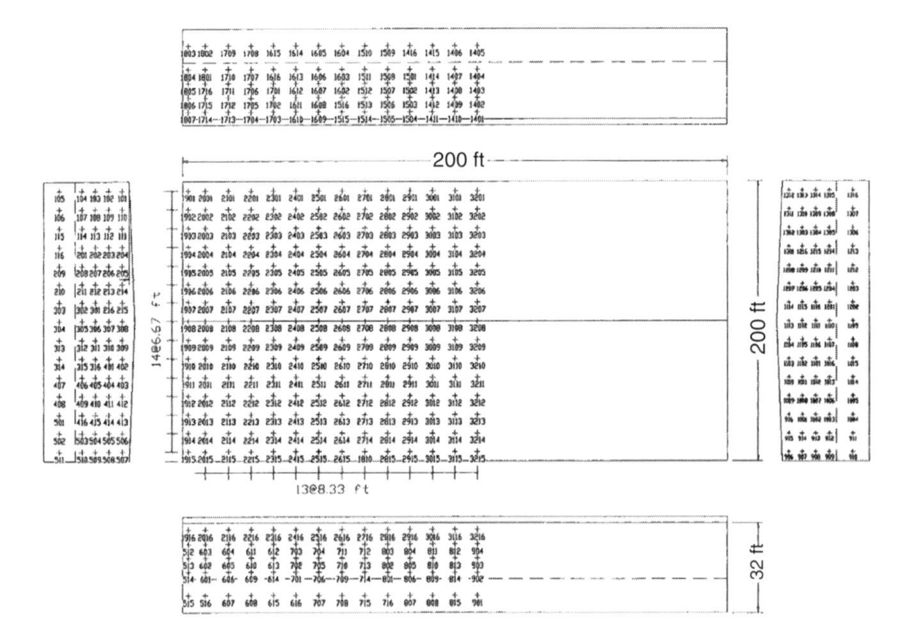

Figure 18.1.2. Building model with locations of pressure taps (from C. E. Ho, D. Surry, and D. Morrish, *NIST/TTU Cooperative Agreement—Windstorm Mitigation Initiative: Wind Tunnel Experiments on Generic Low Buildings*, Alan G. Davenport Wind Engineering Group, The University of Western Ontario, 2003 [18-12]).

Figure 18.1.3. Building model in wind tunnel (from C. E. Ho, D. Surry, and D. Morrish, *NIST/TTU Cooperative Agreement—Windstorm Mitigation Initiative: Wind Tunnel Experiments on Generic Low Buildings*, Alan G. Davenport Wind Engineering Group, The University of Western Ontario, 2003 [18-12]).

The use of DAD procedures is permitted by the ASCE 7-10 Standard (ASCE Commentary Sect. C31.4.2), and allows the estimation of wind effects with specified mean recurrence intervals by accounting for the directional properties of (1) the extreme wind speeds and (2) the building aerodynamics. DAD is evolving into an effective, user-friendly tool (for DAD software and data for rigid buildings, see www.nist.gov/wind). For standard development purposes or for special projects, DAD can also be used for nonlinear analyses providing information on ultimate capacities under wind and gravity loads (Sect. 16.4).

When subjected to wind blowing from any one direction, buildings experience external pressures on all faces. The variation in time and space of external pressures induced by wind from one specific direction is illustrated in the animation, constructed from actual wind tunnel measurements, shown on www.nist.gov/wind.[2] The pressures produce wind loads on components and cladding, and wind forces and torsional moments acting on the main wind force resisting system (MWFRS).

Section 18.2 presents the DAD approach. Section 18.3 briefly discusses the estimation of wind effects with specified mean recurrence intervals by accounting for the dependence upon direction of both the aerodynamic pressures and the wind speeds. Section 18.4 presents material on estimates of design wind effects that account for knowledge errors.

[2]The animation was created by NIST guest researcher Dr. Andrea Grazini, from data provided by the University of Western Ontario. For a still picture from the animation see Fig. 13.2.8.

18.2 DATABASE-ASSISTED DESIGN (DAD)

DAD is an *integrated* wind engineering/structural engineering methodology. The wind engineering input is provided in specified formats by electronic records of aerodynamic and wind climatological data. The structural engineering input consists of information on gravity loads, a preliminary structural design, and influence coefficients that transform wind (and, if present, inertial) forces into wind effects. The output of DAD software consists of peak demand-to-capacity indexes (DCIs) at various cross sections of the structure's members, inter-story drift, and peak accelerations with the respective mean recurrence intervals.[3] The DAD procedure is transparent, that is, fully auditable and traceable by wind engineers, structural engineers, and building officials.

DAD differs from simultaneous pressure integration, which is largely limited to providing information on wind loading, rather than being focused on estimating wind effects used in design and checking the adequacy of the strength and serviceability performance of individual members and of the structure as a whole. DAD can automatically produce time series of DCIs and peak values of those time series corresponding to any specified percentile of the peak value (e.g., 50%). The fact that all calculations are performed in the time domain renders all requisite summations of wind effects easy to perform and physically correct, an attribute not possessed by calculations performed in the frequency domain. To check the adequacy of a structural member, the structural engineer is required to check whether the peak DCIs are smaller than unity, in which case the member or cross section being checked is adequate. If the DCI is larger than unity, the cross section must be redesigned.

Section 18.2.1 describes the origins of DAD. Section 18.2.2 provides details on the DAD approach. Sect. 18.2.3 discusses interpolation issues. Sect. 18.2.4 discusses errors and uncertainties in the estimation of wind effects and how they affect the selection of mean recurrence intervals used in design.

18.2.1 Origins of DAD

DAD has evolved from a methodology developed in the late 1970s and early 1980s at the University of Western Ontario. The aim of the methodology was to develop "pseudo-pressures" for the design of low-rise structures (ASCE Standard 7-10, Sect. 26.2, Envelope Procedure; see also Sect. 2.2.1 of this book).

[3] A DCI is the left-hand side of the cross section's design interaction equation. For steel structures it consists of a sum of ratios in which the numerators are the internal forces (e.g., bending moment and axial force) induced at that cross section by the design wind and gravity forces, and the denominators are the respective member capacities as affected by resistance factors smaller than unity [18-3]. For DCIs of reinforced concrete structures, see [19-10, 19-32].

The methodology is presented with reference to Fig. 18.1.1. The spatial variation of the pressures at a fixed instant in time is represented for portions of lines 3 and r_2. At any fixed point on the building envelope, the pressures vary in time. Pressures are measured in the wind tunnel at a number of pressure taps, of which those affecting Frame 2 are represented in Fig. 18.1.1 as white circles.

Consider, for example, the wind-induced bending moment at time t induced at a cross section j of Frame 2 by wind blowing from a specified direction. The moment is caused by wind forces acting on the frame at time t. To show how these forces are approximated for each of the taps, we refer for specificity to the taps located at the intersections of lines 2 and w_1 and lines 2 and w_2. The following steps are required:

1. Multiply the pressure measured at each tap at time t by the area tributary of the tap. This yields the approximate total wind force acting at time t at the center of that tributary area. For the pressures at taps 2-w_1 and 2-w_2, $p_{2\text{-}w1}(t)$ and $p_{2\text{-}w2}(t)$, the tributary areas are $A_{2\text{-}w1} = (d_{13}/2)(h_2/2)$ and $A_{2\text{-}w2} = (d_{13}/2)(h_e - h_1)/2$, where d_{13} is the distance between lines 1 and 3, h_1 and h_2 are the elevations of taps 2-w_1 and 2-w_2, respectively, and h_e is the elevation of the eave. The corresponding approximate total wind forces are $F_{2\text{-}w1}(t) = p_{2\text{-}w1}(t)A_{2\text{-}w1}$ and $F_{2\text{-}w2}(t) = p_{2\text{-}w2}(t)A_{2\text{-}w2}$, respectively.
2. The roof and wall systems distribute to the frame the forces calculated as in Step 1. For example, assuming that girts are simply supported on the frames, the calculated wind forces associated with taps 1-w_1, 2-w_1, and 3-w_1 acting on Frame 2 are $F_{1\text{-}w1}(t) \times 0.5$, $F_{2\text{-}w1}(t) \times 1.0$, and $F_{3\text{-}w1}(t) \times 0.5$, respectively.
3. The forces $0.5F_{1\text{-}w1}(t)$, $F_{2\text{-}w1}(t)$, and $0.5F_{3\text{-}w1}(t)$ (or similar forces corresponding to the support conditions for the purlins and girts) induce at the cross section j of Frame 2 the bending moments $0.5F_{1\text{-}w1}(t)m_{2\text{-}w1,j}$, $F_{2\text{-}w1}(t)m_{2\text{-}w1,j}$, and $0.5F_{3\text{-}w1}(t)m_{2\text{-}w1,j}$, respectively, where the influence coefficient $m_{2\text{-}w1,j}$ is the moment induced at cross section j by a unit load acting at the intersection of Frame 2 with line w_1.
4. The total bending moment at time t at cross section j of Frame 2 is equal to the sum of contributions similar to those discussed in Step 3 from all the taps affecting the frame.
5. The preceding steps allow the calculation of bending moments and other internal forces induced by pressures measured in the wind tunnel at any time t within an interval t_f corresponding to the duration of a typical storm ($t_f = 20$ min to 1 hr, say). For any response obtained as indicated in this section, the value of interest is the *peak value* during time t_f.

The application of this methodology in the 1970s was affected by limitations of the computational and measurement technology capabilities available at the time. The steps described in Steps 1 through 5 were carried out online following a spatial averaging of pressures by a now obsolete averaging technique. The density of the taps was typically of the order of 1 tap/10 m^2, that is, about one order of magnitude less than current technology allows. Typically, tests were conducted for a total of eight wind directions (45° increments), that is, at a directional resolution almost one order of magnitude lower than in current wind tunnel practice. As was noted earlier, the internal forces being obtained were based on generic sets of influence coefficients, that is, influence coefficients obtained for a few building types purported to be most common, but that, in practice, could differ substantially from the buildings being designed. Similarly, a distance between frames purported to be typical was considered in the online calculations, even though the distance between frames, which depends on the particular design being considered, affects the mean wind loads as well as the extent to which fluctuating loads tributary to the frames are spatially coherent. Finally, the "pseudo-pressures" were developed, largely by eye, for use in the ASCE 7 Standard. These were intended to result in peak moments, horizontal thrusts, and uplift forces comparable to those obtained by using the five steps listed earlier.

The adoption of the methodology just described was an innovative and useful step forward in the codification of wind loads. However, even assuming that the wind tunnel measurements and the performance of the pneumatic averaging technique were satisfactory, errors inherent in that methodology arise, owing to: (1) the relatively small number of taps and small number of wind directions for which measurements were made, (2) the choice of frame properties and locations, which were deemed by the code writers to be typical but can in fact differ widely from those of the building being designed, (3) the use for codification purposes of observed peaks, which, owing to the random nature of the peak response, can differ widely from test to test, and (4) the use of "pseudo-loads," that is, of "eye-ball" envelopes of measured wind effects for the various wind directions considered in the tests. As was mentioned earlier, such errors can exceed 50%.

18.2.2 Description of the DAD Approach

DAD is fundamentally derived from the methodology described in Sect. 18.2.1 [18-1, 18-2]. However, in its practical application, DAD differs from that methodology in that it takes advantage of the following advances [18-11]:

1. A far larger number of pressure taps—by a factor of 10 or more—can be accommodated on a building envelope than was possible in the 1970s.
2. Pressure time histories can be measured simultaneously at all the pressure taps affecting the loading of components and main wind force resisting systems.
3. The simultaneous pressure time histories can be stored electronically for use in calculations.

4. The influence coefficients required to calculate the internal forces can be obtained by standard structural analysis programs for the structure of interest, rather than from generic structures purported to be typical.
5. Time histories of internal forces can be obtained in a routine, user-friendly manner by using software whose input consists of pressure tap locations, stored pressure data, frame locations, and influence coefficients obtained for the structure being designed.
6. DAD software includes a subroutine for calculating *statistics of peak values* by using the information contained in the entire time history of the internal force of interest. Such statistics may consist of the mean and coefficients of variation of the peaks, which can be used in the estimation of the dispersion of the overall uncertainty in the wind effect. Statistics of peak values are preferable to observed values, which vary from record to record. The subroutine based on the Rice method [13-7] allows the estimation of any desired percentile of the peak (Sect. 13.3.5).
7. DAD software also includes software that effects interpolations based on data for buildings with geometric characteristics close to those of the building being designed (Sect. 18.2.3).
8. Database-assisted design databases, software, and examples of its application are available at www.nist.gov/wind. See also [18-13 to 18-15] and www.wind.arch.t-kougei.ac.jp/system/eng/contents/code/w_it.

The software is aimed at implementing the following sequence of operations. A preliminary structural design is developed and used to obtain influence coefficients. Spatial coordinates of the pressure taps, and measured pressure time series, are provided by the wind tunnel laboratory or are available on file. They are used in conjunction with information on the positions of the structural frames and the requisite structural influence coefficients to obtain peak values of the response of interest for each wind direction.

DAD allows the calculation of the time series of internal forces and/or DCIs and their peaks, which are used for the design of the main wind force resisting system and of components such as girts and purlins.

For rigid structures, it is convenient to evaluate peak responses (e.g., peak DCIs) corresponding to a unit wind speed. The resulting direction-dependent response quantities are referred to as directional influence factors (DIFs) [18-4]. Since the linear structural responses being sought are unaffected by dynamic effects, the peak response to the load induced by any arbitrary wind speed can be obtained via multiplication of the corresponding DIF by the square of that wind speed. In this manner, the DIFs can be combined with databases of directional extreme wind speeds from a sufficiently large number of storm events to compute values of the DCIs or forces having specified mean recurrence intervals (Sect. 18.3). These values can then be used to redesign the structure. Influence coefficients are computed for the revised structural design, and the analysis procedure can be repeated as necessary, until satisfactory convergence is achieved.

The ASCE Standard provisions for low-rise buildings were developed on the basis of tests on a modest number of basic configurations—about 10—that do not represent all the building configurations covered by the provisions. Also, interpolations had to be performed by eye. To date, a larger number of building configurations have been covered for DAD development purposes by tests conducted at the University of Western Ontario. For DAD to be a routine tool resulting in safer and more economical design of buildings of a wide variety of types, it is necessary that the acquisition, certification, and archiving of aerodynamic data be expanded in the future. Whenever warranted by economic considerations, ad hoc wind tunnel tests can be performed for the building being designed.

Numerical Example 18.2.1. *Internal forces in a portal frame of an industrial metal building.* Consider a rectangular building with dimensions 200 ft × 100 ft in plan, eave height 20 ft, and gable roof with slopes 1/24, located in terrain with Exposure B in the Miami, Florida, area. Pressures taps were placed on the model in accordance with the schematic of Fig. 18.1.2. (Fig. 18.1.3 is a view of a wind tunnel model ready for the placement of pressure transducers.) The main wind force resisting system consists of portal frames, shown in elevation in Fig. 18.2.1. Steps similar to those described in Sect. 18.2.1, Steps 1 through 5, resulted in time histories of internal forces in the frame. For use in conjunction with wind tunnel data, 10 m 3-s peak wind speeds over open terrain were converted to mean hourly speeds at the wind tunnel reference height. Calculations described in Sect. 18.2.2, item 6, yielded the results of Table 18.2.1.

Table 18.2.1 suggests that statistical inferences based on 30-min or 20-min records differ from inferences based on 1-hr records by relatively small amounts that contribute negligibly to the overall uncertainty in the estimation

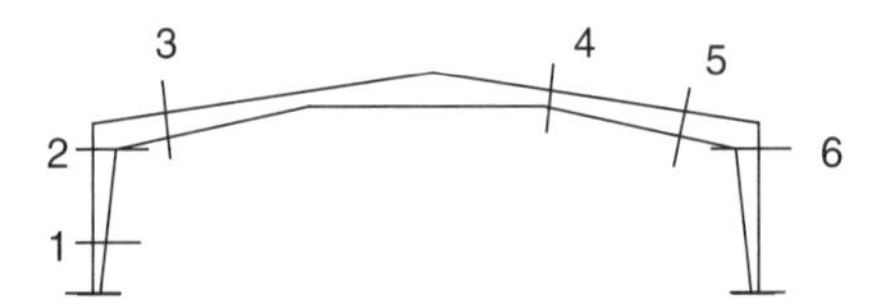

Figure 18.2.1. Elevation of frame for building of Fig. 18.1.2.

TABLE 18.2.1. Mean Values, Standard Deviations, and Percentiles of 1-Hour Record Peak Bending Moments (*kN-m*) Estimated from 1-Hour, 30-Min, and 20-Min Time Histories

Record length	Mean	Standard Deviation	84 percentile	97.5 percentile
1 hour	643	55	695	776
30 min	639	54	691	770
20 min	639	54	690	769

of the peak wind effects. According to these results, it would be possible to use for peak estimation purposes records of, say, 20-min length (about 20 s at model scale), instead of 1 hour, as is common practice at present.

18.2.3 Interpolation

In practice, it is not possible to provide aerodynamic databases for all possible sets of building dimensions corresponding to a given building configuration. Therefore, if pressure records are available for two buildings having the same configuration but somewhat different dimensions, internal forces on a similar building with intermediate dimensions can be obtained by interpolation. We now describe a simple and effective interpolation procedure [18-5].

Assume that pressure data are available for two buildings, denoted by A and C, whose dimensions (and/or roof slopes) are relatively close to and, respectively, larger and smaller than those of the building of interest, denoted by B. The coordinates of the pressure taps of buildings A and C are scaled to match the dimensions of building B. Using these scaled coordinates, DIFs are computed as described in Sect. 18.2.2, treating the measured pressures on buildings A and C as if they had been recorded on a model having the same dimensions as building B. This procedure can be applied using measurements from only one wind tunnel model (i.e., A or C), but improved accuracy can generally be achieved by making use of measurements from several models, having dimensions that bound those of the structure of interest. For each of the models, the tap coordinates are scaled and DIFs are computed. The DIFs for the building of interest are then estimated by taking a weighted average of the DIFs corresponding to each of the wind tunnel models, with greater weight being given to results from models that more closely match the dimensions of the building of interest. The structure is designed on the basis of those DIFs. An example of this interpolation procedure is given in [18-5], in which the results of the interpolation match remarkably well the results based on data obtained directly for the building of interest. For interpolations between buildings with different roof slopes, the extent to which this procedure yields satisfactory results remains to be tested. The requisite software is provided at www.nist.gov/wind.

Alternative interpolation procedures are based on rescaling of the measured pressure time histories of tested buildings [18-6], or the creation of time histories on the unmeasured building via stochastic simulation based on models inferred from buildings A and C [18-7].

18.2.4 Errors and Uncertainties Associated with Laboratory Simulations

Errors and uncertainties associated with wind tunnel simulations are due to the violation of the Reynolds number in the laboratory (Sect. 13.3) and to flow simulation problems [13-5, 13-6]. For this reason, wind tunnel simulations of pressures on low-rise buildings can underestimate the prototype pressures (Fig. 13.3.2). An accessible and transparent compendium of corrections to wind tunnel pressures to approximately match full-scale measurements would be helpful, but remains to be developed.

Section 13.4 describes an alternative wind tunnel simulation method for buildings with relatively small dimensions, such as individual homes. Over distances equal to or smaller than the dimensions of such buildings, the spatial coherence of the low-frequency fluctuations in the incoming flow velocities is close to unity. Therefore, unlike in the case of tall buildings, peak aerodynamic effects on small buildings can be simulated in shear flows with no low-frequency fluctuation components; a commensurate increase in the mean wind speed compensates for the absence of such components. One advantage of this approach is that it reduces errors due to the difficulty of simulating low-frequency fluctuations. A second advantage is that, since the integral turbulence scale is no longer a factor in the simulation, it is no longer necessary to reproduce in the laboratory the ratio between integral turbulence lengths and building dimensions. This allows the tests to be conducted at larger scales than would be possible in conventional simulations, an important advantage given the typically very small dimensions of wind tunnel models of individual homes.

18.3 WIND DIRECTIONALITY EFFECTS

For rigid structures with linearly elastic behavior and orientation defined by an angle φ between a principal axis and the north direction, the wind effect $Q_i(\theta_j, \varphi)$ (e.g., a bending moment, or an axial load) induced by a wind speed $V_i(\theta_j)$ blowing from direction θ_j may be written as

$$Q_i(\theta_j, \varphi) = \kappa(\theta_j, \varphi) V_i^2(\theta_j) \tag{18.3.1}$$

The directional effect coefficient $\kappa(\theta_j, \varphi)$ is the wind effect induced in the building with orientation φ by a unit mean wind speed at the top of the building from direction θ_j, and is called the directional influence factor (DIF—see Sect. 18.2.2). The mean wind speed at the top of the building blowing from direction θ_j in storm event i is denoted by $V_i(\theta_j)$. This section discusses several approaches used in standards or by wind engineering consultants to account for the directionality of both the aerodynamic effects and the extreme wind speeds.

The approach specified in the analytical procedure of the ASCE 7 Standard consists of using a building *wind directionality factor* $K_d = 0.85$ applied to the wind effect calculated by disregarding wind directionality (Sect. 4.2.3). The wind effect induced by wind event i is thus defined as

$$Q_i = 0.85 \max_j[\kappa(\theta_j, \varphi)] \max_j[V_i^2(\theta_j)] \tag{18.3.2}$$

This approach is the simplest, but can either overestimate or underestimate the response, and is therefore typically not used by wind engineering laboratories for estimating wind effects on special structures. The angle φ is omitted from the left-hand side of Eq. 18.3.2 because the building orientation does not affect the maxima of the directional effect coefficients $\kappa(\theta_j, \varphi)$. ASCE 7 Standard estimates of wind effects, other than estimates for MWFRS based on ASCE Chapter 28, are therefore independent of building orientation.

The approach based on *non-parametric statistics* introduced in Sect. 12.7 incorporates the directional aerodynamic and wind climate effects in a simple, transparent, and rigorous manner. This approach requires the development of a matrix of wind speeds $V_i\,(\theta_j)$ in which $j = 16$ or 36, say; $i = 1, 2, \ldots, m$; and m is sufficiently large for the number of storm events or the years of record to cover a time interval longer than the mean recurrence interval of the design wind effect (see Sects. 12.5 and 12.6). A vector with m components is then created, the ith component of which is the largest of the directional wind effects in the ith storm, $\max_j [Q_i\,(\theta_j, \varphi)]$. The components of the vector are rank-ordered, and the wind effect with an $\overline{N}$-yr mean recurrence interval is obtained as in Sect. 12.7.

In principle, *parametric statistics* (Sect. 12.3) may be employed by fitting an extreme value probability distribution to the quantities $\{\max_j [Q_i\,(\theta_j, \varphi)]\}^{1/2}$, which are proportional to wind speeds. This approach is applicable to the case where m represents the number of storm events for which wind speed measurements are available, rather than the larger number of synthetic storm events obtained by Monte Carlo simulation from the measured wind speeds. Whether extreme value statistics applied to the quantities $\{\max_j [Q_i\,(\theta_j, \varphi)]\}^{1/2}$ would yield estimates of wind effects comparable to estimates based on simulated wind speeds remains to be determined.

The *outcrossing* approach is, to our knowledge, currently used by only one or two laboratories. The models include in the data samples non-extreme wind speed data, such as speeds recorded at one-hour intervals [18-8, p. 167] or low speeds occurring in peripheral hurricane zones. This typically causes the underestimation of wind effects, a concern noted in [18-9]. A second drawback is the perception by structural engineers that the approach is opaque (see, e.g., Appendix A5). For details on the outcrossing approach, see Appendix A4.1.

The *sector-by-sector* approach [18-10] used by some wind engineering consultants is described in Appendix A4.2. This approach typically yields unconservative results.

18.4 UNCERTAINTIES IN THE ESTIMATION OF WIND EFFECTS

Owing to the existence of knowledge uncertainties, the estimate of a wind effect with a specified MRI is affected by an uncertainty characterized by a probability distribution. Therefore, a wind effect is determined by (1) its specified MRI $\overline{N}$ (or, equivalently, by its probability of non-exceedance), and

(2) the probability p that the error in the estimation of the wind effect with MRI $\overline{N}$ is not exceeded (Sect. 16.7). ASCE Standard 7-10 specifies MRIs of the design wind speeds that take into account typical knowledge uncertainties. For some buildings, the uncertainties can be larger than those accounted for in the Standard, for example owing to the small size of the available set of wind speed measurements and the consequent large sampling errors inherent in that set. A calibration procedure is described in Sect. 16.7 that takes into account those larger uncertainties, and estimates on their basis design MRIs that ensure that risks inherent in the design do not exceed the risks inherent in Standard provisions for typical structures.

CHAPTER 19

TALL BUILDINGS

19.1 INTRODUCTION

Wind blowing from any given direction induces in tall, flexible buildings fluctuating forces along the principal axes of the building, as well as torsional moments due to the non-coincidence of the point of application of the aerodynamic loads and the building's elastic center. Since inertial forces act at the mass center, additional torsional deformations occur if the building's mass center and elastic center do not coincide.

Assume that the building has a rectangular shape in plan and that the wind blows in a direction normal to a building face. The response in the wind direction and the response normal to the wind direction are then called the *along-wind* and *across-wind* response, respectively. The along-wind response is due to the mean and the fluctuating drag force (Sect. 13.2.2), which in turn are due to positive pressures on the building's windward face, and negative pressures (suctions) on the building's leeward face. The across-wind force is due predominantly to the fluctuating lift force induced by vorticity shed in the building's wake (Sect. 13.2.3). If the wind direction is not normal to one of the building faces, distinguishing between along-wind and across-wind response is no longer meaningful in practice. Rather, for each wind direction, it is necessary to determine the responses along the principal axes of the building and in torsion. If the building's wind-induced motions are sufficiently large, the fluid motions and the building motions interact, and the building can experience *aeroelastic effects*.

Analytical methods are available for estimating the along-wind response of flexible buildings in terrain with uniform upwind surface roughness (see, e.g.,

ASCE Sect. 26.9.5 and Sect. 19.6.1). However, estimates of the along-wind response are not sufficient for design purposes, since the across-wind response is typically comparable to or larger than the along-wind response; for some types of building, torsional effects are significant as well. Dependable analytical methods for estimating across-wind response are not available, except for rough, preliminary estimation methods that cannot be used for structural design. These include the use of an interactive database [19-1] mentioned in ASCE 7-10 Commentary C26.9, and the even rougher estimates by equations included in Sect. 19.6.2. Equations for the preliminary estimation of torsional effects are included in Sect. 19.6.3. For wind skewed with respect to a building face, the wind loading cannot be defined analytically.

For design purposes, structural and serviceability response estimates are performed by using ad hoc wind tunnel test results and extreme wind speed data. In all cases, the mean recurrence intervals of the design wind effects need to conform to the applicable standard requirements. However, if warranted by an uncertainty analysis, they need to be calibrated against the requirements discussed in Sect. 16.7.

The following approaches to wind tunnel testing are presented in this chapter:

1. The high-frequency force balance (HFFB) approach, which makes use of rigid models and strain measurements of base moments and shears.
2. Aeroelastic testing, in which the aeroelastic effects are estimated from measurements of strains in the model's flexible balance.
3. The database-assisted design (DAD) approach, which makes use of rigid models for buildings known to exhibit negligible aeroelastic response, or of flexible models for buildings that exhibit aeroelastic response.

Wind tunnel testing is only one of the elements required to perform response estimates. To obtain the response estimates needed for strength and serviceability design, wind tunnel test results must be used in conjunction with extreme wind speed data sets and dynamic response estimates. As noted in Appendix A5, various wind engineering consultants' procedures differ from each other in ways that can lead to significantly different response estimates. The DAD approach improves upon the current state of the art embodied in the approaches based on strain measurements at the base of the model by:

- Allowing for a clear and effective division of responsibilities between the wind engineering consultant and the structural design engineer.
- Allowing for the provision of full, auditable, traceable, and transparent records of measured aerodynamic data and of the requisite observed

and/or synthetic wind climatological data and their micrometeorological conversions.

- Accounting in a rigorous manner for wind directionality effects.
- Considering actual fundamental modal shapes, rather than modal shapes that vary linearly with height.
- Including effects of higher modes of vibration.
- Combining wind effects in a convenient and physically correct manner.
- The development of transparent and effective algorithms and software that use the wind climatological, aerodynamic, and structural information to yield demand-to-capacity indexes, and estimates of inter-story drift and top-floor accelerations.

The chapter is organized as follows. Section 19.2 is concerned with the high-frequency force balance approach. Section 19.3 briefly discusses the aeroelastic balance approach. Section 19.4 describes the database-assisted design approach, applicable to buildings that do or do not exhibit aeroelastic response. Section 19.5 briefly discusses response mitigation devices. Section 19.6 provides simple algorithms or formulas, and references software for the preliminary estimation of tall building response.

19.2 HIGH-FREQUENCY FORCE BALANCE APPROACH (HFFB)

In the HFFB approach, the test models are rigid. They are supported at the base by a high-frequency force balance that allows measurements of strains

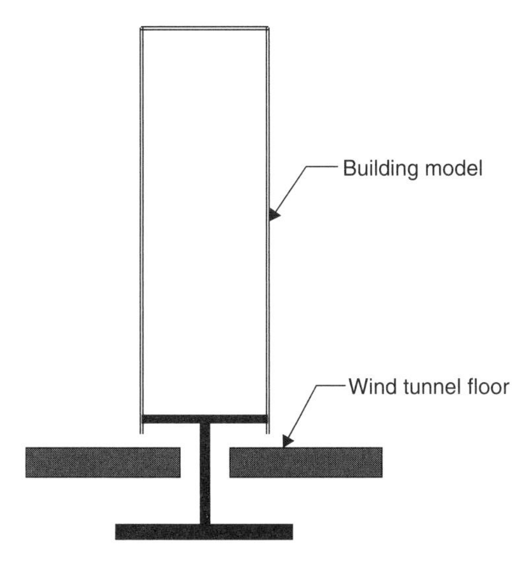

Figure 19.2.1. Schematic of force balance model.

proportional to the base bending moments, shears, and torsional moments, but allows only very small deformations that render the model motions negligibly small (Fig. 19.2.1). The HFFB approach is designed primarily for buildings with approximately straight-line fundamental modal shapes in sway along the principal axes of the building.

For the fundamental mode in the x-direction, assuming that the modal shape is a straight line, the expression for the generalized force is

$$Q_{x1}(t) = \int_0^H w_x(z,t)(z/H)\,dz \tag{19.2.1}$$

(see Eq. 14.3.4, with minor changes of notation), where H = building height, $w_x(z,t)$ = wind loading parallel to the x-direction per unit height, and z/H = fundamental modal shape. Note that the base moment induced by the wind loading $w_x(z,t)$ is proportional to the integral of Eq. 19.2.1. Owing to this coincidence, the base moment is proportional to the generalized force $Q_{x1}(t)$. A similar statement applies to the generalized force in the y-direction, provided that the fundamental modal shape is also a straight line. The HFFB approach takes advantage of this coincidence.

On the other hand, while the generalized aerodynamic torsional moment has the expression

$$Q_{\varphi 1}(t) = \int_0^H T(z,t)\varphi_{T1}(z)\,dz \tag{19.2.2}$$

where $T(z,t)$ is the aerodynamic torsional moment per unit height, and $\varphi_{T1}(z)$ is the fundamental mode of vibration in torsion, the base aerodynamic torsional moment measured in the wind tunnel is

$$Q_{\varphi 1,\mathrm{HFFB}}(t) = \int_0^H T(z,t)\,dz. \tag{19.2.3}$$

Since $\varphi_{T1}(z) \neq 1$ (e.g., $\varphi_{T1}(z) = z/H$), the measured base torsional moment is a poor substitute for the fundamental generalized torsional moment $Q_{\varphi 1}(t)$.

If the fundamental modes of vibration in the x- and y-directions do not vary linearly with height, the measured base bending moments are also inadequate substitutes for the expressions of the respective modal generalized forces. Corrections accounting for the actual modal shapes can be effected [see, e.g., 7-1, p. 340], but they depend upon the distribution of the wind pressures, which is in general unknown, especially for buildings strongly affected by aerodynamic interference effects. The corrections, and the corresponding approximations of the generalized torques and moments, then depend upon assumptions—guesses—on the wind pressure distribution. It is commonly assumed that the pressure distribution is the same as for a building in terrain with uniform roughness, but aerodynamic interactions between buildings in urban settings can render this assumption inadequate.

19.2.1 Advantages and Drawbacks of the HFFB Approach

The HFFB procedure has the advantages that it is relatively inexpensive and fast, and that it is compatible with the presence of architectural details that may render difficult the use of pressure taps. The procedure is convenient for use in studies of aerodynamic alternatives. One drawback of the HFFB approach is that its results can be inaccurate, depending upon the structure's dynamic and aerodynamic characteristics. Also, if, as is typically the case, local pressures are needed for cladding design, separate testing must be performed that adds to the cost of the overall wind tunnel testing.

19.3 AEROELASTIC EFFECTS. TESTING BASED ON STRAIN MEASUREMENTS

Multi-degree-of-freedom models for aeroelastic testing reproduce to the appropriate scale the distribution with height of the building's mass distribution and flexibility. If properly designed and constructed, such models replicate approximately the dynamics of the prototype, including the modal shapes. However, they are expensive, their construction is difficult and time-consuming, and therefore they are rarely used. For details of an elaborate aeroelastic model, see [15-2, Fig. 5.3.20].

A simpler type of aeroelastic model, called the "stick" model, is therefore more commonly used. The "stick" model is similar to the HFFB model, except that its force balance device (1) has the flexibility required for the approximate simulation of the building's aeroelastic motions, and (2) allows the simulation of damping. Various designs of the force balance, as well as analytical correction procedures, are employed to reduce the discrepancies between motions corresponding to linear fundamental modal shapes on the one hand and nonlinear modal shapes on the other [19-2, 19-3, 19-30], and to account, to the extent possible, for the simplified replication in the wind tunnel of the prototype's mass distribution. Even simplified aeroelastic modeling of flexible buildings can provide useful approximate, or at least qualitative, representations of the extent to which aeroelastic effects are significant.

For many tall buildings, those effects are minor or negligible. Tests have shown, for example, that this was true of the World Trade Center twin towers. Moreover, aeroelastic tests indicate that models commonly experience average positive aerodynamic damping. The explanation for such damping lies in the change of the relative velocity of the oncoming flow with respect to the building, as the latter experiences fluctuating motions. A numerical procedure for estimating positive aerodynamic damping for along-wind motions is presented in [19-4]. No analytical procedure is available, however, for aeroelastic effects due to (1) galloping (Sect. 15.3), (2) vortex-induced motions with Strouhal numbers equal or close to the reduced frequency n_1D/V (see Sect. 13.3.1), or (3) vortex-induced motions with reduced frequencies larger than n_1D/V but amplified by increases in the correlations of the pressures along the building height due to the building motions (Sect. 15.2.2).

Rough tentative criteria applicable to untapered buildings with a square shape in plan suggest that aeroelastic tests may be necessary if, for wind normal to a building face, the calculated root-mean-square of the across-wind response is $\sigma_y > \alpha B$, where B denotes the width of the building, and α is a factor that depends upon the upwind terrain roughness. For open, suburban, and urban terrain, $\alpha \approx 0.015$, $\alpha \approx 0.025$, and $\alpha \approx 0.045$, respectively [19-5]. According to results reported in [15-4], for 500 m tall buildings with ratio of height to width $H/D = 10$, square plane cross section, damping ratio $\zeta \approx 1.5\%$, and fundamental natural frequency in the x- and y-motions $n_1 = 0.1$ Hz, these criteria correspond to mean reduced velocities at the top of the building $V_r \approx 9$ and $V_r \approx 11$ in open and urban terrain, respectively, where $V_r = V/(n_1D)$ and $V =$ mean wind speed at the top of the building. Rough calculations based on the use of the Strouhal number suggest that significant vortex-induced vibrations do not occur if the height of typical untapered buildings with a rectangular shape in plan is less than, say, 7 times the smaller dimension in plan.

There are several reasons why aeroelastic effects in tall buildings tend to be small for practical purposes. First, to a greater extent than for chimneys or stacks, vortex-induced forces on buildings tend to have broad frequency distributions, so that only a relatively small fraction of the total vortex-induced force is available for producing oscillations of the lock-in type (Sect. 15.2). This reduces the aeroelastic import of the vorticity. Second, for slightly tapered or untapered buildings, three-dimensional flow effects impede the formation of vortices near the building top, where they would be most effective. Third, nonuniform mean wind speed profiles tend to affect the theoretical vortex-shedding frequencies along the building height, and therefore tend to reduce the extent to which the action of the vorticity is coherent in the vertical direction. A similar favorable effect can be achieved by designing the building to be strongly tapered and with plane cross sections that exhibit discontinuities in the vertical direction. The design of the 828 m tall Burj Khalifa ("Khalifa Tower") (Fig. 19.3.1) was influenced by this fact. The tower's concrete structure is Y-shaped in plan. Each wing has its own high-performance concrete core and perimeter columns, and buttresses the other wings via a six-sided central core, resulting in a torsionally very stiff structure [19-6].

The building configuration can play an important role in reducing or suppressing vortex excitation and otherwise improving aerodynamic and aeroelastic behavior. Chamfering of the corners can be effective in this

Figure 19.3.1. Burj Khalifa, Dubai (by permission of Alfred Molon Photo Galleries).

regard [15-2, Fig. 4.3.6]. Openings or slits in the building can perform a useful aerodynamic function by inhibiting vortex formation, to a degree that needs to be determined by tests.

The role of aeroelastic effects should be checked for responses exceeding the 1,700-yr MRI specified in the ASCE 7 Standard, or the possibly longer MRIs determined as indicated in Sect. 16.7.

Figure 19.3.2 shows the effect of the 1926 Miami hurricane on the 17-story Meyer-Kiser Building, which experienced large torsional moments [19-7]. The 60-story John Hancock building in Boston (Fig. 19.3.3), which also has

Figure 19.3.2. Damage to 17-story Meyer-Kiser Building due to 1926 Miami hurricane [19-7].

an elongated shape in plan, appears to also have experienced large torsional motions, likely amplified by its flexibility. Torsional tuned mass dampers, which alleviate only resonant vibrations (Sect. 19.5), were likely provided for this reason. It appears that structural measures were required as well. However, for legal or other reasons, details concerning the building's aeroelastic response are not publicly available.

The famous *Endless Column* created by Constantin Brancusi (Fig. 19.3.4) has an uncommonly slender shape. The column was found to be "aeroelastically indifferent," meaning that it inhibits coherent vortex shedding as well as galloping. The exceptionally high aspect ratio of the 29.3 m column was made possible by both its aerodynamic properties and its large mass and damping [19-24, 19-25, 19-31].

Figure 19.3.3. John Hancock tower, Boston (by permission of Fotosearch).

19.4 DATABASE-ASSISTED DESIGN

The database-assisted design (DAD) approach is a time domain approach [19-8, 19-9, 19-10, 19-32]. Note the use of the term "design" in its designation. DAD is not aimed merely at providing the structural engineer with wind loads associated with spatially averaged pressures. Rather, DAD is an integrated design methodology that includes member sizing. The member sizing is dictated by (1) the building's aerodynamic and structural properties, (2) the structure's wind environment and its directional interaction with those properties, and (3) design criteria for strength and serviceability under wind and gravity loads. The DAD approach enables the accurate and automated design of each structural member for strength, and of the entire structure for serviceability.

As indicated subsequently in this section, DAD requires the wind engineer to produce (1) the directional pressure time histories from wind tunnel testing recorded simultaneously at a sufficient number of taps, and (2) wind climatological directional data recorded at a nearby weather station or developed by Monte Carlo simulations (Sects. 12.5, 12.6). In addition, the wind engineer must produce the ratio between directional wind speeds at the weather station and the reference directional mean wind speeds at the top of the

Figure 19.3.4. *Endless Column* by Constantin Brancusi, Târgu Jiu, Romania, 1938. By permission of the World Monuments Fund. Copyright Kael Alford/World Monuments Fund.

building, given the building's exposure. This ratio enables the transformation of wind tunnel pressure measurements into prototype pressures on the building envelope.

Once the aerodynamic and wind climatological data produced by the wind engineer are available, the structural engineer can use them for accurately determining individual member demand-to-capacity indexes (DCIs), interstory drift, and top-floor accelerations, corresponding to any specified mean recurrence interval (MRI). The DCI is an indicator of structural strength and adequacy based on structural interaction equations. It incorporates relevant AISC [18-3] or ACI 318 [19-11] and ASCE 7 Standard requirements [2-1], and is briefly discussed in Sect. 19.4.3.

19.4.1 Advantages and Drawbacks of the DAD Approach

The DAD approach has a number of advantages from the point of view of:

- *Accuracy of physical modeling*. DAD can account naturally for higher modes and any modal shape. It also can account automatically for noncoincident elastic centers and centers of mass [19-23] (Sect. 14.4.1.1).

The advantage of the time domain approach used in DAD is that superpositions of wind effects, including summations effected in the expressions for the DCIs, are easily performed as simple algebraic operations, whereas they would pose insurmountable problems in a frequency domain approach. Some wind engineering consultants use methodologies requiring designers to consider as many as 20 separate combinations of global wind effects due to motion in the x- and y-directions and in torsion. These combinations are obtained by using a variety of generic combination factors developed largely by eye. In contrast, a differentiated and accurate approach is made possible by DAD, in which combined wind effects are determined by making use of the influence coefficients specific to each structural member. The automated member-by-member design inherent in the DAD approach is feasible owing to the computing power currently available.

- *Estimation of the wind effects by accounting for wind directionality*. The DAD estimation of wind directionality effects is both convenient and physically correct, and eliminates the possibility of errors similar to those that occurred in the design of the Citigroup Center building in New York City (http://en.wikipedia.org/wiki/Citigroup_Center). That structure is supported by four first-floor columns (legs) located at the centerlines of the structure's four facades, rather than at the structure's corners. Accounting incorrectly for directional winds resulted in inadequate design that required strengthening of the structure following its completion.
- *Convenience*. DAD requires only one test for both structural and cladding design.
- *Transparency*. Every step of the approach can be followed, clearly understood, and audited by wind engineers, structural engineers, and building officials. That wind and structural engineering approaches should satisfy the requirement of transparency is obvious. Nevertheless, this requirement is typically not met satisfactorily by conventional approaches (see Appendix A5). The database-assisted design approach has been developed with this requirement in mind.
- *Clarity in the division of responsibilities between the wind engineer and the structural engineer*. The wind engineer must provide (1) records of time histories of pressures at large numbers of taps on the building surface for the requisite number of wind directions, and (2) records of observed or synthetic directional wind speeds and their conversions to the requisite elevation (top of building) and time average (10 min, 20 min, or 1 hr, say). The conversions are needed so that the data provided to the structural engineer will be consistent with the reference wind speeds for pressure measurements in the wind tunnel. Having access to the full wind loading information, the structural engineer can perform all the requisite calculations required for the design of the structural members. Specialized software is described in [19-9, 19-10] and is listed at www.nist.gov/wind.

The disadvantage of the DAD approach is that it is more expensive and time-consuming than the HFFB approach insofar as manual operations need to be performed in the laboratory to install pressure taps and tubing connections (Fig. 13.3.3). As was pointed out in Sect. 19.2.1, in some instances architectural details of the facade may render difficult the use of pressure taps.

19.4.2 Basic Steps in the DAD Approach

The DAD approach as applied to high-rise buildings entails the steps represented in Fig. 19.4.1 for the particular case of reinforced concrete buildings

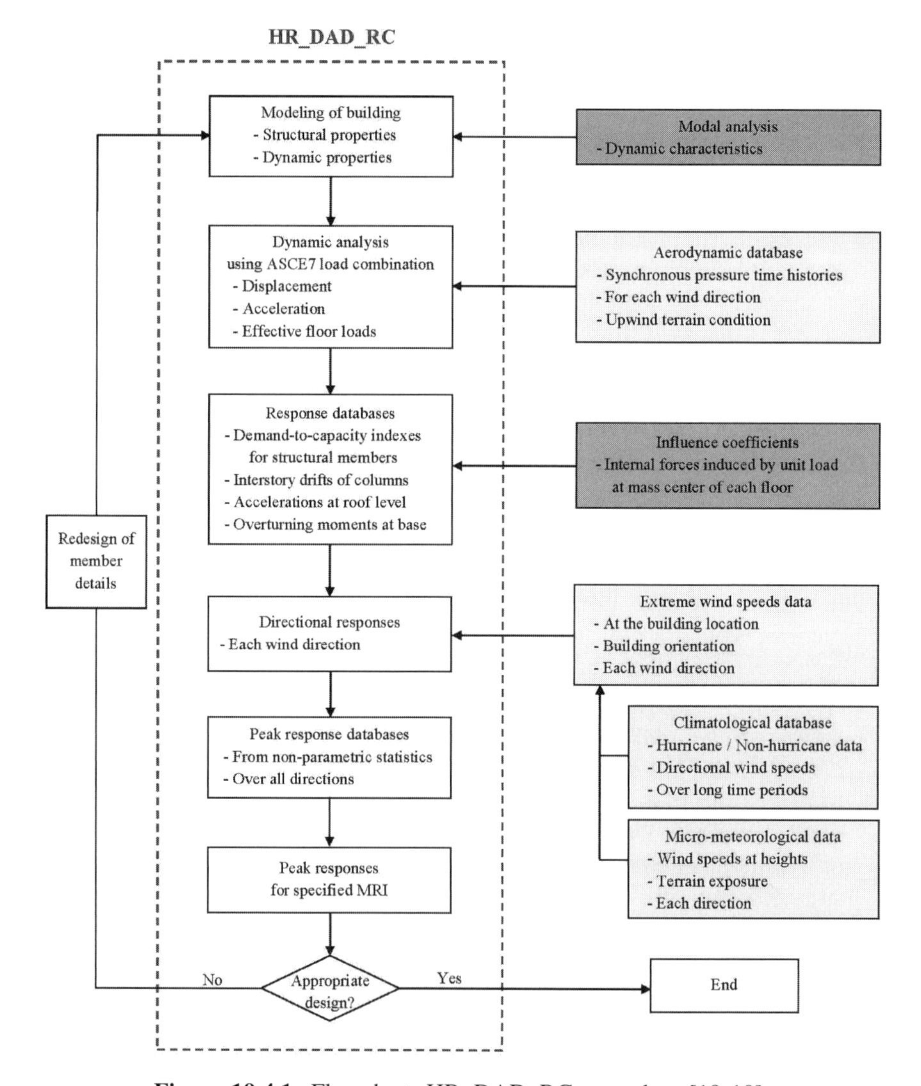

Figure 19.4.1. Flowchart, HR_DAD_RC procedure [19-10].

(The approach is similar for steel structures [19-9]). The processes within the dotted box constitute the main algorithm of the High-Rise Database-Assisted Design for Reinforced Concrete structures (HR_DAD_RC) software [19-10]. The processes outside the box describe information provided by the wind engineer and the structural engineer. The light grey blocks and the dark grey blocks correspond to the wind engineer's and the structural engineer's tasks, respectively. The DAD approach entails the following steps:

1. *Preliminary design.* A preliminary design, based on wind speeds specified in the relevant code or by the wind engineering consultant, is performed by the structural engineer, for example by using the algorithm of ASCE Sect. 26.9.5 [2-1]. This yields an *initial set of building member dimensions*. The fundamental natural frequencies of vibration for the preliminary design can be obtained by modal analysis using a finite element analysis program. The damping ratios are specified by the structural engineer (see [2-1], [19-21, 19-22]).
2. *Pressure coefficient time histories.* The wind engineer provides time histories of nondimensional pressure coefficients for a sufficient number of wind directions (e.g., 16 or 36). An arrangement of pressure taps on the facades of a tall building moment is shown in Fig. 19.4.2.
3. *Dynamic analyses under specified sets of wind speeds.* The building with the member dimensions determined in Step 1 is analyzed by considering the wind forces acting at the mass center of each floor (or group of floors), induced by winds with mean speeds at the building top of, say, 20 m/s, 30 m/s, . . . , 80 m/s, blowing from each of the wind directions. This step is performed by the structural engineer, using as input the directional aerodynamic pressures database (Step 2). The outputs of this step are the effective (aerodynamic plus inertial) lateral forces at each floor (or group of floors) corresponding to each of those wind speeds. Three-dimensional effects are automatically accounted for through the incorporation in the algorithm of the methodology of Sect. 14.4.
4. *Influence coefficients.* The requisite influence coefficients yield internal forces in any member, displacements at each floor, and top-floor accelerations, due to a unit load with specified direction, or a unit torsional moment, acting at the mass center of any floor or group of floors. Influence coefficients are determined by using standard structural software.
5. *Response database.* For each direction and wind speed (e.g., 20 m/s, 30 m/s, . . . , 80 m/s) of Step 3, internal forces are calculated, using summations of effective lateral loads at the mass center of each floor (or group of floors) (Step 3) times the appropriate influence coefficients (Step 4). The forces are added to the respective internal forces induced by factored gravity loads. This allows the calculation of demand-to-capacity indexes indicating whether a member is safe (Sect. 19.4.3). The effective lateral forces and the appropriate influence coefficients

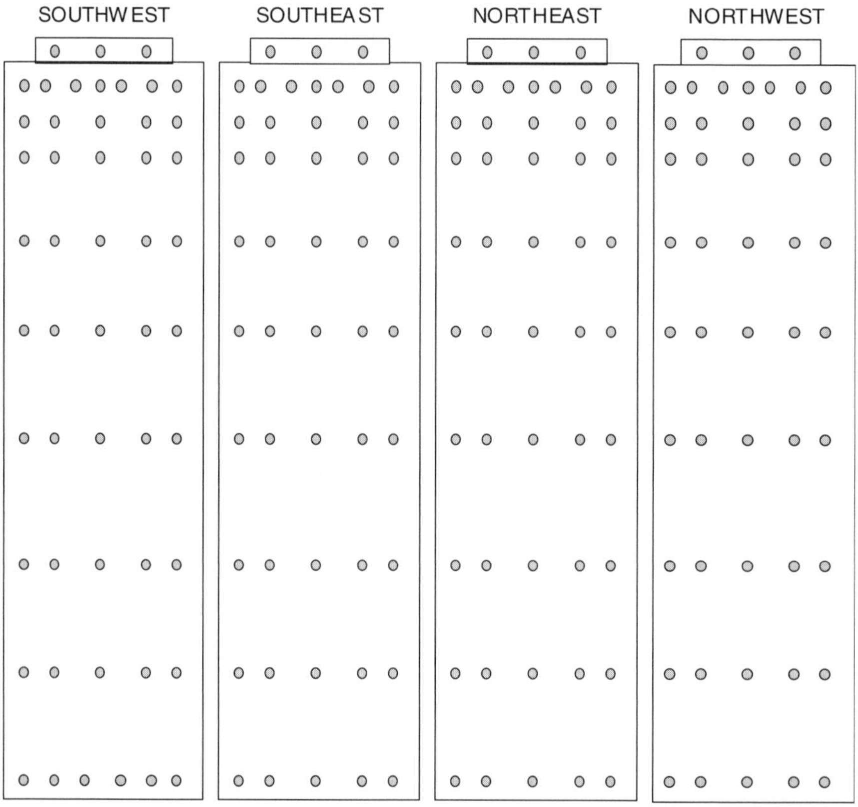

SOUTHWEST SOUTHEAST NORTHEAST NORTHWEST

Figure 19.4.2. Arrangement of pressure taps on the facades of a tall building model.

are similarly used for calculations of inter-story drift along the building height and of top-floor accelerations. The output of this step is a response database, a property of the structure that incorporates its aerodynamic and mechanical characteristics and is independent of the wind climate (Fig. 19.4.3).

6. *Directional wind speed matrix.* Wind speeds at 10 m above ground in open exposure (i.e., a *wind climatological database*) are provided by the wind engineer for a location close to the building. Where necessary, a sufficiently large matrix of wind speeds for each of 36, 16, or 8 directions being considered is developed from measured or simulated wind speed data as indicated in Sects 12.5 and 12.6. Each line of the matrix corresponds to one storm event (if a peaks-over-threshold estimation procedure is used) or to the largest yearly speed (if an epochal estimation procedure is used). The columns of the matrix correspond to the specified wind directions. For hurricane winds, a similar matrix of wind speeds is used. Using micrometeorological relations and/or wind tunnel data, the wind engineer also provides a counterpart to this matrix

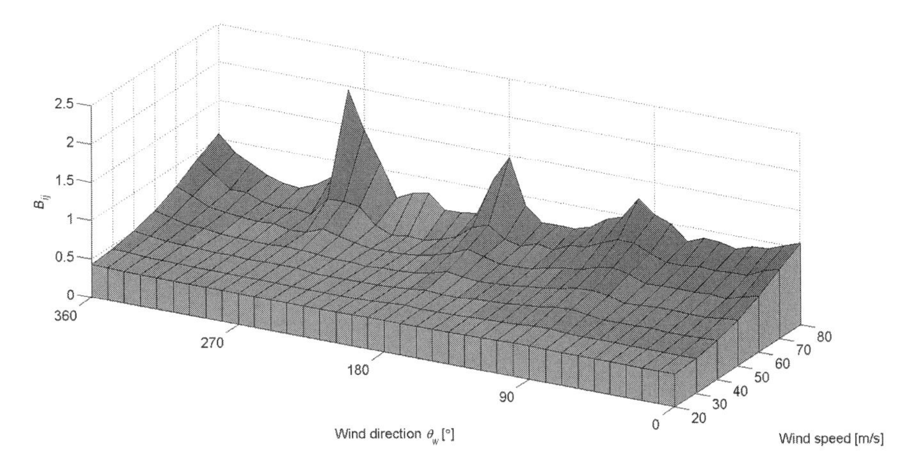

Figure 19.4.3. Response database: dependence on wind speed and direction of a structural member's demand-to-capacity index (denoted by B_{ij}) [19-10].

whose elements are the directional mean hourly wind speeds at the top of the building, in lieu of the directional wind speeds at 10 m above ground in open exposure.

7. *Use of response database and directional wind speeds matrix to obtain the requisite design wind effects for specified mean recurrence intervals.* Design wind effects are calculated by interpolation from the response database of Step 5 and the wind climatological database of Step 6. The wind effects are calculated for winds blowing from each direction in each storm event (or year). A matrix of wind effects is thus developed that is a one-to-one mapping of the directional wind speed matrix. However, for each storm event (or year) only the largest of the directional responses is of interest from a design viewpoint and is therefore retained. A vector of the maximum response induced by each storm event is thus created (Sect. 12.7). This vector is then rank-ordered, and the peak responses corresponding to the required mean recurrence intervals are obtained using non-parametric estimation methods (Sect. 12.7). The peak response of interest can consist of: the demand-to-capacity indexes for any member, inter-story drift, and the peak acceleration.
8. *Repeat the procedure outlined in Steps 1–7* as needed until the results obtained satisfy the design criteria.

Software for performing DAD estimates of the wind-induced response of flexible structures is provided at www.nist.gov/wind. User's manuals are provided in [19-9] for steel buildings and [19-10] for reinforced concrete buildings. The main difference between the two cases resides in the definitions of the demand-to-capacity indexes (Sect. 19.4.3). An example of the application of the DAD approach is presented in [9-10, 19-32] for a structure

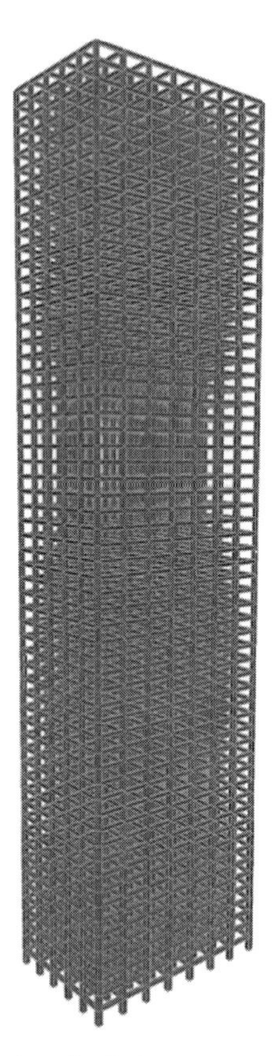

Figure 19.4.4. Schematic of reinforced concrete CAARC building.

with the dimensions of the CAARC building [19-26, 19-27] and a preliminary design reported in [19-28] (Fig. 19.4.4).

The length of the pressure time series measured in the wind tunnel usually corresponds to a prototype time of about one hour. The relation between model and prototype sampling rates is governed by the relation $(nD/V)_{\text{mod}} = (nD/V)_{\text{prot}}$. For example, if the model scale is $D_{\text{mod}}/D_{\text{prot}} = 1/500$, and the velocity scale is $V_{\text{mod}}/V_{\text{prot}} = 1/5$, then the ratio of model to prototype sampling rates is $n_{\text{mod}}/n_{\text{prot}} = 100$. If $n_{\text{mod}} = 400$ Hz, the corresponding prototype sampling rate n_{prot} is 4 Hz. For detailed aerodynamics databases, see [19-33].

19.4.3 Demand-to-Capacity Indexes

For strength design purposes, it is not individual internal forces that need to be considered, but rather the demand-to-capacity indexes (DCIs), that is, the left-hand sides of the applicable interaction equations. For a given steel member cross section, the DCIs are functions of internal forces, and each internal force is in turn a sum of contributions due to (1) wind forces acting along the principal axis x of the building, (2) wind forces acting along the principal axis y of the building, and (3) factored gravity loads. As was pointed out earlier, the fact that all the calculations are performed in the time domain allows the DCI calculations to be performed by simple algebraic operations, without involving the unmanageable combinations of cross-spectra and the multiple guessed-at load effect combination factors that arise in frequency domain calculations.

For example, for steel structures, time series of DCIs, denoted by B_{ij}, where the subscript i identifies the member and j defines the cross section, have expressions of the type [18-1]:

$$\frac{P_{ij}^{f}(t)}{\phi P_j} \geq 0.2:$$

$$B_{ij}^{f}(t) = \frac{P_{ij}^{f}(t)}{\phi P_j} + \frac{8}{9}\left(\frac{M_{ijX}^{f}(t)}{\phi_b M_{jX}} + \frac{M_{ijY}^{f}(t)}{\phi_b M_{jY}}\right) \tag{19.4.1}$$

$$\frac{P_{ij}^{f}(t)}{\phi P_j} < 0.2:$$

$$B_{ij}^{f}(t) = \frac{P_{ij}^{f}(t)}{2\phi P_j} + \left(\frac{M_{ijX}^{f}(t)}{\phi_b M_{jX}} + \frac{M_{ijY}^{f}(t)}{\phi_b M_{jY}}\right) \tag{19.4.2}$$

In Eqs. 19.4.1 and 19.4.2, P_j, M_{jX}, and M_{jY} are the nominal axial and flexural strengths at cross section j; ϕ and ϕ_b are axial and flexural resistance factors; and the quantities in the numerators are the total axial load and bending moments. The indexes X and Y pertain to the principal axes of the cross section (while the indexes x and y mentioned earlier pertain to the principal axes of the building). The superscript f indicates that in the load combinations the gravity loads are factored. For Allowable Stress Design, similar equations hold. Strength design load combinations specified by the ASCE 7-10 Chapter 2 [2-1] include:

$$1.2D + 1.0L + W \tag{19.4.3a}$$

$$0.9D + W \tag{19.4.3b}$$

where D = dead load, L = live load, and W = wind load. Note that the load factors from previous versions of the Standard are absorbed in the wind speeds with longer MRIs specified therein.

The quantities B_{ij} are among the responses of interest in Step 7 of Sect. 19.4.2. For details on expressions for DCIs for reinforced concrete structures, see [19-10, 19-32]. The vector of the DCIs obtained as indicated in Sect. 19.4.2 allows the estimation of the DCI with the mean recurrence interval specified for design. For the cross section j of member i to be acceptable, its DCI must not be greater than unity. For the design of the structural members to be acceptable, it is also necessary that the inter-story drift and the peak acceleration conform to serviceability requirements.

Efforts are currently in progress to develop optimization algorithms based on realistic estimates of the wind effects, aimed at meeting strength and serviceability criteria at the lowest possible cost, with the least possible consumption of material, or with the least possible carbon footprint.

19.4.4 Rigid and Aeroelastic Wind Tunnel Models

Rigid models are used to obtain aerodynamic pressure time histories for the dynamic analysis of buildings that do not exhibit significant aeroelastic effects. For buildings that do exhibit aeroelastic effects, pressure time histories can be measured on aeroelastic models, using the same technique employed for rigid models. The implementation of this approach was reported in [19-12].

19.4.5 Veering Effects

Veering—the phenomenon depicted in Figs. 10.1.5 and 11.2.1—should be taken into account in the design of tall buildings, since winds associated with a specified direction at the ground level act at higher elevations not only with increased speeds, but also with modified directions. Veering effects are usually not large, but need to be checked, especially for buildings with height above approximately 200 m, and at higher latitudes, where Coriolis effects are larger (Sect. 10.1). An approach to accounting for veering in the analysis and design of tall buildings is described in [19-13; www.nist.gov/wind].

19.5 SERVICEABILITY REQUIREMENTS

Wind-related serviceability issues are of concern in two areas: (1) building envelope performance under wind-induced deformations, and (2) occupant discomfort due to building motion.

For the performance of the building envelope to be adequate, the *peak inter-story drift* with specified mean recurrence interval (say, 20 years) must not exceed 1/300 to 1/400 of the story height under unfactored loads. This criterion may be modified, depending upon type of cladding or glazing (brittle cladding may require smaller ratios), cladding attachment details, and insurance considerations. In absolute terms, inter-story drift should not exceed 10 mm unless special details allow nonstructural partitions, cladding, or glazing to accommodate larger drift. However, this criterion may also be modified depending upon specific building features. Total drift ratios with a 20-year MRI should, according to some practitioners, not exceed 1/600 to 1/400 of the building height. Some practitioners believe larger values may be acceptable. For additional material on drift, see the Commentary to [2-1].

Occupant comfort is affected by the visual perception of building oscillations, especially torsional oscillations, which can be particularly unsettling to occupants, and by strong and frequent accelerations of the upper floors, which can produce effects not unlike those associated with seasickness, in more extreme cases. It has been hypothesized that occupant comfort is affected by rapid changes of acceleration (i.e., jerks), but, to our knowledge, no criteria based on such changes have been developed so far. On the other hand, because occupant perception of accelerations is highly variable, criteria on acceptable accelerations differ among codes and practitioners. For example, in North America, proposed ranges of acceptable peak accelerations with 10-year MRI at top floors are 0.15–0.20 m/s^2 for offices and 0.10–0.15 m/s^2 for residential buildings. However, it has been determined that acceptable acceleration levels decrease as the oscillation frequency increases. It has therefore been suggested that these limits be reduced for frequencies of vibration exceeding 0.1 Hz to about half of those values for frequencies of 1 Hz. The International Organization for Standardization (ISO) and the Architectural Institute of Japan (AIJ) criteria similarly reflect the decrease of acceptable acceleration levels as the frequencies of vibration increase.

Compliance with comfort criteria can control structural design. To reduce the resonant response, special damping systems are used, including various types of tuned mass dampers (TMDs) [7-1, p. 356; 19-14 to 19-18]. The effect of the TMDs, the mass of which can be as high as 5% of the building mass, is essentially to add a degree of freedom to the structure. This addition results in the replacement of the original structure, which possesses a mechanical admittance function with a sharp peak (Sect. 14.2.1), with a combined structure/TMD system that has two mechanical admittance functions with lower peaks, one of which occurs at a frequency lower than the original fundamental frequency of the building, while the other occurs at a higher frequency.

For a set of four instructive case studies on tall building serviceability, see [19-29].

19.6 PRELIMINARY ESTIMATES OF FLEXIBLE BUILDING RESPONSE

The ASCE 7 Standard includes a procedure for calculating the along-wind response of flexible buildings in terrain with uniform roughness. However, it provides no indication of effects due to skew winds, or of across-wind and torsional effects. Nor does it provide any indication of how to account for aerodynamic interference from nearby tall buildings. An alternative procedure for calculating along-wind response included in the National Building Code of Canada (NBC) [11-1] has the same limitations. However, unlike the ASCE 7 Standard, NBC includes a procedure that yields estimates—albeit crude—of across-wind response.

Results yielded by ASCE 7, NBC, or similar generic procedures applied to flexible buildings, are only useful for preliminary design purposes. This is also true of generic estimates of tall building torsional response. Having clearly stated these limitations, we present in Sects. 19.6.1, 19.6.2, and 19.6.3, respectively, simple procedures for estimating along-wind, across-wind, and torsional response.

19.6.1 Along-Wind Response

Procedures for calculating along-wind response incorporate (1) a model of the wind profile upwind of the building and of the turbulence in the oncoming wind, (2) a relation between the oncoming wind speeds and the along-wind fluctuating pressures on the building faces normal to the wind, and (3) a method for the calculation of the dynamic response to these pressures. Internally, the procedures for calculating the fluctuating along-wind response make use of spectral methods, which means: (1) representing wind-induced fluctuating pressures at various points of the building faces by their spectra and cross-spectra, which are representations of large numbers of superposed sinusoidal components, each with a distinct frequency n, (2) relating these spectra and cross-spectra to those of the oncoming flow fluctuations, and (3) calculating the spectral densities, r.m.s. values, and peaks of the building responses, by using the relation between the spectra and cross-spectra of the pressures, on the one hand, and the spectra of the response, on the other (Sect. 14.3.7). In all cases, it is assumed that the wind direction is normal to a building face, and that the terrain roughness upwind of the building is fairly uniform, with no significant obstacles near the building.

Several procedures for calculating along-wind response have been proposed. Some use the logarithmic law description of the wind profile, and others use the power law description. Section 19.6.1.1 provides closed-form expressions for the along-wind response based on the logarithmic description of the wind profile and on the use of mean hourly wind speeds near the top of the building. The ASCE 7-10 procedure is a modified version of these expressions that uses the power law description of the wind profile. We do not recommend the use of the NBC procedure, which does not account for the imperfect correlation between windward and leeward pressures or for the dependence of wind spectra on height above ground. All closed-form expressions are based on a straight-line fundamental modal shape, and cannot accommodate curved fundamental modal shapes and higher modes of vibration. Software that can do so is available at www.nist.gov/wind (II B).

19.6.1.1 Closed-Form Expressions for the Along-Wind Response. Consider a building with rectangular shape in plan, in terrain with surface roughness length z_0 uniform over a sufficiently large distance upwind. The building has height H, width B (normal to the wind direction), and depth D (parallel to the wind direction) (Fig. 14.3.2), average pressure coefficient C_w on the windward face and C_l on the leeward face, average specific mass ρ_b, fundamental frequency n_1, and damping ratio ζ_1. The fundamental mode of vibration is assumed to be a straight line. The mean hourly wind speed at elevation H is $\overline{V}(H)$. The following expressions may be used to calculate the peak along-wind deflection at elevation z [19-19]:

$$x_{pk}(z) \approx \frac{0.5\rho u_*^2 C_D B z}{M_1(2\pi n_1)^2}\left[J + 3.75(\mathrm{B} + \mathrm{R})^{1/2}\right] \qquad (19.6.1a)$$

where ρ = specific mass of air, M_1 = fundamental modal mass (if the building's average specific mass ρ_b is independent of elevation z, then $M_1 = BLH\rho_b/3$),

$$u_* = \frac{\overline{V}(H)}{2.5 \ln \dfrac{H}{z_0}}, \qquad (19.6.1b)$$

$$J = 0.78Q^2, \qquad (19.6.1c)$$

$$Q = 2 \ln \frac{H}{z_0} - 1, \qquad (19.6.1d)$$

$$\mathrm{B} = \frac{6.71Q^2}{1 + 0.26\dfrac{B}{H},}, \qquad (19.6.1e)$$

$$\mathrm{R} = \frac{0.59Q^2 N_1^{-2/3}}{\varsigma_1} \frac{C_{Df}^2}{C_D^2} \frac{\mathrm{C}(\eta_1)}{1 + 3.95N_1\dfrac{B}{H}}, \qquad (19.6.1f)$$

$$N_1 = \frac{n_1 H}{u_* Q}, \qquad (19.6.1g)$$

$$C_{Df}^2 = C_w^2 + C_\ell^2 + 2C_w C_\ell \mathrm{C}(\eta_2), \quad C_D = C_w + C_\ell, \qquad (19.6.1h,i)$$

$$\mathrm{C}(\eta) = \frac{1}{\eta} - \frac{1 - e^{-2\eta}}{2\eta^2}, \qquad (19.6.1j)$$

$$\eta_1 = 3.55N_1, \eta_2 = 12.32N_1\Delta/H, \qquad (19.6.1k)$$

Δ = smallest of dimensions H, B, and D. The quantities J, B, and R are measures of the mean, quasi-static, and resonant response, respectively.

The peak acceleration at elevation z is

$$\ddot{x}_{pk}(z) \approx 4.0\frac{0.5\rho u_*^2 C_D B z}{M_1}\mathrm{R}^{1/2}. \qquad (19.6.2)$$

The ratio p between the along-wind response calculated by taking the dynamic amplification into account and the response calculated by neglecting it can be written in the form

$$p \approx \frac{J + 3.75[\mathrm{B} + \mathrm{R}]^{1/2}}{J + 3.75\mathrm{B}^{1/2}}. \tag{19.6.3}$$

If the fundamental modal shape is not a straight line, corrections may be applied to the calculated deflections and accelerations to account for the nonlinearity of the modal shape [7-1, p. 340].

Numerical Example 19.6.1. *Along-wind response estimated by Eqs. 19.6.1 and 19.6.2*. Consider a building with square shape in plan and dimensions $H = 656$ ft, $B = D = 115$ ft, in terrain with uniform terrain roughness length $z_0 = 3.28$ ft (1 m) over a sufficicntly long fetch. The modal shape is assumed to be linear, so Eqs. 19.6.1 and 19.6.2 are applicable. The fundamental frequency of vibration is $n_1 = 0.175$ Hz, and the damping ratio is $\zeta_1 = 0.01$. The average pressure coefficients are $C_w = 0.8$ and $C_l = 0.5$, so the drag coefficient is $C_D = C_w + C_l = 1.3$. The mean hourly wind speed is $\overline{V}(H) =$ 88.3 mph = 129.54 ft/s. It is assumed that the specific mass of air and the average specific mass of the building are, respectively, $\rho = 0.00238$ slugs/ft^3 (1.25 kg/m^3) and 0.381 slugs/ft^3(200 kg/m^3).

Equations 19.6.1a–k yield: $u_* = 9.78$ ft/s, $Q = 9.59$; $J = 71.83$; $\mathrm{B} =$ 590.9; $N_1 = 1.22$; $\eta_1 = 4.35$; $\eta_2 = 2.63$; $\mathrm{C}(\eta_1) = 0.203$, $\mathrm{C}(\eta_2) = 0.308$, $C_{Df}^2 = 1.14$, $\mathrm{R} = 353$, $x_{pk}(z) \approx 0.0024\ z$. For $z = H = 656$ ft, $x_{pk}(H) =$ 1.57 ft, and $\ddot{x}_{pk}(H) \approx 0.76$ ft/s^2.

Numerical Example 19.6.2. *Classification of a building as rigid or flexible.* For design purposes, buildings are classified as rigid if their wind-induced resonant effects are negligible. For buildings with regular shapes in plan, it is possible to determine whether those effects are negligible by performing an approximate dynamic analysis of the *along-wind response* (i.e., the building response in the wind direction induced by wind normal to a building face).

Consider a building with height $H = 150$ ft, flat roof, and a rectangular shape in plan (60 ft × 35 ft). Assume that the basic wind speed (i.e., the peak 3-s gust at 10 m above ground in terrain with Exposure C) is $V = 100$ mph, and that the building has Exposure B in all directions, fundamental natural frequency $n_1 = 0.9$ Hz, and damping ratio $\zeta_1 = 0.02$. The aerodynamic coefficients are $C_D = 1.3$, $C_w = 0.8$, and $C_\ell = 0.5$. The wind is assumed to be normal to the long face of the building (i.e., to the dimension $B = 60$ ft).

First calculate the mean hourly wind speed at height H of the building. The mean hourly speed at 33 ft (10 m) above ground in Exposure C (roughness length $z_{0open} = 0.066$ ft, see Table 11.2.1) is $\overline{V}(z = 33\text{ ft}) =$ $100/1.52 = 65.8$ mph = 96.5 ft/s (Sect. 11.1.2). The mean hourly wind speed at 150 ft over terrain with exposure B ($z_0 = 0.5$ ft, Table 11.2.1) is $\overline{V}(z =$ 150 ft, $z_0 = 0.5$ ft) $= 96.5\,(0.5/0.066)^{0.07}\,[\ln(150/0.5)]\,/[\ln(33/0.066)] =$ 102.0 ft/s (Eq. 11.2.8). We obtain from Eqs. 19.6.1: $u_* = 6.77$ ft/s^2; $Q = 10.41$, $J = 84.49$, $\mathrm{B} = 658.65$; $N_1 = 1.92$; $\eta_1 = 6.80$; $\eta_2 = 5.51$;

$C(\eta_1) = 0.136$; $C(\eta_2) = 0.165$; $C^2_{Df} = 0.64 + 0.25 + 2 \times 0.8 \times 0.5 \times 0.165 = 1.02$; $R = 73.0$. From Eq. 19.6.3, $p \approx 1.03$. It follows that if the building is designed as rigid, its response will be underestimated by 3%. Note that for this building $n_1 < 1$ Hz, and the ratio $H/D = 150/35 = 4.39 > 4.0$.

19.6.2 Across-Wind Response

Unlike the along-wind response, the across-wind response cannot be determined on the basis of approximate representations of the wind-induced pressures. However, empirical expressions based on wind tunnel tests of generic building models have been proposed. The expressions are applicable, provided that: (1) no neighboring tall buildings strongly disturb the oncoming wind flow, and (2) for buildings with a square shape in plan, the calculated r.m.s. across-wind deflection σ_y at the top of the building is such that σ_y/D is less than about 0.015 for open terrain, 0.025 for suburban terrain, and 0.045 for centers of large cities, where D = building depth. It is emphasized that the criteria pertaining to the ratio σ_y/D offer at best a rough indication of tall building behavior.

The expressions for the peak deflections and accelerations at the elevation z of a building with height H are

$$y_{pk}(z) = \mathrm{C}\left[\frac{\overline{V}(H)}{n_1\sqrt{A}}\right]^{\mathrm{p}} \frac{\sqrt{A}}{\varsigma_1^{1/2}} \frac{\rho}{\rho_b} \frac{z}{H} \tag{19.6.4}$$

$$\ddot{y}_{pk}(z) = (2\pi n_1)^2 y_{pk}(z) \tag{19.6.5}$$

where A = area of horizontal section of the building, ρ = specific mass of air, ρ_b = mass of building per unit volume, and C and p are empirical constants. According to [19-20], p = 3.3. Various authors suggest somewhat different values for the constant C. According to [19-20], C = 0.00065. The r.m.s. response may be assumed to be approximately 4 times smaller than the peak response.

Numerical Example 19.6.3. *Across-wind response estimation by Eqs. 19.6.4 and 19.6.5*. For the building, terrain, and wind speed of Numerical Example 19.6.1, for $z = H$ Eqs. 19.6.4 and 19.6.5 yield

$$y_{pk}(H) = 0.00065\left[\frac{129.54}{0.175 \times \sqrt{115 \times 115}}\right]^{3.3} \frac{115}{0.01^{1/2}} \frac{0.00238}{0.381} = 2.18 \text{ ft},$$

$$\ddot{y}_{pk}(H) = 2.39 \text{ ft/s}^2$$

Note that that the calculated across-wind response exceeds the calculated along-wind response, and that $y_{pk}/B = 2.18/115 = 0.019 < 0.025$, suggesting that Eqs. 19.6.4 and 19.6.5 are indeed applicable.

19.6.3 Torsional Response

Like the across-wind response, the torsional response due to the eccentricity of the aerodynamic center with respect to the elastic center can only be estimated by using empirical expressions, with parameters based on wind tunnel tests conducted on buildings with various shapes in terrain with uniform upwind roughness.[1] The equations that follow are not applicable to buildings with non-coincident mass and elastic centers. The peak base torsional moment may be written as

$$T_{pk}[\overline{V}(H)] \approx 0.9\{\overline{T}[\overline{V}(H)] + 3.8T_{\text{rms}}[\overline{V}(H)]\}, \qquad (19.6.6)$$

where $\overline{T}$ = mean base torsional moment, T_{rms} = root mean square value of the fluctuating base torsional moment,

$$\overline{T}[\overline{V}(H)] = 0.038\rho L^4 H n_T^2 [\overline{V}(H)/(n_T L)]^2, \qquad (19.6.7)$$

$$T_{\text{rms}}[\overline{V}(H)] = 0.0017\frac{1}{\zeta_T^{1/2}}\rho L^4 H n_T^2 [\overline{V}(H)/(n_T L)]^{2.68}, \qquad (19.6.8)$$

n_T and ζ_T are the natural frequency and the damping ratio in the fundamental torsional mode of vibration, and A is the area of the horizontal cross section of the building. The peak torsional moment is approximately

$$T_{pk} = 0.9(\overline{T} + 3.8T_{rms}) \qquad (19.6.9)$$

The peak horizontal acceleration induced by torsion at the top of the building at a distance v from the elastic center is

$$\ddot{\varphi}_{pk} v \approx \frac{7.6T_{\text{rms}} v}{\rho_b A H r_m^2} \qquad (19.6.10)$$

where $\ddot{\varphi}_{pk}$ is the peak angular acceleration, and r_m is the radius of gyration of the horizontal cross section of the building.

For a building with a rectangular shape in plan (dimensions B and D),

$$L = (2 \times 2 \times B^2/8)/\sqrt{BD} + (2 \times 2 \times D^2/8)/\sqrt{BD} \qquad (19.6.11)$$

$$r_m = (B^2 + D^2)^{1/2}/\sqrt{12}. \qquad (19.6.12)$$

Numerical Example 19.6.4. *Torsional response of a tall building.* For the building, terrain, and wind speed of Numerical Example 19.6.1, the natural frequency and the damping ratio in the fundamental torsional mode are

[1]The material in this section was kindly communicated by Dr. N. Isyumov.

assumed to be $n_T = 0.3$ Hz and $\zeta_T = 0.01$, respectively. Equations 19.6.11 and 19.6.12 yield $L = 115$ ft and $r_m = 46.95$ ft. We have $\overline{V}(H)/(n_T L) = 129.54/(0.3 \times 115) = 3.75$, $d = 115\sqrt{2}/2 = 81.3$ ft. Equations 19.6.6 to 19.6.10 yield

$$\begin{aligned}
\overline{T}[129.54 \text{ ft/s}] &= 0.038 \times 0.00238 \times 115^4 \times 656 \times 0.3^2 \times 3.75^2 \\
&= 13130 \text{ kip ft}, \\
T_{\text{rms}}[129.54 \text{ ft/s}] &= 0.0017 \frac{1}{0.01^{1/2}} \times 0.00238 \times 115^4 \times 656 \times 0.3^2 \times 3.75^2 \\
&= 14400 \text{ kip ft}, \\
T_{pk} &= 0.9[13130 + 3.8 \times 14400] = 57200 \text{ kip ft}, \\
\ddot{\varphi}_{pk} d &\approx \frac{7.6 \times 14400 \times 10^3 \times 81.3}{0.381 \times 115 \times 115 \times 636 \times 46.95^2} = 1.26 \text{ ft/s}^2.
\end{aligned}$$

PART V

APPENDICES

APPENDIX A1

RANDOM PROCESSES

Consider a process the possible outcomes of which form a collection (or an *ensemble*) of functions of time $\{y(t)\}$. A member of the ensemble is called a *sample function* or a *random signal.* The process is called a *random process* if the values of the sample functions at any particular time constitute a random variable.[1]

A time-dependent random process is *stationary* if its statistical properties (e.g., the mean and the mean square value) do not depend upon the choice of the time origin and do not vary with time. A stationary random signal is thus assumed to extend over the entire time domain. The *ensemble average*, or *expectation*, of a random process is the average of the values of the member functions at any particular time. A stationary random process is *ergodic* if its time averages equal its ensemble averages. Ergodicity requires that every sample function be typical of the entire ensemble.

A stationary random signal may be viewed as a superposition of harmonic oscillations over a continuous range of frequencies. Therefore, some basic results of harmonic analysis are reviewed first (Sects. A1.1 and A1.2). Definitions are then presented of the spectral density function (Sect. A1.3), the autocovariance function (Sect. A1.4), the cross-covariance function, the co-spectrum, the quadrature spectrum, and the coherence function (Sect. A1.5). Mean upcrossing and outcrossing rates are introduced in Sect. A1.6. The

[1]Let a numerical value be assigned to each of the events that may occur as a result of an experiment. The resulting set of possible numbers is defined as a *random variable.* Examples: (1) If a coin is tossed, the numbers zero and one assigned to the outcome heads and to the outcome tails constitute a discrete random variable. (2) To each measurement of a quantity a number is assigned to the result of that measurement. The set of all possible results of the measurements constitutes a continuous random variable.

estimation of peaks of Gaussian random signals is considered in Sect. A1.7, and their application to estimating peaks of non-Gaussian processes is presented in Sect. A1.8.

A1.1 FOURIER SERIES AND FOURIER INTEGRALS

Consider a *periodic* function $x(t)$ with zero mean and period T. It can be easily shown that

$$x(t) = C_0 + \sum_{k=1}^{\infty} C_k \cos(2\pi k n_1 t - \varphi_k) \tag{A1.1.1}$$

where $n_1 = 1/T$ is the *fundamental frequency* and

$$C_0 = \frac{1}{T}\int_{-T/2}^{T/2} x(t)\,dt \tag{A1.1.1a}$$

$$C_k = (A_k^2 + B_k^2)^{1/2} \tag{A1.1.1b}$$

$$\varphi_k = \tan^{-1}\frac{B_k}{A_k} \tag{A1.1.1c}$$

$$A_k = \frac{2}{T}\int_{-T/2}^{T/2} x(t)\cos 2\pi k n_1 t\,dt \tag{A1.1.1d}$$

$$B_k = \frac{2}{T}\int_{-T/2}^{T/2} x(t)\sin 2\pi k n_1 t\,dt \tag{A1.1.1e}$$

Equation A1.1.1 is the *Fourier series expansion* of the periodic function $x(t)$.

If a function $y(t)$ is *nonperiodic*, it is still possible to regard it as periodic with infinite period. It can be shown that if $y(t)$ is piecewise differentiable in every finite interval, and if the integral

$$\int_{-\infty}^{\infty} |y(t)|\,dt \tag{A1.1.2}$$

exists, the following relation holds:

$$y(t) = \int_{-\infty}^{\infty} C(n)\cos[2\pi n t - \varphi(n)]\,dn \tag{A1.1.3}$$

In Eq. A1.1.3, called the *Fourier integral* of $y(t)$ in real form, n is a continuously varying frequency, and

$$C(n) = (A^2(n) + B^2(n))^{1/2} \tag{A1.1.3a}$$

$$\varphi(n) = \tan^{-1}\frac{B(n)}{A(n)} \tag{A1.1.3b}$$

$$A(n) = \int_{-\infty}^{\infty} y(t) \cos 2\pi nt \, dt \tag{A1.1.3c}$$

$$B(n) = \int_{-\infty}^{\infty} y(t) \sin 2\pi kt \, dt \tag{A1.1.3d}$$

From Eqs. A1.1.3a through A1.1.3d and the identities

$$\sin \varphi = \frac{\tan \varphi}{(1 + \tan^2 \varphi)^{1/2}} \tag{A1.1.4a}$$

$$\cos \varphi = \frac{1}{(1 + \tan^2 \varphi)^{1/2}} \tag{A1.1.4b}$$

it follows that

$$\int_{-\infty}^{\infty} y(t) \cos[2\pi nt - \varphi(n)] \, dt = C(n) \tag{A1.1.5}$$

The functions $y(t)$ and $C(n)$, which satisfy the symmetrical relations A1.1.3 and A1.1.5, form a *Fourier transform pair*.

Successive differentiation of Eq. A1.1.3 yields

$$\dot{y}(t) = \int_{-\infty}^{\infty} 2\pi nC(n) \cos[2\pi nt - \varphi(n)] \, dn \tag{A1.1.6a}$$

$$\ddot{y}(t) = \int_{-\infty}^{\infty} 4\pi^2 n^2 C(n) \cos[2\pi nt - \varphi(n)] \, dn \tag{A1.1.6b}$$

A1.2 PARSEVAL'S EQUALITY

The mean square value of the periodic function $x(t)$ with period T (Eq. A1.1.1) is

$$\sigma_x^2 = \frac{1}{T} \int_{-T/2}^{T/2} x^2(t) \, dt \tag{A1.2.1}$$

Substitution of Eq. A1.1.1 into Eq. A1.2.1 yields

$$\sigma_x^2 = \sum_{k=0}^{\infty} S_k \tag{A1.2.2}$$

where $S_0 = C_0^2$ and $S_k = \frac{1}{2} C_k^2 (k = 1, 2, \ldots)$. *The quantity S_k is the contribution to the mean square value of $x(t)$* of the harmonic component with frequency kn_1. Equation A1.2.2 is a form of Parseval's equality.

For a nonperiodic function for which an integral Fourier expression exists, Eqs. A1.1.3 and A1.1.5 yield

$$\begin{aligned}
\int_{-\infty}^{\infty} y^2(t)\,dt &= \int_{-\infty}^{\infty} y(t) \int_{-\infty}^{\infty} C(n)\cos[2\pi nt - \varphi(n)]\,dn\,dt \\
&= \int_{-\infty}^{\infty} C(n) \int_{-\infty}^{\infty} y(t)[2\pi nt - \varphi(n)]\,dn\,dt \\
&= \int_{-\infty}^{\infty} C(n)^2\,dn \\
&= 2\int_{0}^{\infty} C^2(n)\,dn
\end{aligned} \tag{A1.2.3}$$

Equation A1.2.3 is the form taken by Parseval's equality in the case of a nonperiodic function.

A1.3 SPECTRAL DENSITY FUNCTION OF A RANDOM STATIONARY SIGNAL

A relation similar to Eq. A1.2.2 is now sought for functions generated by stationary processes. The spectral density of such functions is defined as the counterpart of the quantities S_k.

Let $z(t)$ be a stationary random signal with zero mean. Because it does not satisfy the condition A1.1.2, $z(t)$ does not have a Fourier transform. An auxiliary function $y(t)$ is therefore defined as follows:

$$y(t) = z(t) \left(-\frac{T}{2} < t < \frac{T}{2}\right) \tag{A1.3.1a}$$

$$y(t) = 0 \text{ elsewhere} \tag{A1.3.1b}$$

The function $y(t)$ so defined is nonperiodic, satisfies Eq. A1.1.2, and thus has a Fourier integral. From the definition of $y(t)$ it follows that

$$\lim_{T\to\infty} y(t) = z(t) \tag{A1.3.2}$$

By virtue of Eqs. A1.2.3 and A1.3.1, the mean square value of $z(t)$ is

$$\begin{aligned}
\sigma_y^2 &= \lim_{T\to\infty} \frac{1}{T}\int_{-T/2}^{T/2} y^2(t)\,dt \\
&= \frac{1}{T}\int_{-\infty}^{\infty} y^2(t)\,dt \\
&= \frac{2}{T}\int_{0}^{\infty} C^2(n)\,dn
\end{aligned} \tag{A1.3.3}$$

The mean square of the function $y(t)$ is then

$$\begin{aligned}\sigma_z^2 &= \lim_{T\to\infty} \sigma_y^2 \\ &= \lim_{T\to\infty} \frac{2}{T}\int_0^\infty C^2(n)\,dn \qquad \text{(A1.3.4)}\end{aligned}$$

With the notation

$$S_z(n) = \lim_{T\to\infty} \frac{2}{T} C^2(n) \qquad \text{(A1.3.5)}$$

Equation A1.3.4 becomes

$$\sigma_z^2 = \int_0^\infty S_z(n)\,dn \qquad \text{(A1.3.5)}$$

The function $S_z(n)$ is defined as *the spectral density function of* $z(t)$. To each frequency $n(0 < n < \infty)$ there corresponds an elemental contribution $S(n)\,dn$ to the mean square value σ_z^2; σ_z^2 is equal to the area under the spectral density curve $S_z(n)$. Because in Eq. A1.3.5 the spectrum is defined for $0 < n < \infty$ only, $S_z(n)$ is called the *one-sided* spectral density function of $z(t)$. This definition of the spectrum is used throughout this text. A different convention may be used where the spectrum is defined for $-\infty < n < \infty$, and the integration limits in Eq. A1.3.5 are $-\infty$ to ∞. This convention yields the *two-sided* spectral density function of $z(t)$.

From Eqs. A1.1.6a and b, following the same steps that led from Eq. A1.1.3 to Eq. A1.3.5, there result the expressions for the spectral density of the first and second derivative of a random process :

$$S_{\dot z}(n) = 4\pi^2 n^2 S_z(n) \qquad \text{(A1.3.6a)}$$

$$S_{\ddot z}(n) = 16\pi^4 n^4 S_z(n) \qquad \text{(A1.3.6b)}$$

A1.4 AUTOCORRELATION FUNCTION OF A RANDOM STATIONARY SIGNAL

From Eqs. A1.1.3a–d, it follows that

$$\begin{aligned}\frac{2}{T}C^2(n) &= \frac{2}{T}[A^2(n) + B^2(n)] \\ &= \frac{2}{T}[A(n)A(n) + B(n)A(n)] \qquad \text{(A1.4.1)}\end{aligned}$$

Using the notations $\tau = t_2 - t_1$ and

$$\tilde{R}(\tau) = \frac{1}{T}\int_{-\infty}^{\infty} y(t_1)y(t_1+\tau)\,dt_1 \qquad \text{(A1.4.2)}$$

Equation A1.4.1 can be written as

$$\frac{2}{T}C^2(n) = 2\int_{-\infty}^{\infty} \tilde{R}(t)\cos 2\pi n\tau\, d\tau \tag{A1.4.3}$$

Equations A1.4.3, A1.3.2, and A1.3.5 thus yield

$$S_z(n) = \int_{-\infty}^{\infty} 2R_z(\tau)\cos 2\pi n\tau\, d\tau \tag{A1.4.4}$$

$$R_z(\tau) = \lim_{T\to\infty}\frac{1}{T}\int_{-T/2}^{T/2} z(t)z(t+\tau)\, dt \tag{A1.4.5}$$

The function $R_z(\tau)$ is defined as the *autocovariance function* of $z(t)$ and provides a measure of the interdependence of the variable z at times t and $t+\tau$. From the stationarity of $z(t)$, it follows that

$$R_z(\tau) = R_z(-\tau) \tag{A1.4.6}$$

Since $R_z(\tau)$ is an even function of τ,

$$\int_{-\infty}^{\infty} 2R_z(\tau)\sin 2\pi n\tau\, d\tau = 0 \tag{A1.4.7}$$

A comparison of Eqs. A1.1.5 and A1.4.4 shows that $S_z(n)$ and $2R_z(\tau)$ form a Fourier transform pair. Therefore,

$$R_z(\tau) = \frac{1}{2}\int_{-\infty}^{\infty} S_z(n)\cos 2\pi n\tau\, dn \tag{A1.4.8a}$$

Since, as follows from Eq. A1.4.4, $S_z(n)$ is an even function of n, Eq. A1.4.8a may be written as

$$R_z(\tau) = \int_{0}^{\infty} S_z(n)\cos 2\pi n\tau\, dn \tag{A1.4.8b}$$

Similarly, by virtue of Eqs. A1.4.4 and A1.4.6,

$$S_z(n) = 4\int_{0}^{\infty} R_z(\tau)\cos 2\pi n\tau\, d\tau \tag{A1.4.9}$$

The definition of the autocovariance function (Eq. A1.4.5) yields

$$R_z(0) = \sigma_z^2 \tag{A1.4.10}$$

For $\tau > 0$ the products $z(t)z(t+\tau)$ are not always positive, as is the case for $\tau = 0$, so

$$R_z(\tau) < \sigma_z^2 \tag{A1.4.11}$$

For large values of τ, the values $z(t)$ and $z(t+\tau)$ bear no relationship to each other, so

$$\lim_{\tau\to\infty} R_z(\tau) = 0 \tag{A1.4.12}$$

The nondimensional quantity $R_z(\tau)/\sigma_z^2$ is the *autocorrelation function* of the function $z(t)$.

A1.5 CROSS-COVARIANCE FUNCTION, CO-SPECTRUM, QUADRATURE SPECTRUM, COHERENCE

Consider two stationary signals $z_1(t)$ and $z_2(t)$ with zero means. The function

$$R_{z_1z_2}(\tau) = \lim_{T\to\infty} \frac{1}{T}\int_{-T/2}^{T/2} z_1(t)z_2(t+\tau)\,dt \tag{A1.5.1}$$

is defined as the *cross-covariance function* of the signals $z_1(t)$ and $z_2(t)$. From this definition and the stationarity of the signals, it follows that

$$R_{z_1z_2}(\tau) = R_{z_2z_1}(-\tau) \tag{A1.5.2}$$

However, in general, $R_{z_1z_2}(\tau) \neq R_{z_1z_2}(-\tau)$. For example, the reader can verify that if $z_2(t) \equiv z_1(t-\tau_0)$,

$$R_{z_1z_2}(\tau_0) = R_{z_1}(0) \tag{A1.5.3}$$

$$R_{z_1z_2}(-\tau_0) = R_{z_1}(2\tau_0) \tag{A1.5.4}$$

The *co-spectrum* and the *quadrature spectrum* of the signals $z_1(t)$ and $z_2(t)$ are defined, respectively, as

$$S^C_{z_1z_2}(n) = \int_{-\infty}^{\infty} 2R_{z_1z_2}(\tau)\cos 2\pi n\tau\,d\tau \tag{A1.5.5}$$

$$S^Q_{z_1z_2}(n) = \int_{-\infty}^{\infty} 2R_{z_1z_2}(\tau)\sin 2\pi n\tau\,d\tau \tag{A1.5.6}$$

It follows from Eq. A1.5.2 that

$$S^C_{z_1z_2}(n) = S^C_{z_2z_1}(n) \tag{A1.5.7a}$$

$$S^Q_{z_1z_2}(n) = -S^Q_{z_2z_1}(n) \tag{A1.5.7b}$$

The *coherence* function is a measure of the correlation between components with frequency n of two signals $z_1(t)$ and $z_2(t)$, and is defined as

$$\text{Coh}_{z_1 z_2}(n) = \frac{[S^C_{z_1 z_2}(n)]^2 + [S^Q_{z_1 z_2}(n)]^2}{S_{z_1}(n) S_{z_2}(n)} \tag{A1.5.7c}$$

A1.6 MEAN UPCROSSING AND OUTCROSSING RATE FOR A GAUSSIAN PROCESS

Let $z(t)$ be a stationary differentiable process with mean zero. The process crosses with positive slope a level k at least once in a time interval $(t,\ t + \Delta t)$ if $z(t) < k$ and $z(t + \Delta t) > k$. If $z(t)$ has smooth samples and Δt is sufficiently small, $z(t)$ will have a single k-crossing with positive slope, (i.e., a single k-upcrossing). The probability of occurrence of the event $\{z(t) < k,\ z(t + \Delta t) > k\}$ can be approximated by the probability of the event $\{z(t) < k,\ z(t) + \dot{z}(t)\Delta t > k\}$. The mean rate of k-upcrossings of $z(t)$ is

$$\begin{aligned} \nu(k) &= \lim_{\Delta t \to 0} \frac{1}{\Delta t} P(k - \dot{z}(t)\Delta t < z(t) < k) = \int_0^\infty \dot{z} f_{\dot{z},\, z}(\dot{z},\ k) d\dot{z} \\ &= E[\dot{z}(t)_+ | z(t) = k] f_z(k) \end{aligned} \tag{A1.6.1}$$

where $f_{\dot{z},\, z}$ and f_z denote the joint probability density function of $(\dot{z}(t),\ z(t))$ and the probability density function of $z(t)$, respectively, and $E[\dot{z}(t)_+ | z(t) = k]$ denotes the expectation of the positive part of $\dot{z}(t)$ conditional on $z(t) = k$.

If $z(t)$ is a stationary Gaussian process with mean zero, then $z(t)$ and $\dot{z}(t)$ are independent Gaussian variables, so that

$$f_{\dot{z},z}(\dot{z}, z) = \frac{1}{2\pi \sigma_{\dot{z}} \sigma_x} \exp\left[-\frac{1}{2}\left(\frac{\dot{z}^2}{\sigma_{\dot{z}}^2} + \frac{z^2}{\sigma_z^2}\right)\right] \tag{A1.6.2}$$

and the mean k-upcrossing rate is

$$\nu(k) = E[\dot{z}(t)_+] f(k) = \frac{\sigma_{\dot{z}}}{\sqrt{2\pi}} \frac{1}{\sqrt{2\pi}\sigma_x} \exp\left(-\frac{k^2}{2\sigma_x^2}\right) \tag{A1.6.3}$$

where σ_z and $\sigma_{\dot{z}}$ denote the standard deviations of z and $\dot{z}(t)$.

Equation A1.6.2 can be extended to the case where the random process is a vector $\mathbf{x}$. Let ν_D denote the mean rate at which the random process (i.e., the tip of the vector with specified origin O) crosses in an outward direction the boundary F_D of a region containing the point O. The rate ν_D has the expression

$$\nu_D = \int_{F_D} d\mathbf{x} \int_0^\infty \dot{x}_n\, f_{\mathbf{x}, \dot{x}_n}(\mathbf{x}, \dot{x}_n)\, d\dot{x}_n \tag{A1.6.4}$$

where $\dot{x}_n$ is the projection of the vector $\dot{\mathbf{x}}$ on the normal to F_D, and $f_{\mathbf{x},\ \dot{x}_n}(\mathbf{x},\ \dot{x}_n)$ is the joint probability distribution of $\mathbf{x}$ and $\dot{x}_n$. Equation A1.6.3 can be written as

$$\nu_D = \int_{F_D} \left\{ \int_0^\infty \dot{x}_n \, f_{\dot{x}_n}[\dot{x}_n | \mathbf{X} = \mathbf{x}] \, d\dot{x}_n \right\} f_{\mathbf{X}}(\mathbf{x}) \, d\mathbf{x}$$

$$= \int_{F_D} \mathbf{E}_0^\infty[\dot{X}_n | \mathbf{X} = \mathbf{x}] f_{\mathbf{X}}(\mathbf{x}) \, d\mathbf{x} \qquad \text{(A1.6.5)}$$

where $f_{\mathbf{X}}(\mathbf{X})$ = probability density of the vector $\mathbf{X}$, and $\mathbf{E}_0^\infty[\dot{X}_n | \mathbf{X} = \mathbf{x}]$ is the average of the positive values of $\dot{X}_n$ given that $\mathbf{X} = \mathbf{x}$. If $\dot{X}_n$ and $\mathbf{X}$ are independent, $\mathbf{E}_0^\infty[\dot{X}_n | \mathbf{X} = \mathbf{x}] = \mathbf{E}_0^\infty[\dot{X}_n]$.

Equation A1.6.4 has been used in wind engineering in an attempt to estimate mean recurrence intervals of directional wind effects that exceed (outcross) a limit state defined by a boundary F_D defined by the limit state being considered. A main difficulty in implementing this approach is the practical estimation of the distribution of the vector of wind effects, or of the images of that vector and of the boundary F_D in the space of directional wind speeds. Such estimation has been attempted by using sets of data that include non-extreme wind speeds. However, because non-extreme speeds are typically not representative of extremes, such use can lead to incorrect estimates of extreme values.

A1.7 PROBABILITY DISTRIBUTION OF THE PEAK VALUE OF A NORMALLY DISTRIBUTED RANDOM SIGNAL

Since

$$\sigma_z^2 = \int_0^\infty S_z(n) \, dn \qquad \text{(A1.7.1)}$$

$$\sigma_{\dot{z}}^2 = 4\pi^2 \int_0^\infty n^2 S_z(n) \, dn \qquad \text{(A1.7.2)}$$

(Eq. A1.3.6a), denoting

$$\nu = (1/2\pi)(\sigma_{\dot{x}}/\sigma_x), \qquad \text{(A1.7.3)}$$

$$\kappa = k/\sigma_x \qquad \text{(A1.7.4)}$$

it follows from Eq. A1.6.3 that the upcrossing rate of the level κ (in units of standards deviations of the process) is

$$E(\kappa) = \nu \exp\left(-\frac{\kappa^2}{2}\right) \qquad \text{(A1.7.5)}$$

where

$$\nu = \left\{ \frac{\int_0^\infty n^2 S_z(n) \, dn}{\int_0^\infty S_z(n) \, dn} \right\}^{1/2} \qquad \text{(A1.7.6)}$$

is the mean zero upcrossing rate, that is,

$$\nu = E(0) \tag{A1.7.7}$$

Peaks greater than $k\sigma_z$ may be regarded as rare events. Their probability distribution may therefore be assumed to be of the Poisson type. The probability that in the time interval T there will be no peaks equal to or larger than $k\sigma_z$ can therefore be written as

$$p(0, T) = \exp\{-E(k)T\} \tag{A1.7.8}$$

The probability $p(0, T)$ can be viewed as the probability that, given the interval T, the ratio K of the largest peak to the r.m.s. value of $z(t)$ is less than κ, that is,

$$P(K < \kappa|T) = \exp\{-E(\kappa)T\} \tag{A1.7.9}$$

The probability density function of K, that is, the probability that $\kappa < K < \kappa + d\kappa$, is obtained from Eq. A1.7.9 by differentiation:

$$p_K(\kappa|T) = \kappa TE(\kappa) \exp\{-\mathrm{E}(\kappa)T\} \tag{A1.7.10}$$

The expected value of the largest peak occurring in the interval T may then be calculated as

$$\overline{K} = \int_0^\infty kp_k \ (k\,|T)\ dk \tag{A1.7.11}$$

The integral of Eq. A1.7.11 is, approximately,

$$\overline{K} = (2 \ln \nu T)^{1/2} + \theta \frac{0.577}{(2 \ln \nu T)^{1/2}} \tag{A1.7.12}$$

[A1-1], where ν is given by Eq. A1.7.6.

A1.8 PROBABILITY DISTRIBUTION OF THE PEAK VALUE OF A NON-GAUSSIAN RANDOM SIGNAL

In many applications, the distribution of the random process of interest is not Gaussian. Pressures on low-rise buildings are an example. For such applications it is necessary to estimate statistics of the peak value of the process during a time interval T. In particular, such statistics are needed in order for comparisons between peaks measured in different experiments to be meaningful. Since the peaks are random variables, they can take on a value lower than the mean in one experiment, and a value larger than the mean in another experiment. For consistency, the comparison should be made between mean values of the peaks, or between quantiles that are the same for both experiments.

The calculation of the probability distribution of the peaks allows such comparisons to be performed. The calculation is performed by using the information contained in the entire record, rather than—as has been common in full-scale measurements practice—between the measured values of the peaks themselves. The approach to the calculation is the following. A probability distribution is fitted to the values of the random variable at sufficiently small consecutive time intervals. This distribution is called the marginal distribution of the process. A process $z(t)$ with a non-Gaussian marginal distribution is called a non-Gaussian process. Denote the distribution of that process by $F_Z(z)$. First, we define a Gaussian variable $y(t)$ by the following transformation:

$$F_Z(z) = \Phi(y) \tag{A1.8.1}$$

that is,

$$y(t) = \Phi^{-1}\{F_Z[z(t)]\} \tag{A1.8.2}$$

We then calculate the distribution of the peak $y_{pk,T}$ of the variate $y(t)$ during the interval T by using the results of Sect. A1.7, and denote that distribution by $F_{Ypk,T}(y_{pk,T})$. Note first that, since any peak $y_{pk,T}$ belongs to the Gaussian process y, there exists an ordinate $\Phi(y_{pk,T})$ of the Gaussian cumulative distribution of $y(t)$. Second, in view of the monotonic nature of the transformation A1.7.12, the image in the process $z(t)$ of the peak value of the process $y(t)$ during the time interval T, $y_{pk,T}$, is the peak value of $z(t)$ during that interval, which we denote by $z_{pk,T}$. We have

$$F_{Z(z_{pk,T})} = \Phi(y_{pk,T}) \tag{A1.8.3}$$

so that

$$z_{pk,T} = F_Z^{-1}[\Phi(y_{pk,T})] \tag{A1.8.4}$$

The distribution of $z_{pk,T}$ is obtained by performing the mapping A1.56 for a sufficient number of points. Once the distribution of $z_{pk,T}$ is available, its mean and quantiles are readily obtained.

We consider the gamma distribution, for which the probability density function is

$$f(z) = \frac{\left(\frac{z-\mu}{\beta}\right)^{\gamma-1} e^{-(z-\mu)/\beta}}{\beta\Gamma(\gamma)} \tag{A1.8.5}$$

where μ, β, γ, and Γ are, respectively, the location parameter, the scale parameter, the shape parameter, and the gamma function. The gamma distribution is sufficiently versatile to cover adequately the entire range of marginal distributions that may be encountered in practice for processes $z(t)$. The procedure for generating the probability $F_Z(z_{pk,T})$ described in this section has been automated, and the software for its implementation is available at www.nist.gov/wind (scroll down to III. Special-Purpose Software, select B).

APPENDIX A2

MEAN WIND PROFILES AND ATMOSPHERIC BOUNDARY LAYER DEPTH

A2.1 EQUATIONS OF BALANCE OF MOMENTA WITHIN THE ATMOSPHERIC BOUNDARY LAYER

It may be assumed that in large-scale storms, within a horizontal site of uniform roughness over a sufficiently long fetch, a region exists within which the flow is horizontally homogeneous. It follows from Eq. 10.1.3 that if the curvature of the isobars is negligible (i.e., if $1/r = 0$), within that region the geostrophic flow along the orthogonal axes x and y is governed by the equations

$$\frac{1}{\rho}\frac{\partial p}{\partial x} = fV_g \tag{A2.1.1a}$$

$$\frac{1}{\rho}\frac{\partial p}{\partial y} = -fU_g \tag{A2.1.1b}$$

where the x-axis coincides with the direction of the friction force (i.e., of the shear stress) at the surface (Fig. 10.1.4), the x- and y-axes are orthogonal; ρ, p, and f are the air density, the mean pressure, and the Coriolis parameter;[1] and U_g, V_g are the components of the geostrophic velocity vector G along the x- and y-axes. The boundary conditions for Eqs. A2.1.1a and b are the following: At the ground level, the flow velocity vanishes, and at the elevation

[1]Note that the notation f designates here the Coriolis parameter, not the Monin coordinate used in Chapter 11 in expressions for the spectral density. The same symbol is employed in both cases to conform to standard usage.

from the ground equal to the boundary layer thickness, the shear stresses due to friction vanish and the wind flows with the gradient velocity.

Within the boundary layer, the equations of balance of momenta for an elemental volume of air are

$$\frac{1}{\rho}\frac{\partial p}{\partial x} - fV - \frac{1}{\rho}\frac{\partial \tau_u}{\partial z} = 0 \tag{A2.1.2a}$$

$$\frac{1}{\rho}\frac{\partial p}{\partial y} + fU - \frac{1}{\rho}\frac{\partial \tau_v}{\partial z} = 0 \tag{A2.1.2b}$$

where τ_u and τ_v are shear stresses parallel to the x- and y-axes, respectively. It follows from Eqs. A2.1.1 and A2.1.2 that

$$V_g - V = \frac{1}{\rho f}\frac{\partial \tau_u}{\delta z} \tag{A2.1.3a}$$

$$U_g - U = -\frac{1}{\rho f}\frac{\partial \tau_v}{\delta z}. \tag{A2.1.3b}$$

A2.2 THE TURBULENT EKMAN LAYER

Let the boundary layer be divided into two regions: a surface layer and an outer layer. In the surface layer, the surface shear τ_0 depends upon the flow velocity at some distance z from the ground, the roughness of the terrain (i.e., a roughness length parameter z_0), and the density ρ of the air, that is,

$$\tau_0 = F(U\mathbf{i} + V\mathbf{j}, z, z_0, \rho) \tag{A2.2.1}$$

where $\mathbf{i}$ and $\mathbf{j}$ are unit vectors in the x and y directions, respectively. Equation A2.2.1 may be written in nondimensional form as follows:

$$\frac{U\mathbf{i} + V\mathbf{j}}{u_*} = f_1\left(\frac{z}{z_0}\right) \tag{A2.2.2}$$

where the quantity

$$u_* = \left(\frac{\tau_0}{\rho}\right)^{1/2} \tag{A2.2.3}$$

is known as the *shear velocity* of the flow, and f_1 is a function of the ratio z/z_0 to be defined. Equation A2.2.2 is a form of the well-known "law of the wall" and describes the flow in the surface layer.

In the outer layer, the reduction of the velocity $[(U_g\mathbf{i} + V_g\mathbf{j}) - (U\mathbf{i} + V\mathbf{j})]$ at height z depends upon the surface shear τ_0; the height to which the effect of the wall stress has diffused in the flow, that is, the boundary-layer thickness δ; and the density ρ of the air. The expression of this dependence in nondimensional form is known as the "velocity defect law":

$$\frac{U\mathbf{i}+V\mathbf{j}}{u_*}=\frac{U_g\mathbf{i}+V_g\mathbf{j}}{u_*}+f_2\left(\frac{z}{\delta}\right) \tag{A2.2.4}$$

where f_2 is a function to be defined.

If it is postulated that a gradual change occurs from the conditions near the ground to conditions in the outer layer, it may be assumed that a region of overlap exists in which both laws are valid. Let Eq. A2.2.2 be written in the form

$$\frac{U\mathbf{i}+V\mathbf{j}}{u_*}=f_1\left[\left(\frac{z}{\delta}\right)\left(\frac{\delta}{z_0}\right)\right] \tag{A2.2.5}$$

From the form of Eqs. A2.2.4 and A2.2.5, and the condition that their right-hand sides be equal in the overlap region, it follows that a multiplying factor inside the function f_1 must be equivalent to an additive quantity outside the function f_2. In the case of the analogous two-dimensional problem, it is well known that the two functions must be logarithms [A2-1]. This requirement will be satisfied if f_1 and f_2 are defined as follows [A2-2]:

$$f_1(\xi)=(\ln \xi^{1/k})\mathbf{i} \tag{A2.2.6}$$

$$f_2(\xi)=(\ln \xi^{1/k})\mathbf{i}+\frac{B}{k}\mathbf{j} \tag{A2.2.7}$$

where B and k are constants [11A-2]. Substituting Eqs. A2.2.6 and A2.2.7 into Eqs. A2.2.5 and A2.2.4, respectively,

$$\frac{U\mathbf{i}+V\mathbf{j}}{u_*}=\frac{1}{k}\left(\ln\frac{z}{\delta}+\ln\frac{\delta}{z_0}\right)\mathbf{i} \tag{A2.2.8}$$

$$\frac{U\mathbf{i}+V\mathbf{j}}{u_*}=\frac{U_g\mathbf{i}+V_g\mathbf{j}}{u_*}+\frac{1}{k}\left(\ln\frac{z}{\delta}\right)\mathbf{i}+\frac{B}{k}\mathbf{j} \tag{A2.2.9}$$

It follows from Eq. A2.2.8 that the mean wind speed along axis x can be written as

$$U(z)=\frac{u_*}{k}\ln\frac{z}{z_0} \tag{A2.2.10}$$

The constant k is known as the *von Kármán constant*. Measurements have determined that $k=0.4$. Equation A2.2.10 is known as the *logarithmic law.* By applying Eq. A2.2.10 to two elevations, the logarithmic law can be written in the form of Eq. 11.2.1, in which the notation $\overline{V}\equiv U$ is used. It may be argued on the basis of Eq. A2.2.8 that the veering angle does not change within the surface layer; nevertheless, according to [11-12], measurements suggest that the veering does affect wind velocities down to lower elevations than inicated by the theory (see Sect. 11.2.3).

If Eqs. A2.2.8 and A2.2.9 are equated in the overlap region, there results

$$\frac{U_g}{u_*} = \frac{1}{k}\ln\frac{\delta}{z_0} \tag{A2.2.11a}$$

$$\frac{V_g}{u_*} = -\frac{B}{k} \tag{A2.2.11b}$$

from which there follows

$$G = \left(B^2 + \ln^2\frac{\delta}{z_0}\right)^{1/2}\frac{u_*}{k} \tag{A2.2.12}$$

We now show that the *boundary layer thickness* δ may be expressed as

$$\delta = c\frac{u_*}{f} \tag{A2.2.13}$$

where c is a constant. To prove this relation, let Eqs. A2.1.3a,b be multiplied by the unit vectors $\mathbf{j}$ and $\mathbf{i}$, respectively. From the expressions thus obtained, and remembering that at the surface $\tau_u = \tau_0$ and that, at $z = \delta$, $\tau_u = \tau_v = 0$, it follows that

$$\int[U\mathbf{i} + V\mathbf{j} - (U_g\mathbf{i} + V_g\mathbf{j})]\,dz = \frac{\tau_0}{\rho f}\mathbf{i} \tag{A2.2.14}$$

where the integration is carried out over the boundary layer depth. Since the bulk of the mass transport takes place in those parts of the boundary layer where Eq. A2.2.4 holds—which include the overlap part of the surface layer down to presumably a very small height—the velocity profile in Eq. A2.2.14 may be approximately described by Eq. A2.2.4. If Eq. A2.2.13 is now substituted into Eq. A2.2.4 and Eq. A2.2.3 is used, the left-hand side of Eq. A2.2.14 becomes

$$\begin{aligned}\int u_* f_2\left(\frac{zf}{cu_*}\right)dz &= \frac{cu_*^2}{f}\int f_2(\varsigma)\,d\varsigma \\ &= const\frac{\tau_0}{\rho f}\end{aligned} \tag{A2.2.14a}$$

This result establishes the validity of Eq. A2.2.13. Substitution of Eq. A2.2.13 into Eq. A2.2.12 yields

$$G = \left[B^2 + \left(\ln\frac{u_*}{fz_0} - A\right)^2\right]^{1/2}\frac{u_*}{k} \tag{A2.2.15}$$

[A2-2]. An independent derivation was obtained in [A2-3] by using the dynamic equations of motion of the flow (the Navier-Stokes equations) in conjunction with a phenomenological turbulent energy model. The quantities B and A are universal constants determined by measurements to have values

$4.3 < B < 5.3$ and $0 < A < 2.8$, and the coefficient c in Eq. A2.2.13 is of the order of 0.25; see [A2-2]. The height above ground z_l that defines the *depth of the atmospheric surface layer* may be assumed to be

$$z_l = b\frac{u_*}{f} \tag{A2.2.16}$$

where b is of the order of 0.025, that is, $z_l \approx 0.1\ \delta$ [A2-4]. This was confirmed by measurements conducted in a rotating wind tunnel [A2-5] and in the atmosphere [11-4].

APPENDIX A3

SPECTRA OF TURBULENT VELOCITY FLUCTUATIONS, KOLMOGOROV HYPOTHESES

The variance of the turbulent fluctuations may be viewed as a sum of squares of amplitudes of elemental harmonic contributions with frequencies n. The harmonic fluctuations at a fixed point have periods $T = 1/n$. In accordance with *Taylor's hypothesis* a flow disturbance – and, in particular, a harmonic fluctuation – travels, more or less "frozen," with the mean flow velocity $\overline{V}(z)$ [A3-1]. Therefore, for fixed time t, a disturbance with time period T is described by a spatially periodic function with wavelength $\lambda = \overline{V}(z)\,T$. Alternatively, that function can be described by its *wave number* $K = 2\pi/\lambda$, or

$$K = \frac{2\pi n}{\overline{V}(z)} \tag{A3.1}$$

Turbulent energy is dissipated by small eddies for which shear deformations and, therefore, viscous stresses, are large. For larger eddies, the viscous stresses and viscous dissipation are small, and the decay time of the fluctuating motions is correspondingly large. The contribution of those larger eddies to the total variance of the turbulent velocity fluctuations may therefore be considered to be approximately steady, which can be the case only if the energy transfer that causes their motion is approximately balanced by the energy lost through viscous dissipation. A range of eddy sizes thus exists for which the motion is determined solely by (1) the rate of energy transfer (or, equivalently, by the rate of energy dissipation, denoted by ϵ, which is approximately equal to the rate of energy transfer) and (2) the viscosity. This statement is known as Kolmogorov's first hypothesis. Since, for that range of eddy sizes, the motion is governed solely by internal parameters of the flow (the rate of energy dissipation and the viscosity), it may be assumed that it is independent of external conditions such as boundaries and, therefore, that local isotropy—the absence of preferred directions of eddy motion—obtains.

The second Kolmogorov hypothesis states that the energy dissipation is produced almost entirely by the very smallest eddies of the flow. Thus, in the inertial subrange—that is, for the range of frequencies for which the eddy motions are locally isotropic, but for which the energy dissipation is small—the influence of viscosity is negligible. In this subrange, the eddy motion may therefore be assumed to be independent of viscosity and, therefore, to be determined solely by the rate of energy dissipation ϵ. It follows that a relation involving $E(K)$ and ϵ holds for sufficiently high K, that is,

$$F[E(K), K, \varepsilon] = 0 \tag{A3.2}$$

where $E(K)$ is the contribution to the variance of the fluctuating velocity per unit wave number.

The dimensions of the quantities between brackets in Eq. A3.2 are $[L^3T^{-2}]$, $[L^{-1}]$, and $[L^2T^{-3}]$, respectively. If we write

$$[E(K)]^\alpha = a_1[K]^\beta[\varepsilon]^\gamma \tag{A3.3a}$$

in which a_1 is a universal constant, it follows from dimensional considerations that, in the inertial subrange,

$$3\alpha = -\beta + 2\gamma \tag{A3.3b}$$

$$-2\alpha = -3\gamma \tag{A3.3c}$$

or

$$\gamma = (2/3)\alpha \tag{A3.3d}$$

Substituting Eq. A3.3.d into Eq. A3.3.b,

$$3\alpha = -\beta + (4/3)\alpha \tag{A3.3e}$$

or

$$\beta = -5\alpha/3 \tag{A3.3f}$$

$$E(K) = a_1K - 5/3\varepsilon^{2/3} \tag{A3.3g}$$

On account of isotropy, the expression for the spectrum of the longitudinal velocity fluctuations, denoted by $S(K)$, is similar to within a constant to Eq. A3.2, Thus,

$$S(K) = a\varepsilon^{2/3}K^{-5/3} \tag{A3.4}$$

where it has been established by measurements that $a \approx 0.5$ [A3-2, A3-3].

From equations of balance of momenta, it can be shown that the rate of energy transfer can be written as [11-24]

$$\varepsilon = \frac{\tau_0}{\rho}\frac{d\overline{V}(z)}{dz} \tag{A3.5}$$

Using Eqs. A3.5, A2.6, and A2.13,

$$\varepsilon = \frac{u_*^3}{kz} \tag{A3.6}$$

The fact that the logarithmic law (Eq. A2.13) was used implies that Eq. A3.6 is only valid in the surface atmospheric boundary layer. Substituting Eq. A3.6 into Eq. A3.4 and using Eq. A3.1 yields

$$\frac{nS(z,n)}{u_*^2} = 0.26\, f^{-2/3} \tag{A3.7}$$

where the nondimensional frequency (the Monin coordinate) f is defined by Eq. 11.3.8, and

$$S(z,n)\, dn = S(z,K)\, dK \tag{A3.8}$$

Note that

$$S(\mathrm{z},0) = \frac{4\overline{u^2}L_u^x}{\overline{V}} \tag{A3.9}$$

Equation A3.9 is obtained by setting $n = 0$ in Eq. A1.25, applying to the resulting equation the relation $R_u(t) = R_{u_1u_2}(x)/\overline{V}$ (implicit in Taylor's hypothesis), and using the definition of L_u^x (Eq. 11.3.3). In addition, it follows from Eq. A1.25 that the derivative of $S(z, 0)$ with respect to n vanishes at $n = 0$.

Some wind tunnel workers use the assumption that

$$L_u^x(z) = \frac{1}{2\pi}\frac{\overline{V}(z)}{n_{\mathrm{peak}}} \tag{A3.10}$$

where n_{peak} is the frequency at which the non-dimensional spectrum of the longitudinal velocity fluctuations attains its peak. Because the shape of the spectrum at low frequencies, where that peaks typically occurs, is poorly known, the use of Eq. A3.10 can cause the estimation of L_u^x to be in error severalfold [11-18].

The development of Eq. A3.7 was preceded by a model [A3-4] that does not account for the variation of the spectrum with height above ground and should therefore not be used in calculations of tall building response to wind. For details, see [7-1, Fig. 2.3.3b].

APPENDIX A4

WIND DIRECTIONALITY EFFECTS, OUTCROSSING AND SECTOR-BY-SECTOR APPROACHES

An approach to accounting for wind directionality effects in a simple, rigorous, and transparent manner is described in Chapter 12. Nevertheless, some wind engineering consultants use approaches developed in the 1970s or 1980s, which are discussed in this Appendix.

A4.1 APPROACH BASED ON THE OUTCROSSING OF THE LIMIT-STATE BOUNDARY

This approach is applied by some wind engineering consultants within the framework of the wind tunnel procedure. As was mentioned earlier, the approach typically includes in the statistical samples non-extreme wind speed data, such as largest speeds recorded during successive 1-hour intervals [18-8, p. 167], or low speeds occurring in peripheral hurricane zones. We do not recommend its use unless detailed documentation is provided in a transparent manner so that it can be understood and clearly assessed by the structural engineer in charge of the project (for the need for such transparency, see Appendix A5). As was noted earlier, the outcrossing approach tends to underestimate wind effects corresponding to specified MRIs, a concern noted in [18-9].

A4.1.1 Statement of Outcrossing Problem for Structures Experiencing Directional Wind Effects

Let us represent the wind velocity as a vector **v** with origin at a point O, and let the wind speed (i.e., the magnitude of the vector **v** corresponding to the

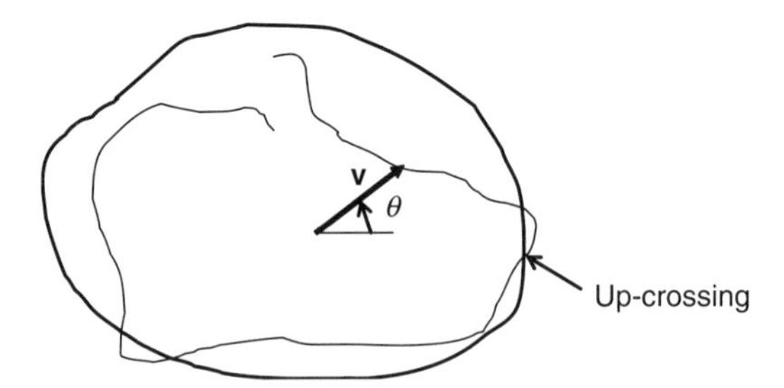

Figure A4.1.1. Limit state boundary of wind effect, and plot of the wind effect induced by wind vector v *as* a function of azimuth angle θ. Outcrossing of the boundary occurs twice in the figure.

wind direction θ) be denoted by $v(\theta)$. To every angle θ there corresponds a wind speed $v_R(\theta)$ that induces in the structure or member being considered the limit state R of interest. The limit-state boundary is the curve D described by the tip of the velocity vector **v** with magnitude $v_R(\theta)$ as θ varies from $0^\circ$to 360° (Fig. A4.1.1).

In particular, if

$$Q(\theta) = k(\theta)v^2(\theta) \tag{A4.1.1}$$

holds, where $Q(\theta)$ is the wind effect induced by the wind speed $v(\theta)$, then

$$v_R(\theta) = [R/k(\theta)]^{1/2}. \tag{A4.1.2}$$

The wind velocity vector **v** varies continuously, that is, its tip describes a curve that evolves in time. If that vector were statistically stationary (see Sect. 11.1.1 for the definition of this term), the theory of random processes would allow the estimation of the mean time interval T between successive *outcrossings* of the curve D by the vector **v**. (The term "outcrossing" denotes crossing from inside the boundary, as indicated in Fig. A4.1.1). The structure should be designed so that T is sufficiently large. For example, if R were the allowable stress in the member of interest, for non-hurricane regions T should be 50 years, say.

A4.1.2 Mean Time Interval *T* Between Successive Outcrossings

The procedure for estimating the mean time interval T between outcrossings is similar for the vector process **v** and for a scalar random process; see Sect. A1.6. The theory governing the estimation of T is mathematically straightforward. Elements of the theory are reproduced here only for the general information of the reader.

For the time-dependent vector $\mathbf{v}(t)$ with origin O, it can shown that

$$T = \frac{1}{\displaystyle\int_D E_0^\infty(\dot{v}_n|v_R)f_v(v_R)dD}, \tag{A4.1.3}$$

where v_n is the projection of the vector $\mathbf{v}$ on the normal to line D; $E_0^{\infty}(\dot{v}_n|\mathbf{v}_R)$ is the average of the positive values of the time derivative of v_n, conditional on the tip of the vector $\mathbf{v}$ being on line D; $f_\mathrm{v}(\mathbf{v})$ is the probability density of the vector $\mathbf{v}$; and dD is the elemental length along line D.

If polar coordinates are used, Eq. A4.1.3 can be written as

$$T \approx \frac{1}{\sum_{\theta_i} E_0^{\infty}[\dot{v}_n(\theta_i)|v(\theta_i)=v_R(\theta_i)] f_{v(\theta_i)}[v_R(\theta_i)] \left\{ \left[v_R^2 + \left(\left. \frac{dv(\theta)}{d\theta} \right|_{v(\theta_1)=v_R(\theta_1)} \right)^2 \right]^{1/2} \Delta\theta \right\}} \tag{A4.1.4}$$

where v_R is the magnitude of the vector $\mathbf{v}$ extending from origin O to the line D, θ is the angle defining the direction of $\mathbf{v}$, and $\{v_R^2(\theta) + (\frac{dv}{d\theta}|_{v=v_R})^2\}^{1/2}\Delta\theta$ is the elemental length dD that corresponds in polar coordinates to the angular increment $\Delta\theta$. Some users assume that $\dot{v}_n(\theta)$ is independent of $v(\theta)$ (this would imply $E_0^{\infty}[\dot{v}_n(\theta)|v(\theta) = v_R(\theta)]$ $E_0^{\infty}[\dot{v}_n(\theta)]$). However, given the anisotropy of the wind vector process, it is not established that this assumption is acceptable. The main difficulty in using the limit-state-boundary outcrossing approach lies in obtaining a reliable, unbiased estimate of the averages $E_0^{\infty}[\dot{v}_n(\theta)|v(\theta) = v_R(\theta)]$ (see comments on the outcrossing approach at the beginning of Sect. A4.1).

A4.2 THE SECTOR-BY-SECTOR APPROACH [18-10]

Let the sectorial wind effects

$$Q_i(\overline{N}) = k_i v_i^2(\overline{N})(i = 1,2,..,q) \tag{A4.2.1}$$

[for simplicity we use the notations $Q_i \equiv Q(\theta_i)$, $k_i \equiv k(\theta_i)$ and $v_i \equiv v(\theta_i)$] be calculated for each sector $i(i = 1, 2, 3, \ldots n)$, and let the *largest* of those sectorial wind effects be denoted by $Q_m(\overline{N})$, that is,

$$Q_i(\overline{N}) < Q_m(\overline{N}) \tag{A4.2.2}$$

for $i \neq m$. The sector-by-sector design criterion requires that the structure be designed so that the wind effect $Q_m(\overline{N})$ (e.g., a stress) induces in the structure the limit state R (e.g., the allowable stress). Some consultants use this criterion in the belief that it results in a design wherein the MRI of the limit state R is $\overline{N}$.

We show that the MRI of the limit state R for a member designed in accordance with this criterion is *less* than $\overline{N}$. This is the case because the sector-by-sector design criterion does not account for the effects of wind speeds from all sectors $i \neq m$, where m denotes the sector for which $Q_m(\overline{N})$, the largest of the sectorial wind effects $Q_i(\overline{N})$, occurs.

Let $F_Q(Q \le R)$ denote the probability that none of the largest yearly wind effects $Q_i (i = 1, 2, \ldots, n)$ exceeds the limit state R. The probability that $Q \le R$ is equal to the probability that $v_1 \le v_1(\overline{N}_1)$ *and* $v_2 \le v_2(\overline{N}_2), \ldots, v_m \le v_m(\overline{N}), .., v_n \le v_n(\overline{N}_n)$:

$$F_Q(Q \le R) = \text{Prob}[v_1 \le v_1(\overline{N}_1), v_2 \le v_2(\overline{N}_2), \ldots, v_m \le v_m(\overline{N}), .., v_n \le v_n(\overline{N}_n)]. \tag{A4.2.3}$$

where $\overline{N}_i$ is the mean recurrence interval for which the speed $v_i(\overline{N}_i)$ induces the limit state R.

It follows from Eq. A4.2.2 that, for $i \ne m$, $\overline{N}_i > \overline{N}$. If it is assumed that the speeds $v_1, v_{2,}, \ldots, v_n$ are mutually *independent* (this assumption implies that their mutual correlations are zero), then it follows from Eq. A4.2.3 that[1]

$$F_Q(Q \le R) = \text{Prob}[v_1 \le v_1(\overline{N}_1)]\text{Prob}[v_1 \le v_2(\overline{N}_2)] \ldots \text{Prob}[v_m \le v_m(\overline{N})]$$
$$\text{Prob}[v_n \le v_n(\overline{N}_n)] < \text{Prob}[v_m \le v_m(\overline{N})] \tag{A4.2.4}$$

Numerical Example A4.2.1. *Mean recurrence interval of wind effect induced by directional wind speeds under the sector-by-sector design criterion.* Assume, in Eq. A4.2.4, that $n = 8$ directional sectors, $\overline{N} = 50$ years, and $\overline{N}_i = 250$ years. Equation AA4.2.4 yields $F_Q(Q \le R) = (1 - 1/50)(1 - 1/250)^7 = 0.98 \times 0.996^7 = 0.95$, corresponding to a mean recurrence interval of the event $Q > R$ of $1/(1 - 0.95) = 20$ years.

Under the independence assumption, the sector-by-sector design criterion is unconservative. The criterion remains unconservative, even if the mutual correlations of the sectorial speeds do not vanish but are relatively small, as is typically the case for extreme directional wind speeds.

[1]Equation A4.2.4 follows from Eq. A4.2.3 by virtue of the definition of independent events. For example, consider the throw of a die and the throw of a coin. It is easy to see that, since the outcome of the throw of the die and the outcome of the throw of the coin are independent events, the probability $P(1, \text{H})$ that those outcomes are, respectively, “one” *and* “heads” is equal to the product of the probabilities $P(1)$ of the outcomes “one” and $P(\text{H})$ of the outcome “heads”, that is, $P(1, \text{H}) = P(1)P(\text{H}) = (1/6)(1/2) = 1/12$.

APPENDIX A5

REPORT ON ESTIMATION OF WIND EFFECTS ON THE WORLD TRADE CENTER TOWERS

Note. *The material that follows reproduces NIST document NCSTAR1-2, Appendix D, dated 13 April 2004* (http://wtc.nist.gov/NCSTAR1/NCSTAR1-2index.htm), *submitted by Skidmore, Owings & Merrill LLP, Chicago, Illinois* (wtc.nist.gov). *The documents listed in Sections 3.1, 3.2, and 3.3 are not in the public domain, but are believed to be obtainable under the provisions of the Freedom of Information Act. The material illustrates difficulties encountered by practicing structural engineers in evaluating wind engineering laboratory reports, and contains useful comments on the state of the art in wind engineering at the time of its writing.*

1.0 Table of Contents

2.0 Overview

2.1 Project Overview

The objectives for Project 2 of the WTC Investigation include the development of reference structural models and design loads for the WTC Towers. These will be used to establish the baseline performance of each of the towers under design gravity and wind loading conditions. The work includes expert review of databases and baseline structural analysis models developed by others as well as the review and critique of the wind loading criteria developed by NIST.

2.2 Report Overview

This report covers work on the development of wind loadings associated with Project 2. This task involves the review of wind loading recommendations developed by NIST for use in structural analysis computer models. The NIST recommendations are derived from wind tunnel testing/wind engineering reports developed by independent wind engineering consultants in support of insurance litigation concerning the WTC towers. The reports were provided voluntarily to NIST by the parties to the insurance litigation.

As the third party outside experts assigned to this Project, SOM's role during this task was to review and critique the NIST developed wind loading criteria for use in computer analysis models. This critique was based on a review of documents provided by NIST, specifically the wind tunnel/wind engineering reports and associated correspondence from independent wind engineering consultants and the resulting interpretation and recommendations developed by NIST.

3.0 NIST-Supplied Documents

3.1 Rowan Williams Davies Irwin (RWDI) Wind Tunnel Reports, Final Report, Wind-Induced Structural Responses World Trade Center—Tower 1, New York, New York, Project Number: 02-1310A, October 4, 2002; Final Report, Wind-Induced Structural

Responses World Trade Center—Tower 2, New York, New York, Project Number:02-1310B, October 4, 2002.

3.2 Cermak Peterka Petersen, Inc. (CPP) Wind Tunnel Report
Wind-Tunnel Tests—World Trade Center, New York, NY
CPP Project 02-2420
August 2002

3.3 Correspondence

Letter dated October 2, 2002 from Peter Irwin/RWDI to Matthys Levy/Weidlinger Associates, Re: Peer Review of Wind Tunnel Tests

World Trade Center, RWDI Reference #02-1310

Weidlinger Associates Memorandum dated March 19, 2003 from Andrew Cheung to Najib Abboud, Re: Errata to WAI Rebuttal Report

Letter dated September 12, 2003 from Najib N. Abboud/Hart-Weidlinger to S. Shyam Sunder and Fahim Sadek/NIST, Re: Responses to NIST's Questions on "*Wind-Induced Structural Responses, World Trade Center,* Project Number 02-1310A and 02-1310B, October 2002, by RWDI, Prepared for Hart-Weidlinger"

Letter dated April 6, 2004

From: Najib N. Abboud /Weidlinger Associates
To: Fahim Sadek and Emil Simiu
Re: Response to NIST's question dated March 30, 2004 regarding "Final Report, Wind-Induced Structural Responses, World Trade Center—Tower 2, RWDI, Oct 4, 2002"

3.4 NIST Report, Estimates of Wind Loads on the WTC Towers, Emil Simiu and Fahim Sadek, April 7, 2004

4.0 Discussion and Comments

4.1 General

This report covers a review and critique of the NIST recommended wind loads derived from wind load estimates provided by two independent private sector wind engineering groups, RWDI and CPP. These wind engineering groups performed wind tunnel testing and wind engineering calculations for various private sector parties involved in insurance litigation concerning the destroyed WTC Towers in New York. There are substantial disparities (greater than 40%) in the predictions of base shears and base overturning moments between the RWDI and CPP wind reports. NIST has attempted to reconcile these differences and provide wind loads to be used for the baseline structural analysis.

4.2 Wind Tunnel Reports and Wind Engineering

The CPP estimated wind base moments far exceed the RWDI estimates. These differences far exceed SOM's experience in wind force estimates for a particular building by independent wind tunnel groups.

In an attempt to understand the basis of the discrepancies, NIST performed a critique of the reports. Because the wind tunnel reports only summarize the wind tunnel test data and wind engineering calculations, precise evaluations are not possible with the provided information. For this reason, NIST was only able to approximately evaluate the differences. NIST was able to numerically estimate some corrections to the CPP report but was only able to make some qualitative assessments of the RWDI report. **It is important to note that wind engineering is an emerging technology and there is no consensus on certain aspects of current practice.** Such aspects include the correlation of wind tunnel tests to full-scale (building) behavior, methods and computational details of treating local statistical (historical) wind data in overall predictions of structural response, and types of suitable aeroelastic models for extremely tall and slender structures. It is unlikely that the two wind engineering groups involved with the WTC assessment would agree with NIST in all aspects of its critique. This presumptive disagreement should not be seen as a negative, but reflects the state of wind tunnel practice. It is to be expected that well-qualified experts will respectfully disagree with each other in a field as complex as wind engineering.

SOM's review of the NIST report and the referenced wind tunnel reports and correspondence has only involved discussions with NIST; it did not involve direct communication with either CPP or RWDI. SOM has called upon its experience with wind tunnel testing on numerous tall building projects in developing the following comments.

4.2.1 CPP Wind Tunnel Report

The NIST critique of the CPP report is focused on two issues: a potential overestimation of the wind speed and an underestimation of load resulting from the method used for integrating the wind tunnel data with climatic data. NIST made an independent estimate of the wind speeds for a 720-year return period. These more rare wind events are dominated by hurricanes that are reported by rather broad directional sectors (22.5°). The critical direction for the towers is from the azimuth direction of 205–210°. This wind direction is directly against the nominal "south" face of the towers (the plan north of the site is rotated approximately 30 degrees from the true north) and generates dominant cross-wind excitation from vortex shedding. The nearest sector data are centered on azimuth 202.5 (SSW) and 225 (SW). There is a substantial drop (12%) in the NIST wind velocity from the SSW sector to the SW sector. The change in velocity with direction is less dramatic in the CCP 720-year velocities or in the ARA hurricane wind roses included in the RWDI report. This sensitivity to directionality is a cause for concern in trying to estimate a wind speed for a particular direction. However, it should be noted that the magnitude of the NIST interpolated estimated velocity for the 210 azimuth direction is similar to the ARA wind rose. The reduction of forces has been estimated by NIST based on a square of the velocity; however, a power of 2.3 may be appropriate based on a comparison of the CPP 50-year (nominal) and 720-year base moments and velocities.

The NIST critique of the CPP use of sector by sector approach of integrating wind tunnel and climatic data is fairly compelling. The likelihood of some degree of underestimation is high but SOM is not able to verify the magnitude of error (15%) which is estimated by NIST. This estimate would need to be verified by future research, as noted by NIST.

4.2.2 RWDI Wind Tunnel Report

The NIST critique of RWDI has raised some issues but has not directly estimated the effects. These concerns are related to the wind velocity profiles with height used for hurricanes and the method used for up-crossing.

NIST questioned the profile used for hurricanes and had an exchange of correspondence with RWDI. While RWDI's written response is not sufficiently quantified to permit a precise evaluation of NIST's concerns, significant numerical corroboration on this issue may be found in the April 6 letter (Question 2) from N. Abboud (Weidlinger Associates) to F. Sadek and E. Simiu (NIST).

NIST is also concerned about RWDI's up-crossing method used for integrating wind tunnel test data and climatic data. This method is computationally complex and verification is not possible because sufficient details of the method used to estimate the return period of extreme events are not provided.

4.2.3 Building Period Used in Wind Tunnel Reports

SOM noted that both wind tunnel reports use fundamental periods of vibrations that exceed those measured in the actual (north tower) buildings. The calculations of building periods are at best approximate and generally underestimate the stiffness of a building thus overestimating the building period. The wind load estimates for the WTC towers are sensitive to the periods of vibration and often increase with increased period as demonstrated by a comparison of the RWDI base moments with and without P-Delta effects. Although SOM generally recommends tall building design and analysis be based on P-Delta effects, in this case even the first order period analysis (without P-Delta) exceeds the actual measurements. It would have been desirable for both RWDI and CPP to have used the measured building periods.

4.2.4 NYCBC Wind Speed

SOM recommends that the wind velocity based on a climatic study or ASCE 7-02 wind velocity be used in lieu of the New York City Building Code (NYCBC) wind velocity. The NYCBC wind velocity testing approach does not permit hurricanes to be accommodated by wind tunnel testing as intended by earlier ASCE 7 fastest mile versions because it is based on a method that used an importance factor to correct 50-year wind speeds for hurricanes. Because the estimated wind forces are not multiplied by an importance factor, this hurricane correction is incorporated in analytical methods of determining wind forces but is lost in the wind tunnel testing approach of determining wind forces.

4.2.5 Incorporating Wind Tunnel Results in Structural Evaluations

It is expected that ASCE 7 load factors will also be used for member forces for evaluating the WTC towers. Unfortunately, the use of ASCE 7 with wind tunnel-produced loadings is not straightforward. Neither wind tunnel report gives guidance on how to use the provided forces with ASCE 7 load factors.

The ASCE 7 load factors are applied to the nominal wind forces and, according to the ASCE 7 commentary, are intended to scale these lower forces up to wind forces associated with long return period wind speeds. The approach of taking 500-year return period wind speeds and dividing the speeds by the square root of 1.5 to create a nominal design wind speed; determining the building forces from these reduced nominal design wind speeds; and then magnifying these forces by a load factor (often 1.6) is, at best, convoluted. For a building that is as aerodynamically active as the WTC, an approach of directly determining the forces at the higher long return period wind speeds would be preferred. The CPP data did provide the building forces for their estimates of both 720-years (a load factor of 1.6) and the reduced nominal design wind speeds. A comparison of the wind forces demonstrates the potential error in using nominal wind speeds in lieu of directly using the underlying long period wind speeds.

It should also be noted that the analytical method of calculating wind forces in ASCE 7 provides an importance factor of 1.15 for buildings such as the WTC in order to provide more conservative designs for buildings with high occupancies. Unfortunately, no similar clear guidance is provided for high occupancy buildings where the wind loads are determined by wind tunnel testing. Utilizing methods provided in the ASCE 7 Commentary would suggest that a return period of 1800 years with wind tunnel-derived loads would be comparable to the ASCE 7 analytical approach to determining wind loads for a high occupancy building.

It would be appropriate for the wind tunnel private sector laboratories or NIST, as future research beyond the scope of this project, to address how to incorporate wind tunnel loadings into an ASCE 7-based design.

4.2.6 Summary

The NIST review is critical of both the CPP and RWDI wind tunnel reports. It finds substantive errors in the CPP approach and questions some of the methodology used by RWDI. It should be noted that boundary layer wind tunnel testing and wind engineering is still a developing branch of engineering and there is not industry-wide consensus on all aspects of the practice. For this reason, some level of disagreement is to be expected.

Determining the design wind loads is only a portion of the difficulty. As a topic of future research beyond the scope of this project, NIST or wind tunnel private sector laboratories should investigate how to incorporate these wind tunnel-derived results with the ASCE 7 Load Factors.

4.3 NIST Recommended Wind Loads

NIST recommends a wind load that is between the RWDI and CPP estimates. The NIST recommended values are approximately 83% of the CPP estimates and 115% of the RWDI estimates. SOM appreciates the need for NIST to reconcile the disparate wind tunnel results. It is often that engineering estimates must be done with less than the desired level of information. In the absence of a wind tunnel testing and wind engineering done to NIST specifications, NIST has taken a reasonable approach to estimate appropriate values to be used in the WTC study. However, SOM is not able to independently confirm the precise values developed by NIST.

The wind loads are to be used in the evaluation of the WTC structure. It is therefore recommended that NIST provide clear guidelines on what standards are used in the evaluations and how they are to incorporate the provided wind loads.

REFERENCES

2-1 *Minimum Design Loads for Buildings and Other Structures*, ASCE/SEI 7-10, Reston, VA: American Society of Civil Engineers, 2010.

7-1 E. Simiu and R. H. Scanlan, *Wind Effects on Structures*, 3rd ed., John Wiley & Sons, New York, 1996.

10-1 M. D. Powell and T. A. Reinhold, "Tropical Cyclone Destructive Potential by Integrated Kinetic Energy," *Proceedings of the 11th Americas Conference on Wind Engineering*, University of Puerto Rico, San Juan, Puerto Rico, 2009.

10-2 *Standard for Estimating Tornado and Extreme Wind Characteristics at Nuclear Power Sites*, ANSI/ANS-2.3-1983, American Nuclear Society, 1983.

11-1 *Canadian Structural Design Manual*, Supplement No. 4 to the National Building Code of Canada, National Research Council of Canada, Ontario, 1971.

11-2 *Eurocode 1: Actions on Structures – Part 1-4, General Actions – Wind Actions*, CEN TC 250, 2006.

11-3 H. Lettau, "Note on aerodynamic roughness element description," *J. Applied Meteorol.*, **8**, 828–832, 1969.

11-4 M. D. Powell, T. A. Reinhold, and P. J. Vickery, "Reduced drag coefficient for high wind speeds in tropical cyclones," *Nature*, **422**, 279–283, 2003.

11-5 P. A. Irwin, "Exposure categories and transitions for design wind loads," *J. Struct. Eng.*, **132**, 1755–1763, 2006.

11-6 P. S. Jackson, "On the displacement height in the logarithmic velocity profile," *J. Fluid Mech.*, **111**, 15–25, 1981.

11-7 A. G. Davenport, "The Relationship of Wind Structure to Wind Loading," in *Proceedings of the Symposium on Wind Effects on Buildings and Structures*, Vol. 1, National Physical Laboratory, Teddington, U.K., Her Majesty's Stationery Office, London, 1965, 53–102.

11-8 E. Simiu, "Logarithmic profiles and design wind speeds," *J. Eng. Mech.*, **99**, 1073–1083, 1973.

11-9 M. D. Powell, personal communication, 2005.

11-10 S. A. Hsu, "Estimating 3-s and max instantaneous gusts from 1-minute sustained winds during a hurricane." *Electronic Journal of Structural Engineering*, **77–79**, www.ejse.org/Archives/Fulltext/2008/Normal/200821.pdf, 2009.

11-11 B. Yu and A. Gan Chowdhury, "Gust Factors and Turbulence Intensities for the Tropical Cyclone Environment," *J. Appl. Met. Climatol.*, **48**, 534–552, 2009.

11-12 D. H. Wood, "Internal Boundary Layer Growth Following a Step Change in Surface Roughness," *Boundary Layer Meteorol.*, **22**, 241–244, 1982.

11-13 J. L. Franklin, M. L. Black, and K. Valde, "GPS Dropwindsonde Wind Profiles in Hurricanes and Their Operational Implications," *Weather and Forecasting*, Vol 18, 32–44, http://journals.ametsoc.org/doi/pdf/10.1175/1520-0434%282003%29018%3C0032%3AGDWPIH%3E2.0.CO%3B2, 2003.

11-14 M. T. Chay and C. W. Letchford "Pressure distributions on a cube in a simulated thunderstorm downburst. Part A: stationary downburst observations," *J. Wind Eng. Ind. Aerodyn.*, **90**, 711–732, 2002.

11-15 W. Letchford and M. T. Chay, "Pressure distributions on a cube in a simulated thunderstorm downburst. Part B: moving downburst observations," *J. Wind Eng. Ind. Aerodyn.*, **90**, 733–753, 2002.

11-16 K. Butler, S. Cao, A. Kareem, Y. Tamura, and S. Ozono, "Surface pressure and wind load characteristics on prisms immersed in a simulated transient gust front flow field," *J. Wind Eng. Ind. Aerodyn.*, **98**, 2010, 299–316.

11-17 A. Sengupta, F. L. Haan, P. P. Sarkar, and V. Balaramudu, "Transient loads on buildings in microburst and tornado winds," *J. Wind Eng. Ind. Aerodyn.*, **96**, 2008, 2173–2187.

11-18 J. Counihan, "Adiabatic atmospheric boundary layers: A review and analysis of data from the period 1880–1972," *Atmosph. Environ.*, **9**, 1975, 871–905.

11-19 M. Shiotani, *Structure of Gusts in High Winds, Parts 1–4*, The Physical Sciences Laboratory, Nikon University, Furabashi, Chiba, Japan, 1967–1971.

11-20 P. Duchêne-Marullaz, "Effect of High Roughness on the Characteristics of Turbulence in the Case of Strong Winds," *Proceedings of the Fifth International Conference on Wind Engineering*, Vol. 1, Pergamon Press, Oxford, 1980.

11-21 H. V. Theunissen, *Characteristics of the Mean Wind and Turbulence in the Planetary Boundary Layer*, Review No. 32, Institute for Aerospace Studies, University of Toronto, 1970.

11-22 B. Yu, A. Gan Chowdhury, and F. J. Masters, "Hurricane Wind Power Spectra, Cospectra, and Integral Length Scales," *Boundary Layer Meteorol.*, **129**, 411–430, 2008.

11-23 J. C. Kaimal, J. C. Wyngaard, Y. Izumi, and O. R. Coté, "Spectral Characteristics of Surface-Layer Turbulence," *J. Royal Meteorol. Soc.*, **98**, 563–589, 1972.

11-24 J. L. Lumley and H. A. Panofsky, *The Structure of Atmospheric Turbulence*, John Wiley & Sons, Hoboken, NJ, 1964.

11-25 B. J. Vickery, "On the Reliability of Gust Loading Factors," *Proceedings, Meeting Concerning Wind Loads on Buildings and Structures*, Building Science Series 30, National Bureau of Standards, Washington, DC, 1970.

11-26 L. Kristensen and N. O. Jensen, "Lateral Coherence in Isotropic Turbulence and in the Natural Wind," *Boundary Layer Meteorol.*, **17**, 352–373, 1979.

11-27 J. L. Franklin, M. L. Black, and K. V. Eyewall, "Wind Profiles in Hurricanes Determined By GPS Dropwindsondes," National Hurricane Center, Miami, FL, 2001, www.nhc.noaa.gov/aboutwindprofile.shtml.

12-1 L. T. Lombardo, J. A. Main, and E. Simiu, "Automated extraction and classification of thunderstorm and non-thunderstorm wind data for extreme value analysis," *J. Wind Eng. Ind. Aerodyn.*, 97, 120–131, 2009.

12-2 E. Simiu, R. Wilcox, F. Sadek, and J. J. Filliben, "Wind speeds in ASCE 7 Standard Peak-Gust Map: Assessment," *J. Struct. Eng.*, **129**, 427–439, 2003.

12-3 E. Simiu, R. Wilcox, F. Sadek, and J. J. Filliben, Closure to "Wind Speeds in ASCE 7 Standard Peak-Gust Map: Assessment," *J. Struct. Eng.*, **131**, 997, 2005.

12-4 J. A. Peterka, "Database of peak gust wind speeds," in www.nist.gov/wind, I Extreme Winds, Data Sets, 4, Texas Tech/CSU data, Original_Superstation_List.txt, 2001.

12-5 E. Simiu, J. Biétry, and J. J. Filliben, "Sampling Errors in the Estimation of Extreme Winds," *J. Struct. Div.*, ASCE **104**, 491–501, 1978.

12-6 R. Vega, Wind Directionality: A Reliability Based Approach, doctoral dissertation, Texas Tech University, 2008.

12-7 N. A. Heckert, E. Simiu, and T. Whalen, "Estimates of Hurricane Wind Speeds by Peaks over Threshold Method," *J. Struct. Eng.*, **124**, 445–449, 1998.

12-8 E. Castillo, A. S. Hadi, N. Balakrishnan, and J. M. Sarabia, *Extreme Value and Related Models with Applications in Engineering and Science*, John Wiley & Sons, Hoboken, NJ, 2004.

12-9 E. Simiu, N. A. Heckert, J. J. Filliben, and S. K. Johnson, "Extreme Wind Load Estimates Based on the Gumbel Distribution of Dynamic Pressure: An Assessment," *Structural Safety*, **23**, 221–229, 2001.

12-10 Structural Design Actions – Wind Actions – Commentary AS/NZ 1170.2 Supplement 1:2002, Standards Association International, Sydney, and Standards New Zealand, Wellington, 2002.

12-11 J. D. Holmes and W. W. Moriarty, "Application of the generalized Pareto distribution to extreme value analysis in wind engineering," *J. Wind Eng Ind. Aerodyn.*, **83**, 1–10, 1999.

12-12 J. D. Holmes, *Wind Loading of Structures*, 2nd ed., Francis and Taylor, New York, 2007.

12-13 E. Simiu and J. A. Lechner, "Discussion of 'Classical Extreme Value Model and Prediction of Extreme Winds,' by Janos Galambos and Nicholas Macri" (*J. Struct. Eng.*, **125**, 792–794, 1999), *J. Struct. Eng.*, **128**, 271, 2002.

12-14 J. R. M. Hosking and J. R. Wallis, "Parameter and quantile estimation for the generalized Pareto distribution," *Technometrics*, **29**, 339–349, 1987.

12-15 L. de Haan, "Extreme Value Statistics," in *Extreme Values Statistics and Applications*, J. Galambos, J. Lechner, and E. Simiu, eds., Vol. 1, Kluwer, Dordrecht, The Netherlands, 1995, 93–122.

12-16 Y. An and M. D. Pandey, "A comparison of methods of extreme wind speed estimation," *J. Wind Eng. Ind. Aerodyn.*, **93**, 535–545, 2005.

12-17 F. T. Lombardo and E. Simiu, "Discussion of 'A comparison of methods of extreme wind speed estimation,' by Y. An and M. D. Pandey" (*J. Wind Eng.*

Ind. Aerodyn., **93**, 535–545, 2005), *J. Wind Eng. Ind. Aerodyn.*, **96**, 2452–2453, 2008.

12-18 L. R. Russell, "Probability distributions for hurricane effects," *J. Waterways, Harbors, and Coastal Eng. Div.*, ASCE **97**, 139–154, 1971.

12-19 M. E. Batts, L. R. Russell, and E. Simiu, "Hurricane Wind Speeds in the United States," *J. Struct. Div.*, ASCE **100**, 2001–2015, Oct. 1980.

12-20 P. N. Georgiou, A. G. Davenport, and B. J. Vickery, "Design wind loads in regions dominated by tropical cyclones," *J. Wind Eng. Ind. Aerodyn.*, **13**, 139–152, 1983.

12-21 P. J. Vickery and L. A. Twisdale, "Prediction of hurricane wind speeds in the United States." *J. Struct. Eng.*, **121**, 1691–1699, 1995.

12-22 P. J. Vickery, D. Wadhera, L. A. Twisdale Jr., and F. M. Lavelle, "U.S. Hurricane Wind Speed Risk and Uncertainty," *J. Struct. Eng.*, **135**, 310–320, 2009.

12-23 M. Grigoriu, *Algorithms for Generating Large Sets of Synthetic Directional Wind Speed Data for Hurricane, Thunderstorm, and Synoptic Winds*, NIST Technical Note 1626, National Inst. of Standards and Technology, www.nist.gov/wind, 2009.

12-24 M. Grigoriu, *Applied Non-Gaussian Processes: Examples, Theory, Simulation, Random Vibration*, Prentice Hall, Englewood Cliffs, NJ, 1995.

12-25 E. Gumbel, *Statistics of Extremes*, Columbia Univ. Press, New York, 1958.

12-26 S. Coles and E. Simiu, "Estimating Uncertainty in the Extreme Value Analysis of Data Generated by a Hurricane Simulation Model," *J. Eng. Mech.*, **129**, 1288–1294, 2003.

12-27 J. Pickands, "Bayes Quantile Estimation and Threshold Selection for the Generalized Pareto Family," in *Extreme Values Statistics and Applications*, J. Galambos, J. Lechner, and E. Simiu, eds., Vol. 1, 1995, 123–138.

12-28 N. J. Cook and R. I. Harris, "Exact and general FT1 penultimate distributions of extreme wind speeds drawn from tail-equivalent Weibull parents," *Struct. Safety*, **26**, 391–420, 2004.

12-29 S. Coles, *An Introduction to Statistical Modeling of Extreme Values*, Springer-Verlag, London, U.K., 2001.

12-30 E. Simiu, F.T. Lombardo, and D. Yeo, Discussion of "Ultimate Wind Load Design Gust Speeds in the United States for Use in ASCE-7, *J. Struct. Eng,.* **136**, 613-625," *J. Struct. Eng.* (in press).

13-1 W. C. L. Shih, C. Wang, D. Coles, and A. Roshko, "Experiments on flow past rough cylinders at large Reynolds numbers," *J. Wind Eng. Industr. Aerodyn.*, **49**, 351–368, 1993.

13-2 J. D. Ginger, J. D. Holmes, and P. Ilim, "Variation of internal pressure with varying sizes of dominant openings and volumes," *J. Struct. Eng.*, **136**, 1319–1326, 2010.

13-3 O. Mahrenholz, personal communication, 1998.

13-4 G. Schewe and A. Larsen, "Reynolds number effects in the flow around a bluff bridge deck cross section," *J. Wind Eng. Ind. Aerodyn.*, **74–76**, 149–169, 1998.

13-5 R. P. Hoxey, A. M. Reynolds, G. M. Richardson, and J. L. Short, "Observations of Reynolds number sensitivity in the separated flow region on a bluff body," *J. Wind Eng. Industr. Aerodyn.*, **73**, 231–249, 1998.

13-6 F. Long, *Uncertainties in pressure coefficients derived from full and model scale data*, Report to the National Institute of Standards and Technology, Wind Science and Engineering Research Center, Department of Civil Engineering, Texas Tech University, Lubbock, TX, 2005.

13-7 F. Sadek and E. Simiu, "Peak Non-Gaussian Wind Effects for Database-Assisted Low-Rise Building Design," *J. Eng. Mech.*, **128**, 530–539, 2002.

13-8 W. P. Fritz, B. Bienkiewicz, B. Cui, O. Flamand, T. C. E. Ho, H. Kikitsu, C. W. Letchford, and E. Simiu, "International Comparison of Wind Tunnel Estimates of Wind Effects on Low-Rise Buildings: Test-Related Uncertainties," *J. Struct. Eng.*, **134**, 1887–1890, 2008.

13-9 B. Bienkiewicz, M. Endo, and J. A. Main, "Comparative Inter-laboratory Study of Wind Loading on Low-rise Industrial Buildings," ASCE/SEI Structural Congress, April 30–May 2, 2009, American Society of Civil Engineers, Austin Texas.

13-10 D. Surry, T. C. E. Ho, and G. A. Kopp, "Measuring Pressures Is Easy, Isn't It?" *Proceedings of the International Conference on Wind Engineering*, Texas Tech University, Lubbock, TX, 2003, **2**, 2618–2623.

13-11 T. C. E. Ho, D. Surry, D. Morrish, and G. A. Kopp, "The UWO contribution to the NIST aerodynamic database for wind loads on low buildings: Part I. Archiving format and basic aerodynamic data," *J. Wind Eng. Ind. Aerodyn.*, **93**, 1–30, 2005.

13-12 L. M. St. Pierre, G. A. Kopp, D. Surry, and T. C. E. Ho, "The UWO contribution to the NIST aerodynamic database for wind loads on low buildings: Part II. Comparison of data with wind load provisions," *J. Wind Eng. Ind. Aerodyn.*, **93**, 31–59, 2005.

13-13 B. F. Coffman, J. A. Main, D. Duthinh, and E. Simiu, "Wind effects on low-rise metal buildings: Database-assisted design versus ASCE 7-05 Standard Estimates," *J. Struct. Eng.*, **136**, 744–748, 2010.

13-14 P. Huang, A. Gan Chowdhury, G. Bitsuamlak, and R. Liu, "Development of devices and methods for the simulation of hurricane winds in a full-scale testing facility." *Wind and Structures*, **12**, 151–177, 2009.

13-15 E. Simiu, G. Bitsuamlak, A. Gan Chowdhury, E. Tecle, J. Li, and D. Yeo, "Testing of residential structures under wind loads," ASCE *Nat. Haz. Rev.*, in press.

13-16 A. M. Nathan, "The effect of spin on the flight of a baseball," *Am. J. Phys.*, **76**, 119–124, 2008.

13-17 T.-C. Fu, A.M. Aly, A. G. Chowdhury, G. Bitsuamlak, D. Yeo, and E. Simiu, "A proposed technique for determining aerodynamic pressures on residential homes." *Wind and Structures* (in press).

13-18 P. Irwin, K. Cooper, and R. Girard, "Correction of distortion effects caused by tubing systems in measurements of fluctuating pressures," *J. Wind Eng. Ind. Aerodyn.*, **5**, 93–107, 1979.

13-19 G. A. Kopp, M. J. Morrison, E. Gavanski, D. J. Henderson, and H. P. Hong, "Three Little Pigs Projects: Hurricane Risk Mitigation by Integrated Wind Tunnel and Full-Scale Laboratory Tests," *Nat. Haz. Rev.* **11**, 151–161, 2010.

14-1 J. D. Robson, *An Introduction to Random Vibration*, Elsevier, Amsterdam, The Netherlands, 1964.

15-1 R. H. Scanlan and E. Simiu, "Aeroelasticity in Civil Engineering," in *A Modern Course in Aeroelasticity*, E. H. Dowell, ed., Kluwer, Dordrecht, The Netherlands, 2004.

15-2 E. Simiu and T. Miyata, *Design of Buildings and Bridges for Wind*, John Wiley & Sons, Hoboken, NJ, 2006.

15-3 M. Novak, "Aeroelastic galloping of prismatic bodies," *J. Eng. Mech. Div.*, ASCE **95**, 115–142, 1969.

15-4 H. Kawai, "Vortex-induced vibration of tall buildings," *J. Wind Eng. Ind. Aerodyn.*, **41–44**, 117–128, 1992.

15-5 E. Simiu, Buffeting and aerodynamic stability of suspension bridges in turbulent flow, Doctoral dissertation, Department of Civil and Geological Engineering, Princeton University, 1971.

15-6 T. Theodorsen, *General Theory of Aerodynamic Stability and the Mechanisms of Flutter*, NACA Report No. 496, 1935.

16-1 M. K. Ravindra, C. A. Cornell, and T. V. Galambos, "Wind and snow load factors for use in LRFD," *J. Struct. Div.*, ASCE **104**, 1443–1457, 1978.

16-2 P. Hanzlik, S. Diniz, A. Grazini, M. Grigoriu, and E. Simiu, "Building Orientation and Wind Effects Estimation," *J. Eng. Mech.*, **131**, 254–258, 2005.

16-3 S. Jang, L.-W. Lu, F. Sadek, and E. Simiu, "Database-assisted wind load capacity estimates for low-rise steel frames," *J. Struct. Eng.*, **128**, 1594–1603, 2002.

16-4 D. Duthinh and W. P. Fritz, "Safety evaluation of low-rise steel structures under wind loads by nonlinear database-assisted technique," *J. Struct. Eng.*, **133**, 587–594, 2007.

16-5 D. Duthinh, J. A. Main, A. P. Wright, and E. Simiu, "Low-rise steel structures under directional winds: mean recurrence interval of failure," *J. Struct. Eng.*, **134**, 1383–1388, 2008.

16-6 D. Duthinh and E. Simiu, "Safety of structures in strong winds and earthquakes: Multi-hazard considerations," *J. Struct. Eng.*, **136**, 230–233, 2010.

16-7 C. Crosti, D. Duthinh, and E. Simiu, "Risk consistency and synergy in multi-hazard design," *J. Struct. Eng.*, in press.

16-8 L. T. Phan, E. Simiu, M. A. McInerney, A. A. Taylor, and M. D. Powell, *Methodology for the Development of Design Criteria for Joint Hurricane Wind Speed and Storm Surge Events: Proof of Concept*, NIST Technical Note 1482, National Institute of Standards and Technology, Gaithersburg, MD, 2007.

16-9 L. T. Phan, D. N. Slinn, S. W. Kline, *Introduction of Wave Set-up Effects and Mass Flux to the Sea, Lake, and Overland Surges from Hurricanes (SLOSH) Model*, NISTIR 7689, National Institute of Standards and Technology, Gaithersburg, MD, 2010.

16-10 F. Sadek, S. Diniz, M. Kasperski, M. Gioffrè, and E. Simiu, "Sampling errors in the estimation of peak wind-induced internal forces in low-rise structures," *J. Eng. Mech.*, **130**, 235, 2004.

16-11 B. Ellingwood, T. V. Galambos, J. G. MacGregor, and C. A. Cornell, *Development of a Probability Based Load Criterion for American National Standard A58*, NBS Special Publication 577, National Bureau of Standards, Washington, D.C., 1980.

17-1 C. R. Cole, D. A. Macpherson, and K. A. McCullough, "A Comparison of Hurricane Loss Models," *Journal of Insurance Issues*, **33**, No. 1, 31-53, 2010.

17-2 K. Gurley, J. P. Pinelli, C. Subramanian, A. Cope, L. Zhang, J. Murphree, and A. Artiles, *Florida Public Hurricane Loss Model*, Engineering Team Final Report, International Hurricane Research Center, Florida International University, Miami, FL, www.cis.fiu.edu/hurricaneloss/documents/Engineering/Engineering_Volume_I.pdf (similar URL for Volume II and Volume III), 2005.

17-3 J. J. Filliben, K. Gurley, J.-P. Pinelli, and E. Simiu, "Fragility curves, damage matrices, and wind-induced loss estimation," *Proceedings of the 3rd Intern. Conf. Computer Simulation in Risk Analysis and Hazard Mitigation*, June 19–21, 2002, Sintra, Portugal, 2002, 119–126, www.nist.gov/wind (scroll to IV Publications, select 4 NIST Talks and Conf. Proceed).

17-4 J.-P. Pinelli, E. Simiu, K. Gurley, C. Subramanian, L. Zhang, A. Cope, J. Filliben, and S. Hamid, "Hurricane Damage Prediction Model for Residential Structures," *J. Struct. Eng.*, **130**, 1685–1691, 2004.

18-1 E. Simiu, F. Sadek, T. M. Whalen, S. Jang, L.-W. Lu, S. M. C. Diniz, A. Grazini, and M. A. Riley, "Achieving safer and more economical buildings through database-assisted, reliability-based design for wind," *J. Wind Eng. Ind. Aerodyn.*, **91**, 1587–1611, 2003.

18-2 T. M. Whalen, E. Simiu, G. Harris, J. Lin, and D. Surry, "The use of aerodynamic databases for the effective estimation of wind effects in main wind-force resisting systems: Application to low buildings," *J. Wind Eng. Ind. Aerodyn.*, **77-78**, 685–692, 1998.

18-3 American Institute of Steel Construction, *Manual of Steel Construction: Load and Resistance Factor Design*, Vol. I, 3rd ed., 2001.

18-4 A. Rigato, P. Chang, and E. Simiu, "Database-assisted design, standardization, and wind direction effects," *J. Struct. Eng.*, **127**, 855–860, 2001.

18-5 J. A. Main, "Interpolation issues in the estimation of peak internal forces using pressure databases," *Proceedings of the 6th Pacific-Asia Conference on Wind Engineering* (C. K. Choi, ed.), Seoul, Korea, 2005.

18-6 Y. Chen, G. A. Kopp, and D. Surry, "Interpolation of Pressure Time Series in an Aerodynamic Database for Low Buildings," *J. Wind Eng. Ind. Aerodyn.*, **91**, 737–765, 2003.

18-7 F. Masters and K. Gurley, "Multivariate Stochastic Simulation of Wind Pressure over Low-rise Structures through Linear Model Interpolation," *Proceedings of the 10th Americas Conference on Wind Engineering* (M. Levitan et al., eds.), Baton Rouge, LA, June 1–4, 2005.

18-8 *Wind Tunnel Studies of Buildings and Structures*, ASCE Manuals and Reports on Engineering Practice No. 67, American Society of Civil Engineers, Reston, Virginia, 1999.

18-9 N. Isyumov et al., "Predictions of Wind Loads and Responses from Simulated Tropical Storm Passages," *Proceedings of the 11th International Conference on Wind Engineering*, June 2–5, 2003, Lubbock, Texas (D. A. Smith and C. W. Letchford, eds.), Lubbock, Texas Tech University, 2005.

18-10 E. Simiu and J. J. Filliben, "Wind tunnel testing and the sector-by-sector approach," *J. Struct. Eng.*, **131**, 1143–1145, 2005.

18-11 J. A. Main and W. P. Fritz, *Database-assisted design for wind: Concepts, software, and examples for rigid and flexible buildings*. NIST Building Science Series 180, National Institute of Standards and Technology, Gaithersburg, MD, 2006.

18-12 T. C. E. Ho, D. Surry, and D. Morrish, *NIST/TTU Cooperative Agreement – Windstorm Mitigation Initiative: Wind Tunnel Experiments on Generic Low Buildings*, Alan G. Davenport Wind Engineering Group, The University of Western Ontario, www.nist.gov/wind, 2003.

18-13 D.-W. Seo and L. Caracoglia, "Derivation of equivalent gust effect factors for wind loading on low-rise buildings through database-assisted-design approach," *Eng. Struct.*, **32**, 328-336, 2010.

18-14 A. F. Mensah, P. L. Datin, D. O. Prevatt, R. Gupta, and J. W. van de Lind, Database-assisted design methodology to predict wind-induced structural behavior of a light-framed wood building, *Eng. Struct.*, **33**, 674-684, 2011.

18-15 Y. Quan, Y. Tamura, M. Matsui, S. Y. Cao, and A. Yoshida, "TPU aerodynamic database for low-rise buildings," Proceedings, 12th international Conference on Wind Engineering, Cairns, Australia, July 2–6, 2007, 1615-1622, www.wind.arch.t-kougei.ac.jp/system/eng/contents/code/w_it.

19-1 Y. Zhou, T. Kijewski, and A. Kareem, "Aerodynamic loads on tall buildings: interactive database," *J. Struct. Eng.*, **129**, 394–404, http://aerodata.ce.nd.edu/, 2003.

19-2 *Wind tunnel testing: A general outline*, The Boundary Layer Wind Tunnel Laboratory, The University of Western Ontario, London, Ontario, Canada, 1999.

19-3 Y. Zhou and A. Kareem, "Aeroelastic balance," *J. Eng. Mech.*, **129**, 253–292, 2003.

19-4 R. D. Gabbai and E. Simiu, "Aerodynamic Damping in the Along-Wind Response of Tall Buildings," *J. Struct. Eng.*, **136**, 117–119, 2010.

19-5 C. S. Kwok and W. H. Melbourne, "Wind-induced lock-in excitation of tall structures," *J. Struct. Div.*, ASCE **107**, 57–72, 1981.

19-6 W. F. Baker, D. S. Korista, and L. C. Novak, "Structural Design of the World's Tallest Building, The Burj Dubai Tower," www.som.com/content.cfm/burj_khalifa, 2009.

19-7 F. E. Schmit, "The Florida Hurricane and Some of Its Effects," *Eng. News Record*, **97**, 624–627, 1926.

19-8 E. Simiu, R. D. Gabbai, and W. P. Fritz, "Wind-induced tall building response: a time domain approach." *Wind and Structures*, **11**, 427–440, 2008.

19-9 S. M. J. Spence, *High-Rise Database-Assisted Design 1.1 (HR_DAD_1.1): Concepts, Software, and Examples*, NIST Building Science Series 181, National Institute of Standards and Technology, Gaithersburg, MD, www.nist.gov/wind, 2009.

19-10 D. H. Yeo, *Database-Assisted Design for Wind: Concepts, Software, and Example for High-Rise Reinforced Concrete Structures*, NIST Technical Note 1665, National Institute of Standards and Technology, Gaithersburg, MD, www.nist.gov/wind, 2010.

19-11 American Concrete Institute, *Building Code Requirements for Structural Concrete (ACI 318-08) and Commentary*, American Concrete Institute, Farmington Hills, MI, 2008.

19-12 G. Diana, S. Giappino, F. Resta, G. Tomasini, and A. Zasso, "Motion effects on the aerodynamic forces for an oscillating tower through wind tunnel tests," *Proceedings*, 5th European & African Conference on Wind Engineering, Florence, Italy, 2009, 53–56.

19-13 D. H. Yeo and E. Simiu, *Database-Assisted Design for Wind: Veering Effects on High-rise Structures*, NIST Technical Note 1672, National Institute of Standards and Technology, Gaithersburg, MD, 2010.

19-14 J. P. Den Hartog, *Mechanical Vibrations*, 4th ed., McGraw-Hill Book Company, Inc., New York, 1956.

19-15 R. J. McNamara, "Tuned Mass Dampers for Buildings," *J. Struct. Div.*, ASCE **103**, 1785–1798, 1977.

19-16 R. Luft, "Optimal Tuned Mass Dampers for Buildings," *J. Struct. Div.*, ASCE **105**, 2766–2772, 1979.

19-17 A. Kareem and S. Kline "Performance of multiple mass dampers under random loading," *J. Struct. Eng.*, **121**, 348–361, 1995.

19-18 F. Ubertini, "Prevention of suspension bridge flutter using multiple tuned mass dampers," *Wind & Structures*, **13**, 235–256, 2010.

19-19 G. Solari, "Along-Wind Response Estimation: Closed Form Solution," *J. Struct. Div.*, ASCE **108**, 225–244, 1982.

19-20 NRC-CNRC, *User's Guide NBC 1995, Structural Commentaries, Part 4*, Institute for Research in Construction, Ottawa, 1995.

19-21 W. P. Fritz, N. Jones, and T. Igusa, "Predictive Models for the Median and Variability of Building Period and Damping," *J. Struct. Eng.*, **135**, 276–586, 2009.

19-22 Y. Tamura, K. Suda, and A. Sasaki, "Damping in Buildings for Wind-Resistant Designs," Invited Lecture, *Proceedings of the International Symposium on Wind and Structures for the 21st Century*, Cheju, Korea, 2005.

19-23 I. Venanzi, W. P. Fritz, and E. Simiu, "Structural design for wind of tall buildings with non-coincident mass and elastic centers: A database-assisted design approach," *Proceedings of the 6th Asia-Pacific Conference on Wind Engineering* (C.-K. Choi, ed.), Seoul, Korea, 2005.

19-24 D. Lungu, G. Solari, G. Bartoli, M. Righi, R. Vacareanu, and A. Villà, "Reliability under wind loads of the Brancusi Endless Column, Romania," *Int. J. Fluid Mech. Res.*, **29**, 323–328, 2002.

19-25 R. D. Gabbai, "Influence of Structural Design on the Aeroelastic Stability of Brancusi's Endless Column," *J. Eng. Mech.*, **134**, 462–465, 2008.

19-26 W. H. Melbourne, "Comparison of measurements on the CAARC standard tall building model in simulated model wind flows," *J. Wind Eng. Ind. Aerodyn.*, **6**, No. 1-2, 73–88, 1980.

19-27 R. L. Wardlaw and G. Moss, "A standard tall building model for the comparison of simulated natural winds in wind tunnels," *Proceedings*, International Conference on Wind Effects on Buildings and Structures, Tokyo, Japan, 1971, 1245–1250.

19-28 M. Teshigawara, "Structural design principles," Chapter 6 in *Design of modern high-rise reinforced concrete structures*, H. Aoyama, ed., Imperial College Press, London, 2001.

19-29 P. P. Hoogendoorn and R. A. Cabal, "Four tall buildings in Madrid. Study of the wind-induced response in serviceability limit state," http://www.ctbuh.org/LinkClick.aspx?fileticket=wzmKep%2B2rOY%3D&tabid=71&language=en-US, 2009.

19-30 N. Isyumov, "The aeroelastic modeling of tall buildings," in *Wind Tunnel Modeling for Civil Engineering Applications*, T. A. Reinhold, ed., Cambridge University Press, Cambridge, U.K., 1982, 373–403.

19-31 E. Dragomirescu, H. Yamada, and H. Katsuchi, "Experimental investigation of the aerodynamic stability of the 'Endless Column', Romania," *J. Wind Eng. Ind. Aerodyn*., **97**. 475-484, 2009.

19-32 D. Yeo, "High-rise reinforced concrete structures: Database-assisted design for wind." *J. Struct. Eng.* (in press).

19-33 Y. Tamura, "Wind and tall buildings" Proceedings, 5th Europe-African Conference on Wind Engineering (EACWE5), Florence, Italy, 19th–23rd July, 2009. www.wind.arch.t-kougei.ac.jp/system/eng/contents/code/w_it.

A1-1 A. G. Davenport, "Note on the Distribution of the Largest Value of a Random Function with Application to Gust Loading," *J. Inst. Civil Eng.*, **24**, 187–196, 1964.

A2-1 G. B. Schubauer and C. M. Tchen, *Turbulent Flow*, Princeton Univ. Press, Princeton, NJ, 1961.

A2-2 G. T. Csanady, "On the Resistance Law of a Turbulent Ekman Layer," *J. Atmosph. Sci.*, **24**, 467–471, 1967.

A2-3 A. S. Monin and A. M. Yaglom, *Statistical Fluid Mechanics: Mechanics of Turbulence*, MIT Press, Cambridge, MA, 1971.

A2-4 H. Tennekes, "The Logarithmic Wind Profile," *J. Atmosph. Sci.*, **30**, 234–238, 1973.

A2-5 G. C. Howroyd and P. R. Slawson, "The Characteristics of a Laboratory Produced Turbulent Ekman Layer," *Boundary Layer Meteorol.*, **3**, 299–306, 1964.

A3-1 J.O. Hinze, *Turbulence*, Second edition, McGraw-Hill, New York, 1975.

A3-2 H. Tennekes and J. L. Lumley, *A First Course in Turbulence*, MIT Press, Cambridge, MA, 1972.

A3-3 C. Foias, O.P. Manley, R. Rosa, and R. Temam, *Navier-Stokes Equations and Turbulence*, Cambridge University Press, Cambridge, 2001.

A3-4 A.G. Davenport, "The Spectrum of Horizontal Gustiness Near the Ground in High Winds," *J. Royal Meteorol. Soc.*, **87**, 194-211 (1961).

INDEX